DIE ENTSTEHUNG VON STERNEN
DURCH KONDENSATION DIFFUSER MATERIE

VON

G. R. BURBIDGE

F. D. KAHN

R. EBERT, S. v. HOERNER, ST. TEMESVÁRY

MIT 36 ABBILDUNGEN

Springer-Verlag Berlin Heidelberg GmbH

Alle Rechte, insbesondere das der Übersetzung in fremde Sprachen, vorbehalten
Ohne ausdrückliche Genehmigung des Verlages ist es auch nicht gestattet, dieses
Buch oder Teile daraus auf photomechanischem Wege (Photokopie, Mikroskopie)
zu vervielfältigen

ISBN 978-3-662-01329-8   ISBN 978-3-662-01328-1 (eBook)
DOI 10.1007/978-3-662-01328-1

© by Springer-Verlag Berlin Heidelberg 1960

Ursprünglich erschienen bei Springer-Verlag OHG . Berlin · Göttingen · Heidelberg 1960.

Softcover reprint of the hardcover 1st edition 1960

Die Wiedergabe von Gebrauchsnamen, Handelsnamen, Warenbezeichnungen usw.
in diesem Werk berechtigt auch ohne besondere Kennzeichnung nicht zu der
Annahme, daß solche Namen im Sinn der Warenzeichen- und Markenschutz-
Gesetzgebung als frei zu betrachten wären und daher von jedermann benutzt
werden dürfen

Vorwort

Auf der Hamburger Tagung der Gesellschaft Deutscher Naturforscher und Ärzte im September 1956 wurde folgendes Preisausschreiben bekanntgegeben:

„Der Vorstand der Gesellschaft Deutscher Naturforscher und Ärzte hat beschlossen, von Zeit zu Zeit ein Preisausschreiben zu veranstalten, das die Aufmerksamkeit lenken soll auf bestimmte naturwissenschaftliche oder medizinische Probleme, die im jeweiligen Zeitpunkt eine besondere Behandlung verlangen.

In diesem Jahr ist ein Preis von DM 7000.— ausgesetzt für die Behandlung des Themas ‚Die Entstehung von Sternen durch Kondensation diffuser Materie'.

Dabei wird eine Darstellung der verschiedenen gegenwärtig in der Literatur vorhandenen Ansätze und Theorien, nicht die einseitige Propagierung einer bestimmten Theorie, gewünscht.

Im einzelnen ist folgendes zu sagen:

A. Die Preisschrift soll eine zusammenfassende und kritische Übersicht geben über

1. die Gründe für die Annahme

a) daß bestimmten Sterngruppen ein definiertes Alter zugeschrieben werden muß,

b) daß gegenwärtig Sterne aus interstellarer Materie entstehen;

2. die in der Literatur vorhandenen Ansätze zu einer physikalischen Theorie der Entstehung von Sternen und Sterngruppen durch Kondensation interstellarer Materie,

3. die Versuche

a) das Farbenhelligkeitsdiagramm verschiedener Sterngruppen,

b) die räumliche Verteilung und die Geschwindigkeitsverteilung der Sterne von verschiedenem Typus und verschiedener ‚Population' theoretisch zu verstehen.

B. Es bleibt den Bewerbern überlassen zu entscheiden, wieweit sie

1. die Aufsammlung von Materie durch bestehende Sterne,

2. die Entstehung von Doppelsternen und mehrfachen Sternsystemen oder andere ihnen notwendig erscheinende Fragenkomplexe in die Darstellung hineinziehen wollen.

C. Nicht erwartet wird die Darstellung der Sternentstehung unter physikalischen Bedingungen, die von den gegenwärtigen qualitativ wesentlich verschieden sind.

Der Preis wird für die beste Arbeit ausgesetzt. Wenn mehrere gleichgute Arbeiten eingehen, kann der Preis geteilt werden. Wenn keine Bewerbungen eingehen oder die eingegangenen Bewerbungen den Ansprüchen des Preisrichterkollegiums nicht genügen, so fällt die ausgesetzte Summe an die Gesellschaft zurück. Der Kreis der Bewerber ist nicht beschränkt. Die Preisschrift soll zur Erleichterung der Arbeit der Preisrichter in deutscher, englischer oder französischer Sprache abgefaßt sein. Jede Preisschrift ist in drei Exemplaren an die Hamburger Sternwarte in Bergedorf einzusenden. Sie soll durch ein Deckwort gekennzeichnet sein; Name und Anschrift des Verfassers sind in einem verschlossenen Briefumschlag unter dem gleichen Deckwort mitzuteilen. Der letzte Termin für den Eingang der Arbeiten bei der Hamburger Sternwarte in Hamburg-Bergedorf ist der 30. April 1958, 24.00 Uhr.

Die Teilnehmer unterwerfen sich der Entscheidung des Preisgerichts, das unter Ausschluß des Rechtsweges verbindlich entscheidet.

Als Preisrichter wurden vom Vorstand der Gesellschaft die Herren BIERMANN (Göttingen), HECKMANN (Hamburg) und UNSÖLD (Kiel) berufen.

Wir hoffen zuversichtlich, daß das Preisausschreiben dazu verhelfen wird, einen undurchsichtigen Fragenkomplex zu klären, um die künftige Forschung in einem Felde zu erleichtern, das immer stärkere Arbeit erfordert."

Die Entscheidung des Preisgerichtes lautet:

„A. Auf das Preisausschreiben der Gesellschaft Deutscher Naturforscher und Ärzte, das im September 1956 bekanntgemacht wurde, sind 12 Bewerbungen eingegangen, von welchen drei von hohem wissenschaftlichem Wert sind, während neun den gestellten Bedingungen nicht genügen.

Die drei genannten Arbeiten tragen (in alphabetischer Reihenfolge) die Kennworte

1. CANOPUS (in englischer Sprache)
2. Wie Sterne sich bilden, wie Sterne vergehn, das möchten die FACHLEUTE gerne verstehn (in deutscher Sprache)
3. MONDAY (in englischer Sprache).

B. Das Preisgericht bestand aus den Herren BIERMANN (Göttingen/ München), HECKMANN (Hamburg) und UNSÖLD (Kiel). Außerdem wirkten noch dankenswerterweise die Herren OORT (Leiden) und SCHWARZSCHILD (Princeton) mit.

C. Jede der drei genannten Arbeiten war von so hohem Gesamtgewicht, daß man ihr wahrscheinlich — wäre sie allein ohne die beiden anderen eingebracht worden — den vollen Preis zuerkannt hätte. Die Urteile aller Preisrichter machten deutlich, daß jede der Arbeiten ihre spezifischen Vorzüge hat — bei aller Verschiedenheit der Durchführung im einzelnen. Es war daher für die Preisrichter schwierig, zu einer differenzierten Bewertung zu gelangen.

Tatsächlich haben alle Preisrichter Differenzierungen vorgenommen, die aber nicht übereinstimmten. Sie empfanden stark die Problematik, Vor- und Nachteile — die schwer vergleichbar waren — in Zahlenverhältnissen auszudrücken. Sie sind deshalb nach reiflicher Überlegung — unter Berücksichtigung von vielerlei Umständen — zu dem Beschluß gekommen, die ausgesetzte Summe den Autoren in gleichen Teilen zuzusprechen.

D. In Gegenwart des Rechtsanwalts Dr. C. LAMERSDORF (Bergedorf) wurden am 25. 9. 1958 in einer Sitzung der Preisrichter die verschlossenen Umschläge geöffnet, die den Arbeiten beigelegen hatten und die Namen der Autoren enthielten.

Die Arbeit CANOPUS ist verfaßt von Dr. FRANZ D. KAHN (Manchester).

Die Arbeit FACHLEUTE hat die Herren Dr. ROLF EBERT (Hamburg), Dr. SEBASTIAN VON HOERNER (Heidelberg), Dr. STEPHAN TEMESVÁRY (München) zu Verfassern, und

die Arbeit MONDAY stammt von Dr. GEOFFREY R. BURBIDGE (Williamsbay, Wisc.).

E. Unsere Gesellschaft ist glücklich, mitteilen zu können, daß die ursprünglich ausgesetzte Summe von DM 7000.— durch eine Zuwendung der Hamburgischen Wissenschaftlichen Stiftung auf DM 9000.— erhöht werden konnte, so daß auf jede der drei Arbeiten DM 3000.— entfallen.

F. Als unser Preisausschreiben bekanntgemacht wurde, hatten wir die Hoffnung, es möge die Forschung in einem Bereich anregen, der noch in voller Entwicklung begriffen ist. Wir dürfen uns heute freuen, daß unsere Erwartungen übertroffen wurden.

Die Preisrichter möchten mit Nachdruck dafür eintreten, daß die Arbeiten so schnell wie möglich veröffentlicht werden."

Die Gesellschaft Deutscher Naturforscher und Ärzte hat zu danken dem Kuratorium der Hamburgischen Wissenschaftlichen Stiftung für die Erhöhung des ausgesetzten Preises, dem Springer-Verlag, der seine Verbundenheit mit der Gesellschaft erneut bekundet hat durch die bereitwillige Übernahme des Druckes, den Herren J. HARDORP, K. ROHLFS, E. SCHÜCKING, H. H. VOIGT und P. WELLMANN, die sich am Lesen der Korrekturen beteiligten. Außerdem fertigte Herr VOIGT ein Sachwortverzeichnis an, das den Wert des Buches für den Leser erhöht. Wir hielten es für ausreichend, das Verzeichnis nur in englischer Sprache zu geben, obwohl es sich auch auf den deutschen Anteil bezieht.

Im Namen des Vorstandes der
Gesellschaft Deutscher Naturforscher und Ärzte

Hamburg-Bergedorf, im Dezember 1959   O. HECKMANN

Inhaltsverzeichnis

The Formation of Stars by the Condensation of diffuse Matter.
By Dr. G. R. BURBIDGE (Williamsbay, Wisc.) 1

The Formation of Stars by the Condensation of diffuse Matter.
By Dr. F. D. KAHN (Manchester) 104

Die Entstehung von Sternen durch Kondensation diffuser Materie.
Von Dr. R. EBERT (Hamburg), Dr. S. v. HOERNER (Heidelberg)
und Dr. ST. TEMESVÁRY (München) 184

Subject index 325

The Formation of Stars by the Condensation of diffuse Matter

by

G. R. BURBIDGE

with 8 Figures

Contents

	Page
Introduction	2
I. The formation of stars out of gas and dust	3
1. Protostar formation through the concentration of dust	4
2. Protostar formation through the direct influence of other stars	8
3. Star formation through the concentration of pre-stellar nuclei	12
4. The influence of turbulence on protostar formation	13
5. The mass-distribution function in star formation	13
6. Evidence for stars in the process of formation	14
7. The removal of angular momentum	16
8. The formation of binary systems	17
9. Protostar formation in the presence of a magnetic field	19
10. Accretion of matter	22
11. Protostar formation in the absence of stars and dust	27
II. Theoretical evolutionary tracks	32
12. Gravitational contraction of protostars on to the main sequence	32
13. Mixing in main-sequence stars	36
14. Evolution of stars down the main sequence	37
15. Stellar models with inhomogeneities in chemical composition	40
16. Evolutionary tracks of HOYLE and SCHWARZSCHILD	44
17. Evolution of massive stars off the main sequence	49
III. Associations and clusters	51
18. Definitions of O- and T-Associations	51
19. Properties of T Tauri stars	52
20. Expansion of O-Associations	53
21. Expansion of T-Associations	55
22. Rate of star formation in O- and T-Associations	55
23. Early observational work and interpretations of galactic clusters	57
24. Modern observational work and interpretations; Dating of clusters	58
25. The very young Orion Nebula cluster	61
26. The very young clusters NGC 2264 and NGC 6530	62
27. Possible explanations of the H-R Diagrams of the Orion Nebula cluster, NCG 2264 and NCG 6530	63
28. Young clusters (h and χ Persei, Pleiades)	66
29. Clusters of intermediate age	68
30. The old clusters NGC 752 and M 67	68

Die Entstehung von Sternen

31. General conclusions from H-R Diagrams of galactic clusters 69
32. Empirical deductions about evolution of field stars in the solar neighborhood 70
33. Color-Magnitude diagrams of globular clusters; early observations, difference from solar-neighborhood stars 72
34. Modern observations and interpretation: Dating of globular clusters 73
35. Effect of chemical composition; fitting of main sequences of globular and galactic clusters 73
36. Horizontal branch stars in clusters 75
37. Luminosity functions of clusters 77

IV. Properties of stellar populations; problems of star formation and evolution on the galactic scale 78
38. Stellar populations 78
39. Spatial and velocity distribution of different kinds of stars 79
40. Spiral structure in the Galaxy 83
41. Stellar evolution in external galaxies: General remarks 84
42. M 31 84
43. M 33 86
44. The Magellanic Clouds 86
45. Elliptical galaxies in the local group 88
46. Intergalactic star clusters in the local group 90
47. More distant galaxies 91
48. Are there evolutionary sequences of galaxies? 92

V. Conclusion 95

References. 97

Introduction

This paper is concerned

(1) with the processes of formation of stars out of the interstellar gas and dust,

(2) with the properties of clusters and associations of stars,

(3) with the evolution which can take place in these clusters and the theoretical attempts which have been made to explain this evolution, and

(4) with the spatial and velocity distributions of different types of stars, as indicators of the ages and past histories of stars.

The order in which these topics will be discussed is as follows. First we shall describe the arguments which have been put forward concerning the formation of stars out of the interstellar gas and dust. In this section also, observational evidence for star formation will be summarized. In the second chapter we shall describe the attempts which have been made to understand the evolutionary track of a single star. This is necessary at this point in order that we can discuss the evolution of groups of stars.

In chapter three we shall describe the properties of clusters and associations of stars, discussing both their evolutionary history as far as it can be deduced from present theories, and also the semi-empirical approach which has been adopted by some workers.

In chapter four we shall discuss the spatial distribution and velocity distribution of different populations of stars in our Galaxy, and the bearing of these parameters on the evolutionary history and age of various types of stars. We shall also briefly discuss the extrapolation of these ideas to external galaxies.

In chapter five we outline some of the many problems which remain unsolved.

I. The formation of stars out of gas and dust

The presence in the solar neighborhood of O- and B-type stars of high luminosity is the strongest indication that star formation is currently taking place in the disk of our Galaxy. For, as will become clear from our later discussion, these luminous stars are using up their nuclear fuel so rapidly that their total lives can be measured only in millions or tens of millions of years. Thus they must have condensed in times which are recent as compared with the total age of our Galaxy, which we shall take to be of the order of 10^{10} years. On the other hand, the oldest clusters of stars in our Galaxy were probably formed at an epoch close to that at which the galaxy itself first condensed out of the intergalactic medium. Thus it is reasonable to suppose that star formation has been going on continuously ever since the galaxy first formed, though its rate has probably not remained constant.

A successful theory of star formation which starts from assuming a diffuse distribution of gas and dust, must begin by showing that the condition for gravitational instability for a mass of the right order of magnitude, i.e. $10^2-10^4 M_\odot$, since some fragmentation will probably occur thereafter, is fulfilled. Thus the well-known Jeans criterion must be satisfied. We may use the virial relation to express this condition, which is that

$$2U + \Omega < 0,$$

where U is the total internal energy of the mass and Ω is its gravitational potential energy. This is the simplest form of the condition for gravitational instability. Here we have supposed that the gas pressure is zero at the boundary of the configuration. In the case in which the magnetic energy is taken into account, a further term must be added to the left-hand side of the inequality. A fourth term must be added if Coriolis forces are to be taken into account. A fifth term must be added in the case in which it is supposed that a pressure is applied at the boundary of the configuration.

From the condition stated above, we see that in this, the simplest case, very abnormal conditions in the interstellar gas are required for contraction to occur. Thus if we suggest that the temperature of the medium is 100°K, and the critical mass in the initial contraction is

$100\,M_\odot$, then the density of the gas must be $\sim 10^{-19}\,\text{g/cm}^3$, or about 10^5 times the normal value of the gas density in the disk of the galaxy and about 10^4 times the normal values in the interstellar clouds. On the other hand, contraction will occur at the same critical mass, and with a density of $10^{-23}\,\text{g/cm}^3$, if the temperature is reduced to about $2°\text{K}$. This simple example illustrates the problem associated with star formation. The normal conditions in the interstellar gas are such that the formation of groups of stars will never occur. For star formation to take place we need regions of high density and/or low temperature. Also a pressure applied at the boundary of the mass involved, due to external influences, may make the condition for gravitational instability easier to fulfil. However, both the presence of a magnetic field and Coriolis forces will make it harder for contraction to take place, though rotation may give rise to some fragmentation of the original contracting masses, and this is required to produce protostars out of masses which are much greater than normal stellar masses.

Approaches to the problem of star formation, in which all of these different factors have been considered, will be described in the remainder of this chapter. We shall also describe ideas concerning the existence of pre-stellar nuclei and accretion which may take place.

1. Protostar formation through the concentration of dust

Here we shall discuss attempts that have been made to describe the effects of the dust and the stars on the dynamical and temperature conditions in the interstellar gas, which may lead to the formation of new stars, or the rejuvenation of older stars.

Dust is a widespread constituent of the interstellar medium in the spiral arms in our own and other galaxies. It has been shown by LINDBLAD (1), TER HAAR (2) and KRAMERS and TER HAAR (3) that in a gaseous medium most of the atoms other than hydrogen and helium will tend to stick together, forming first molecules and then more complicated assemblies, i. e., dust grains will form. The only condition, therefore, for the formation of dust is that the galaxy has evolved sufficiently so that heavy elements, particularly the metals, are present. When dust is present the inelastic collisions between neutral hydrogen atoms and the grains are powerful cooling agents for the interstellar medium. The equilibrium temperature in the spiral arms where the dust is plentiful is ≤ 100 degrees. In regions where the temperature is below average and the density is greater, dust grains form and the temperature is reduced even further. Since the gas pressure tends to remain constant as the temperature decreases, the density will be further increased, and the rate of formation of dust grains will become even greater. These arguments, originally due to SPITZER (4), (5), suggest that the growth of dust is rapidly accelerated in limited regions, thus eventually forming what SPITZER has described as a "diffuse cloud", an idealized concept, since it is possible that the dust grains start to concentrate together before most of the atoms have condensed into grains.

When dust grains reach the size at which they might begin to absorb radiation effectively, different grains in the galactic plane will be forced towards each other by the radiation pressure of the general galactic field. This process as a basis for star formation was first proposed by SPITZER (4) and WHIPPLE (6). An idealized situation obtains when we consider concentration of the dust in directions parallel to the galactic plane. The diffusion velocity V_d of a dust grain at distance R from the center of a diffuse cloud is given by Whipple's equation

$$V_d = \frac{2 m_d g}{\pi a^2 n_H m_H V_H} = -\frac{dR}{dt}, \tag{1}$$

where m_d and a are the mass and radius of the grain, and where n_H, m_H and V_H are the number density, mass, and root mean square velocity of the neutral hydrogen atoms. The acceleration of a grain at distance R from the center of a diffuse cloud is given by

$$g = \frac{\pi^2 a^4 N_d (1-\gamma) Q^2 U R}{3 m_d},$$

where N_d is the number density of dust inside the cloud minus that outside the cloud, γ is the albedo, U is the energy density of radiation due to stars, and Q is the ratio of the effective scattering area of the particle to its geometrical cross-section. If n_d is the number density of dust outside the cloud we can put

$$N_d = \frac{K}{R^2} - n_d.$$

Substituting in (1) for g and N_d and solving the differential equation we obtain

$$\left(\frac{R}{R_0}\right)^2 = \frac{N_d(0)}{n_d}(1 - e^{t/t_r}) + 1,$$

where $N_d(0) = N_d$ at $t = 0$ and $R = R_0$. Here

$$t_r = \frac{3 n_H m_H V_H}{4 U (1-\gamma) Q \pi a^2} = \frac{3}{2} \tau_r, \tag{2}$$

where τ_r is the time for the mass of the concentration to increase by a factor e. If t_c is the time at which R vanishes, which has been called the time of concentration, then

$$\frac{t_c}{t_r} = \log_e\left(1 + \frac{n_d}{N_d(0)}\right). \tag{3}$$

This equation shows the characteristics of any exponential build-up in that, for example, if $n_d = N_d(0)$, $t_c = 0.7 t_r$, whereas if $0.1 n_d = N_d(0)$, t_c

is only increased to 2.4 t_r. Under reasonable conditions ($T_H \cong 100$ degrees, $n_H = 1$, $U = 5.2 \times 10^{-13}$ erg/cm^3, $(1 - \gamma) Q = 1$, $\pi a^2 Q n_d = 3 \times 10^{-22}$), SPITZER and WHIPPLE have found that for $N_d(0)/n_d = 0.58$ (R.H.S. = 1), $t_c = 3.9 \times 10^7$ years.

This is a short time compared with the age of the Galaxy, and modifications in the numbers due to new estimates of the interstellar density and temperature made since 1946 will not increase the time appreciably. It is a time just short enough for appreciable concentration to take place before a cloud is seriously distorted by galactic rotation. SPITZER has shown that effective limits on the size of the concentration, defined by R_0, can be made. If R_0 is too small, V_d will be less than the thermal velocities of the grains, in which case only a slight concentration will be produced. The partial pressure of the grains will balance the radiative force. Thus

$$\frac{R_0}{t_c} > \frac{3 kT}{m_d},$$

and if $T = 100°$K, and $a = 10^{-5}$ cm, $t_c = 3.9 \times 10^7$ years, then $R_0 \gg 10^{-3}$ parsec. The value 10^{-3} parsec is an extreme lower limit, and SPITZER has concluded that the size of regions which can remain sufficiently cool for dust concentration for $\sim 10^7$ years must be considerably larger than this, so that concentrations with initial radii not much less than a parsec will arise.

On the other hand, if R_0 is too great the high inward velocities of the dust grains become so great that the gas is heated by collisions. Further, if the contraction velocity is greater than the gas kinetic velocity, the gas will be dragged inward and the concentration of the grains will be slowed down. Thus

$$\frac{R_0}{t_c} < \frac{3 kT}{m_H},$$

or, for the values used by SPITZER, $R_0 < 64$ parsec. This is of the same order as the radius of a spiral arm in gas and dust.

For sizes greater than this, the simple theory would break down and the effect might be that the initial concentration would begin to fragment. An upper limit on the density of the dust is about 2×10^{-26} g/cm^3; this corresponds to the situation in which all of the elements that can be concentrated into grains are so concentrated ($\sim 2\%$ of the total mass in hydrogen gas with a mean density of 10^{-24} g/cm^3), and in which it is assumed that the chemical composition of the gas and stars is the same. Thus the total mass of dust which can be concentrated in a single condensation lies in the range $10^{-3} - 200\ M_\odot$.

This theory of dust concentration will hold only up to the point at which other factors, notably the opacity of the cloud and gravitational contraction, become important. Thus different methods must be used when the dust has been concentrated by a factor of about a hundred, so that the dust and gas densities become comparable. At this point both dust and gas begin to move together and continue to contract in this way until gravitational forces take over. In the case in which the gas and dust are mixed together, the usual Jeans condition for the gravitational instability can be used provided that the effective gravitational constant G_{eff} is used. If f_r/f_g is the ratio between the radiative and gravitational attraction of the two grains, and a fraction q of the matter is in the form of dust, then

$$G_{eff} = G\left(1 + q^2 \frac{f_r}{f_g}\right).$$

The usual condition for contraction in the absence of dust is that

$$\frac{GM_0}{R_0} \geq 3\mathfrak{R}T.$$

Now it is found that the relation between the wavelength for instability in the presence and in the absence of dust is

$$L^2 = L_0^2 \frac{(1-q)^2}{(1+q^2 f_r/f_g)}.$$

Thus since

$$L_0^2 = 4\left(\frac{\pi}{3}\right)^2 R_0^2,$$

$$R^2 = R_0^2 \frac{(1-q)^2}{(1+q^2 f_r/f_g)}. \tag{4}$$

Therefore the condition for instability in the presence of dust becomes

$$\frac{GM(1+q^2 f_r/f_g)^{1/2}}{R_0(1-q)} \geq 3\mathfrak{R}T. \tag{5}$$

Since the critical mass $M \propto R^3$, and $M_0 \propto R_0^3$, this can be written

$$\frac{GM_0}{R_0} \frac{(1-q)^2}{(1+q^2 f_r/f_g)} \geq 3\mathfrak{R}T. \tag{6}$$

If $T = 100°$ K and $\varrho_H = 1.6 \times 10^{-24}$ g/cm³, the condition for contraction in the absence of dust is that $M_0 \geq 10^5 M_\odot$. However, if dust is present we may follow SPITZER by putting $f_r/f_g = 1.4 \times 10^4$ for icegrains with radii 10^{-5} cm. Then if $q = 0.1$, 0.5, and 0.9, $(1-q)(1+q^2 f_r/f_g)^{-1/2} = 7.6 \times 10^{-2}$, 8.5×10^{-3}, and 9.4×10^{-4} respectively, and the critical masses become 580, 7.2, and $8.8 \times 10^{-2} M_\odot$ respectively.

Thus as q increases a large cloud becomes unstable and begins to contract as a whole, while local concentrations of dust will rapidly lead to condensation within the cloud, and a kind of fragmentation will proceed. The time for the protostar to contract to half its dimensions (i. e. an increase in density of order 10) is given by $(\varrho G_{eff})^{-1/2}$ which, for $q = 0.5$ and $\varrho = 3 \times 10^{-24}$ g/cm³, is about 10^6 years. Contraction beyond a density of 10^{-23} g/cm³ through gravitational processes is not possible, since at this point the gas pressure becomes comparable with the radiative pressure. Beyond this point the gravitational contraction process will take over, and the rate of contraction will be slowed up. However, it may be supposed that beyond this point the contraction will take place and some fragmentation may occur along the lines considered by HOYLE (to be discussed later), the breaking down into fragments only ceasing when the isothermal contraction goes over to the adiabatic condition and the protostar becomes highly opaque.

This process of protostar formation leads naturally to a chemical composition which is more characteristic of the dust than of the interstellar medium as a whole. Since there are probably not extreme differences in composition between the youngest stars and the interstellar medium out of which they form, this suggests that if the protostars form out of a nucleus of dust they are later able to attract more gas which forms the bulk of the star. We shall later discuss the theory of accretion in this connection. The alternative to this explanation is to try and provide a theory which will give the necessary degree of compression to the gas and dust of the interstellar medium to form a protostar, without concentrating the dust alone.

Despite this difficulty it does appear that there is considerable concentration of dust relative to gas in some dark nebulae. Thus it is usually the case that ϱ_d (compressed)$/\varrho_d$ (original) $>$ ϱ_g (compressed)$/\varrho_g$ (original).

This process of dust concentration will fail if the effect of the radiation pressure on the grains is very much less than that used in this treatment, and there is some evidence that the grains do have a very high albedo. The albedo depends both on the absorption and scattering and on whether the grain is metallic or dielectric. Thus instead of putting $(1-\gamma) Q = 1$ as was done originally, we might put $(1-\gamma) Q \approx 0.01$. Under these conditions t_c becomes about 3.9×10^9 years, so that the whole of the later discussion is invalidated, because any concentration will be dispersed by the effects of galactic shear long before any appreciable density increase is attained. However, at the present stage, our knowledge of grain physics is still very inadequate, and we have felt that a discussion of the older work in which all of the parameters favored rapid concentrations is not out of place.

2. Protostar formation through the direct influence of other stars

In a later section we shall describe the observations of globules which may be protostars in the process of formation. Bok originally proposed that these are dust concentrations, but more recently OORT (7) has suggested that they are mainly composed of gas with only a small admixture of dust. In the latter case the concentration may have been produced by a mechanism of protostar formation proposed by BIERMANN and SCHLÜTER (8) and OORT and SPITZER (9).

The Biermann-Schlüter and Oort-Spitzer mechanism presupposes that a hot O-type star is initially present, associated with a complex of interstellar clouds. The theory is incomplete in that the presence of the O-type star is treated as an observational fact, and no theory concerning its origin is proposed.

It is supposed that the O-star is suddenly born in the middle of a cloud complex, an unreal picture, since the star will take $\sim 10^5$ years to contract on to the main sequence, but one which leads to a simple model. The sequence of events will then be as follows. The newly-born star will start to ionize the hydrogen surrounding it, and it is easily shown that the velocity of the ionization front is so rapid (at least one order of magnitude larger than the velocity of sound in the medium) that it can be said to ionize instantly up to approximately the radius of the Strömgren sphere for the star. (This rapid, almost explosive, ionization is a consequence of the initial assumption.) Provided that the complex is large as compared with the Strömgren sphere, then the temperature inside the Strömgren sphere will be about 10000 degrees, which must be compared with the value of about 100 degrees outside it, and the inner pressure will be increased accordingly. Consequently the hot mass will expand into the cool shell.

Thus a compression region will proceed into the cool shell with a velocity of the order of that of sound in the hot gas (~ 10 km/sec). If the rate of radiation is sufficient to keep the temperature of the compressed region near to that in the cool shell, then the density will be increased by a factor ~ 100. The compression front will continue to expand outward after it has reached the outer edge of the complex. Its velocity of expansion is determined both by the rocket effect of the matter ionized from the inner surface of the shell and the pressure in the inner sphere, while the acceleration finally ceases through the braking action of the interstellar gas outside the cloud complex. OORT and SPITZER have calculated the minimum total masses which will escape complete evaporation through this process, and they amount to $\sim 10^3 - 10^5 \, M_\odot$ for exciting stars ranging from B 0 to O 5. Since the cloud complexes are highly irregular it can be seen that in some cases the material will be completely ionized to its periphery, while in others masses of compressed neutral hydrogen will be ejected, to form separate clouds. This mechanism was developed primarily to explain the generation of motions of the interstellar clouds, and the velocities which are obtained are shown to be in reasonable agreement with the observed velocities of clouds.

For the condensation of protostars the critical masses from the Jeans criterion can be calculated. If we suppose that initially the mean densities lie in the range 3.3×10^{-23} — 1.6×10^{-22} g/cm³ ($n_H = 20$ to 100), that the compression factor is 100, and that $T = 100$ degrees, then the critical masses for gravitational contraction are $900 - 400\,M_\odot$ while the radii (the condensations are supposed to be spherical) are ~ 1 parsec. This simple argument may not be applicable, however, since the initial clouds are probably very far from being spherical (cf. the later discussion in this section). If these critical masses have any validity, it appears that the globules observed in the vicinity of a number of nebulae must be further advanced in protostar evolution than is the case when the clouds first begin to condense, since the majority of them are probably less massive, more dense, and certainly much smaller than they were at this stage. Thus several stages of fragmentation must again take place before the stellar masses are achieved.

This process of protostar formation will give rise to an expanding system of stars, and thus it has been suggested that it is responsible for the birth of expanding O and B associations whose stellar velocity vectors have been shown in a number of cases to be traceable to a very small initial volume of space, as will be discussed in Chapter III.

The formation of the luminous O star which starts this whole process has not been explained, and it is not impossible that an alternative explanation of the whole birth process may be forthcoming. Perhaps a more rapid, or even explosive, mechanism may be worth investigation. The very large amounts of kinetic energy which are found to be present in such complexes as the Orion nebula, amounting to some 10^{50} ergs (10), are reminiscent of the energies released in supernova outbursts. ÖPIK (11) has suggested that the great arc in this region which amounts to some $10^5\,M_\odot$ and is expanding with a velocity of ~ 10 km/sec is the result of a supernova outburst in which the supernova shell collided with the surrounding quiescent interstellar medium, and that a compression region has been driven outward. The momentum balance demands an initial mass for the supernova shell of at least $100\,M_\odot$ if an ejection velocity as great as 10^4 km/sec is assumed. This is a very large supernova mass, and it has led SAVEDOFF (10) to reject the hypothesis and to explain the energy as being due to O-star formation. However, our knowledge of the range of supernova masses and velocities is very limited. If an explosion were responsible for this feature, then it is probable that the compression region was the place of origin of the high-velocity stars AE Aurigae, μ Columbae, and 53 Arietis which have apparently come from this vicinity (see chapter III).

As has just been described, the Biermann-Schlüter-Oort-Spitzer mechanism leads to a compression of cool gas by surrounding hot gas, i. e. a pressure is exerted on the cool gas at its boundary. The gravitational instability of a gas cloud under an external pressure at its boundary has recently been studied by EBERT (12), BONNOR (13) and MCCREA (14). It has been shown by these authors that the general form of the virial can be written

$$2U + \Omega - 3pV = 0,$$

where p is the pressure exerted over the boundary, and V is the volume enclosed by the boundary. If Ω is neglected this is simply Boyle's law. If on the other hand we can neglect the external pressure, then we have the usual statement that for equilibrium

$$2U + \Omega = 0.$$

The system will contract if

$$2U + \Omega - 3pV < 0.$$

Since, by definition, the net pressure on the boundary is everywhere inward, it is clear that the term involving this pressure will make it more easy for the cloud to contract. This has been shown in some detail by EBERT who has given exact solutions for a sphere using Emden's equation, and by MCCREA using more approximate arguments. A convenient way of showing the result has been given by MCCREA as follows. For a uniform sphere the critical mass above which the system will collapse is given by

$$M_c = 8.91 \ T^2 \left(\frac{p}{k}\right)^{-1/2} M_\odot$$

which must be compared with the critical mass in the absence of an external pressure which is

$$M_c = 41.1 \ T^{3/2} \ n_H^{-1/2} \ M_\odot.$$

The corresponding critical radii in the two cases are

$$R_c = 2.85 \ T \left(\frac{p}{k}\right)^{-1/2} \text{parsec}$$

and

$$R_c = 6.4 \ T^{1/2} \ n_H^{-1/2} \text{ parsec}.$$

Some numerical examples have been described by these authors. The effect of an external pressure on the gravitational instability of a cylinder and a plane stratified layer has also been considered. The results show that in the latter case gravitational collapse will not occur in the way that it does for a spherical distribution of matter. This has lead MCCREA to conclude that if stars are formed out of material which originally had elongated form, or which lies in sheets, it must first break up through forces exerted by differential motions, or because of density fluctuations, so that the fragments have all of their dimensions comparable, before gravitational collapse can occur.

3. Star formation through the concentration of pre-stellar nuclei

KRAT (15) has argued that the process of star formation consists of the concentration of many pre-stellar bodies into a larger body. Collisions between these pre-stellar nuclei, which are envisaged as dark bodies with planetary structures (masses $\sim 10^{23}$ g), give rise to gas and dust which then condense through dynamical instability. He has proposed that such processes are going on continually in star clusters which contain a large supply of invisible planetary material.

These pre-stellar nuclei have also been proposed by UREY (16) in considering the origin of the solar system. He has suggested that objects of considerable mass and size ($\sim$ lunar dimensions) accumulate in a dust cloud at low temperature. No detailed theory of the way in which dust grains will be accreted into solid objects of large mass has yet been proposed. Once solid objects, perhaps of lunar size, have been formed, it is proposed that these will tend to be attracted together, and will then proceed rapidly to accrete gas and dust in the cloud. The collisions of these pre-stellar nuclei — the attraction may lead to an implosion — would probably lead to fragmentation and then recondensation into a gaseous star.

These ideas have further been discussed by HUANG (17). He has supposed that all double and multiple stars are formed from pre-stellar nuclei, thus overcoming the angular momentum difficulties which are discussed later in this chapter. He has also supposed that no sharp distinction between the formation of multiple stellar systems, and single stars together with planetary systems, can be made, a point of view which has been put forward by KUIPER (see section 8 of this chapter) and by STRUVE (18).

Huang has noted that a difficulty associated with the formation of pre-stellar nuclei is that if they are formed by random collisions between dust grains, the time scale for their formation is of the order of 10^9 years. He has therefore proposed that their growth will be accelerated if large dust concentrations can be formed by dynamical means, and has suggested that such condensations will be formed in vortices in the boundary layers between colliding clouds. That the appropriate boundary-layer conditions are formed in shearing collisions between dust clouds seems unlikely. These arguments are better translated into terms which apply more directly to the state of dust clouds. These will have large Reynolds numbers and we can suppose that the conditions are those of compressible turbulence. Consequently some compression in collisions between clouds, or the normal density fluctuations to be expected in a turbulent cloud, may lead to some acceleration in the dust concentration. The time scale for such processes remains unknown.

4. The influence of turbulence on protostar formation

VON WEIZSÄCKER (19) has considered protostar formation in the framework of a general cosmogony in which it is supposed that turbulence plays the dominant role. Thus it is supposed that the initial localized compression of the gas out of which the condensation begins is a density fluctuation characteristic of the turbulent gas. No detailed arguments for star formation in the general field have been put forward, but the importance of developing such a cosmogony using a theory of compressible hydromagnetic turbulence, since this will be more characteristic of astrophysical conditions, must be stressed. VON WEIZSÄCKER has considered in some detail the sequence of events in which the protostar out of which the solar system formed was developed.

5. The mass-distribution function in star formation

None of the theories of star formation have been able to predict the distribution of masses among the protostars. However, by using the observed luminosity functions for different clusters and for the solar-neighborhood stars it has been possible to determine this function for stars *after they have arrived* on the main sequence. This problem is discussed in Chapter III. It is found that the mass distribution function has the form

$$\xi(M) = kM^{-1.35}.$$

The $M^{-1.35}$ dependence is approximately followed for all of the clusters in which the function has been determined and also for the solar-neighborhood stars. It may well be universal. The value of k is different for different clusters, and is determined by the star density in the cluster. If it were possible to trace the evolution of the protostars back from the main sequence through their gravitational contraction stages and beyond in an unambiguous way, this observed mass-distribution might be expected to give valuable information concerning the process of star formation. However, a number of complicating factors are apparently at work.

As we shall show in Chapter III, arguments may be advanced which suggest that considerable fragmentation takes place at a fairly early stage of the gravitational contraction of a protostar. Also some astronomers believe that protostars form with their masses lying only within two narrow mass ranges, via the O-associations and the T-associations. In this case the mass-distribution function of stars on the main sequence has little to do with the original mechanism of star formation. However, on any other hypothesis the observed mass-distribution function on the main sequence must be intimately related to the process of star formation, and the constancy of the index -1.35 (perhaps better expressed as constant between the limits of -1.3 and -1.4) suggests that the same processes have occurred in clusters of many different ages.

6. Evidence for stars in the process of formation

Observational evidence for star formation has come from several directions. Objects which may be protostars at an early stage are the globules described by Bok and his associates, and the Herbig-Haro objects. Later stages of a star's contraction towards the main sequence may be represented by the T Tauri stars.

A number of small round dense nebulae or globules have been detected (20), (21), and it has been suggested by Bok that these are protostars being formed by the concentration of dust as we have described previously. A number of these have been found in selected regions, particularly in the Rosette Nebula in Monoceros, the Southern Coalsack itself, and others. Bok has used the measures of the total photographic absorptions and dimensions of known obscuring clouds to estimate the total masses and densities of some of the globules. He has used radii of the dust grains of 10^{-5} cm. For three globules, including the Coalsack, he has made the following estimates:

	Globule I	Globule II	Coalsack
Diameter (parsec).	0.06	0.5	8
Density (g/cm³).	1.3×10^{-21}	5×10^{-23}	3×10^{-24}
Minimum mass ($M_\odot$) . . .	0.002	0.05	13

It has further been argued that these globules are probably still growing, mainly because of the large effective gravity $G_{eff} \approx 50\ G$, so that, for example, the Coalsack could double its mass in a further 3×10^7 years. It is of considerable interest that some globules have been discovered in Messier 8, and these are very close in space to the young cluster NGC 6530 where the stars are mostly still contracting towards the main sequence and where T Tauri stars are found (see Chapter III). It is not impossible, therefore, that this is a region where all stages of star formation are going on together.

It should be emphasized that the densities of the globules given by Bok are lower limits, and very much higher densities and thus larger total masses are entirely possible. Also, in some nebulosities there are large numbers of globules, very much smaller than those originally described by Bok, and there is probably a continuous sequence of sizes going down at least to the limit of resolution of the 200-inch telescope.

A recent study by Bok (22) by means of star counts in dust clouds in the Ophiuchus and Taurus regions has shown that very dense dust concentrations are present. For example, in Ophiuchus a dark cloud with a minimum total mass $\sim 30\,M_\odot$ has been found. The mean dust density in this cloud is about 300 times the average dust density in the interstellar medium. Also, 21-cm observations in some of the large complexes associated with these regions have shown that together with the dust concentrations there are often large volume concentrations of neutral atomic hydrogen (23), so that the total masses may be ~ 10 times the mass of the dust. However these concentrations have been achieved, it is clear that they are regions in which star formation on a large scale can be expected to occur.

The Herbig-Haro objects are small knots of emission nebulosity, found in regions of the sky where heavy obscuration is present (24), (25). For example, many of these objects are found in the region of the Orion Nebula. They often have a star-like nucleus. Their spectra show forbidden lines of [O I], [O II], and [S II], as well as hydrogen emission lines. The continuum radiation between the emission lines is about a factor 10 less strong than the radiation in the emission lines. That the Herbig-Haro objects are related to the T Tauri stars is indicated by the fact that both occur in dense regions of interstellar matter, and also by the presence of the same forbidden emission lines in the nebulosity around T Tauri as appear in the Herbig-Haro objects. AMBARTZUMIAN(26) has suggested that the Herbig-Haro objects may represent an earlier evolutionary stage of the T Tauri stars.

A remarkable event has been described by HERBIG (27). Direct photographs of a Herbig-Haro object in Orion taken in 1954 showed two star-like nuclei which were not present on photographs taken in 1946 and 1947. This may perhaps be an observation of the birth of a star. However, a difficulty associated with this interpretation is the rapidity with which the luminosity of the protostar appears to have grown and the condensation changed in 7 years, as compared with the rate of gravitational contraction.

FESSENKOV (28) has proposed that stars are currently being born in gas filaments. According to these ideas chains of stars like beads on a string are formed, the string being the filament of gaseous nebulosity. FESSENKOV and ROSHKOVSKY (28) have described observational evidence for a number of star chains in the Network Nebula in Cygnus, in NGC 6960, 6992, and 6995. The mean density along such filaments of nebulosity has been estimated by them to be about $10^{-19}\,\text{g/cm}^3$. Now if we suppose that a protostar has already formed, the condition for another condensation to contract near to it is that the tidal force due to the nearby protostar does not disrupt it in the time necessary for it to contract, and this is a function of the distance apart of the two condensations and the density of the material. It has been argued that a separation of ~ 0.1 parsec between the stars in the chain is consistent with the mean density of 10^{-19} g/cm^3. The very existence of real *chains* of stars means that they must be very young since the galactic tidal forces will very rapidly disrupt such a configuration. However, considerable doubt has been cast on this interpretation of the observations. For example, the work of CHAMBERLAIN[1] gives mean densities in the Network Nebula of 10^{-21} g/cm^3. Moreover, Mount Wilson photographs of some of the same regions do not show many of the star chains (*29*). There is also the problem as to whether an apparently regular configuration of stars can be explained by accidental groupings. Thus the observations must be treated with reservation.

7. The removal of angular momentum

Whatever the initial state of the condensation, it is highly probable that it will contain a large amount of angular momentum so that as it contracts the protostar will have a very high rotation. It is clear that as the contraction continues the effect of the angular momentum will be to halt contraction perpendicular to the axis of rotation and lead to the formation of a disk which is unstable in its equatorial region. This situation will develop rather rapidly and certainly will occur before the final fragmentation processes within the condensation have taken place.

The magnitude of this effect of the conservation of angular momentum in a contracting protostar is illustrated by the following extreme example. The angular velocity of rotation of our galaxy amounts to about 0.04 km/sec per parsec. If a stellar mass initially contracted from a gas cloud with a radius of 2 parsecs into a star with a radius $5\,R_\odot$ without loss of angular momentum, the surface rotational velocity of the star would amount to about 10^7 km/sec, an entirely ridiculous result.

Several types of mechanism which will lead to loss of angular momentum from protostars and stars have been proposed.

It has been suggested that a fragmentation process in which a large proportion of the stars form bound multiple systems in which the angular momentum becomes angular momentum of relative motion might be envisaged. The high proportion of stars which are contained in binary systems supports this hypothesis. The best argument in favor of this idea is that it appears to be the only one which is capable of explaining the magnitude of the effect. It will be discussed further later on. However, some other ideas have been proposed.

It has been suggested (*30*) that the effect of compression in interacting gas clouds may reduce the angular momentum rapidly. Thus two clouds with the same rotational sense will, in a collision, form a thin compressed layer which will try to rotate with an angular velocity of

---

[1] CHAMBERLAIN, J. W.: Astrophysic. J. **117**, 399 (1953).

the same order as that of the clouds, while currents will flow in opposite directions on the two surfaces of the layer, and will probably vanish long before the collision is completed. The angular momentum in the layer can be reduced by a factor of the order of $(\varrho_c/\varrho)^{2/3}$ where ϱ_c and ϱ are the compressed layer density and the cloud density, respectively.

Another mechanism which has been proposed is that of non-magnetic braking due to the interaction of the surrounding interstellar medium on the star (*31*), (*32*). This mechanism has been closely related to the formation of protostars in a turbulent medium in which a fast rotating protostar surrounded by a large rotating envelope is formed. It is then supposed that the interaction of a gaseous envelope on the central mass will be such that the central mass will lose angular momentum because of the drag of the envelope. This angular momentum will be transferred to the envelope, and it will be transported away as the envelope is dissipated. Unfortunately it appears that this is a very weak mechanism under almost all astrophysical conditions, since it can transport away only a very small fraction (estimated by TER HAAR to be $\gtrsim 10^{-4}$) of the total initial angular momentum.

A plausible mechanism which has been worked out in some detail is that of the magnetic braking of the rotation of a central mass by a surrounding ionized cloud (*33*), (*32*), (*34*). A star containing a magnetic field will ionize the surrounding medium and will then, through the ionized cloud, be magnetically coupled to the interstellar gas; it will then be slowed down by this coupling. The rate of slowing will depend both on the initial strength and form of the magnetic field assumed for the star, and on the extent of the Strömgren sphere, i. e., on the luminosity of the star. LÜST and SCHLÜTER (*34*) have shown that, if a young star possesses a surface magnetic field of the order of 100 gauss, the effect of this braking is to reduce its rotation from the condition of limiting stability to negligible values in times $\gtrsim 10^6$ years. However, the large amount of angular momentum which the protostar must lose even to reach this condition of limiting stability is not readily explicable in this way. At the very low temperatures which it must have if it originally condensed out of a gas and dust concentration, there will be very little ionization, and the coupling between the concentration and the surrounding medium will be very weak. Thus the condition for contraction in a magnetic field, which will be discussed in Section 9 of this chapter, and the conditions demanded for angular momentum to be lost by magnetic braking are, to a first approximation, mutually exclusive.

8. The formation of binary systems

We have previously pointed out in section 7 that the problems associated with the angular momentum of contracting protostars are so severe that it appears that the solution lies in the idea that stars are formed in general in multiple systems which take up the net angular momentum of the initial clouds. The theories of SPITZER and HOYLE and others have to a large extent ignored the angular momentum problem, and the fact that the angular momentum may be an important factor in limiting the condensation also has not been discussed.

Until quite recently it was believed that the fission theory of binary star formation based on the work of DARWIN and JEANS might still be tenable (35). However, a formidable array of arguments are now available which show that this theory cannot be correct. These are based on the observational data (36), (37), on arguments relating to stellar structure and evolution, and on dynamical grounds directly refuting the arguments of JEANS (38).

Consequently the way appears to be open for the development of a theory based on the observations and theory described earlier, but taking into account the angular momentum problem encountered in the formation of single stars. Such an approach has been described by KUIPER (37), who has proposed that multiple stars are formed by independent condensations in random positions within protostars. He has pointed out that a general model might be expected to explain the two principal statistical properties of binary stars; i. e., the distribution of semi-major axes, a, and the distribution of mass ratios between the two components. The first is equivalent to the frequency distribution of the total angular momentum present in the original protostars, after allowance is made for components subtracted from mutiple stars and from those wide binaries that are destroyed by stellar encounters. The second is just the mass division of the condensations. This, for main sequence stars, is apparently a random function, which is what might be expected. Whether or not this is consistent with the Salpeter mass function discussed in Chapter III remains to be tested.

The transfer of angular momentum in encounters between systems at an early stage of evolution has been considered by HUANG and STRUVE (39). They have suggested that such an exchange might lead to a distribution of angular momentum over a coeval group of stars such that all will have values higher or lower, on the average, than the corresponding values of rotation for stars in the Galaxy as a whole.

KUIPER has considered that the interior region of the protostar, which he conceives as a type of globule similar to that described in Section 6 of this chapter, will be highly turbulent, since the Reynolds number is very high. The angular momentum of the system is then contained in the eddies, the major part in the largest outer eddies. Then, using reasonable assumptions, it is possible to obtain a frequency distribution function for $\log_e(a)$, which is of the form

$$F(\log_e a) = \frac{2}{\sqrt{\pi}} x^3 e^{-x^2},$$

where $x = A/A_p = \alpha(a/a_1)^{1/2}$, $a_1 = 1$ A. U., and α is a pure number, which is a function of β. The initial peripheral velocity is βv_c, v_c being the free circular velocity under attraction by the protostar. It is clearly very difficult to estimate β. KUIPER has shown that $0 < \beta < 1$; since the protostar does not expand and dissolve, $\beta \ll 1$. A is the total angular momentum and A_p is its most probable value. KUIPER has arbitrarily chosen $\beta = 0.1$. Comparison of this formula with the observational results then shows that the observed dispersion is higher than the computed one, i. e., there are more close and more wide binaries than are predicted theoretically. Variation in β alters the position of the maximum but not the dispersion. For $\beta = 0.1$, the maximum falls close to $\log a = 1.4$, or $a = 25$ A. U. KUIPER has given a number of plausible arguments which explain the sense of the departure of theory from observation.

Some evolutionary effects are to be expected for all those binary systems in which the components are quite close (perhaps with $a \leq 1$ A. U.).

KUIPER's arguments are compatible with the observations of regularities among multiple stars. Thus the formation of three condensations will lead to instability and the ejection of one of them, unless two of the systems are very close together as compared with the third. Similarly, a quadruple system will reduce to a triple or a binary, unless the mutual distances are arranged in a hierarchy. Only in associations which have positive total energies, and are therefore dispersing, are such situations not obtained.

9. Protostar formation in the presence of a magnetic field

There is strong evidence for the existence of a galactic magnetic field with an average strength of about 10^{-5} gauss, and doubts have been expressed as to whether an initial instability can finally lead to a condensation of stellar dimensions in the presence of so much magnetic energy. The problem has been discussed by MESTEL and SPITZER (40). In the presence of a magnetic field the virial condition is written

$$2U + \mathfrak{M} + \Omega = 0,$$

where

$$\mathfrak{M} = \frac{1}{8\pi} \int H^2 dv$$

is the magnetic energy. For a uniform sphere with magnetic field H,

$$\mathfrak{M} = \frac{1}{6} H^2 R^3.$$

Thus the critical mass above which a cloud is able to condense (neglecting U) is given by

$$M \geq \frac{H^3}{(6\,G)^{3/2}} \left(\frac{3}{4\,\pi\,\varrho}\right)^2. \tag{7}$$

Thus if $H = 7 \times 10^{-6}$ gauss and $\varrho = 1.6 \times 10^{-23}$ g/cm^3,

$$M \geq 1.7 \times 10^5\, M_\odot,$$

or if $H = 10^{-6}$ gauss (somewhat less than its galactic strength), $M \geq 500\, M_\odot$. This means that, as long as the magnetic field is frozen into the material, a mass smaller than this cannot fragment into stars. For a mass greater than this limit, contraction will lead to amplification of the magnetic field following the law

$$\frac{H}{H_0} = \frac{R_0^2}{R^2},$$

so that the magnetic energy increases at a rate proportional to R^{-1}; i. e., it has the same dependence as the gravitational energy, so that it dilutes gravity by a factor independent of the density.

This argument neglects the fact that the material can contract quite normally along the magnetic lines of force where the magnetic pressure is zero. This might suggest that in a spiral arm in which we suppose that the magnetic field lines lie parallel to the axis of the "cylinder", the condensation along the arms will proceed more rapidly than transverse contraction, and only when the unstable element has formed a spheroid oblate to the direction of the field such that the effective gravity along the field is equal to the effective gravity (normal gravity minus magnetic impedance) across the field, normal condensation could take place.

However, the critical length given by the JEANS criterion for the contraction of a cylinder is

$$L \geq \left(\frac{\pi \mathfrak{R} T}{G\,\varrho}\right)^{1/2},$$

while the condition for the lateral instability is that

$$R > \left(\frac{2 \mathfrak{R} T}{\pi\,G\,\varrho}\right)^{1/2},$$

so that

$$L \approx R \frac{\pi}{\sqrt{2}}.$$

But to allow condensation without magnetic impedance we used condensation of a cylinder of volume $\pi R^2 L$, where $L \gg R$. Thus an initial length very much greater than the critical length is required, and since this would be highly unstable against further break-up this mechanism of protostar formation will not occur.

It is necessary, therefore, to consider in more detail whether the coupling between the magnetic field and the matter is so strong in the interstellar medium, that the conditions described above are completely valid. MESTEL and SPITZER have written down the equations of motion for the plasma and for the neutral gas and dust. These equations for the plasma are

$$\frac{\vec{j} \times \vec{H}}{c} - (n_i F_{iH} + n_e F_{eH}) - \nabla (p_i + p_e) + n_i m_i \nabla \Phi = n_i m_i \frac{d\vec{V_i}}{dt} \quad (8)$$

and for the neutral gas

$$- (n_i F_{iH} + n_e F_{eH}) - \nabla p_H + n_H m_H \nabla \Phi = n_H m_H \frac{d\vec{V_H}}{dt}. \quad (9)$$

Here n_H, n_e, n_i are the number densities of neutral hydrogen, electrons, and ions, respectively, F_{eH} and F_{iH} are the mean forces on electrons and ions due to collisions between them and neutral atoms and molecules, p_H, p_i, and p_e are the partial pressures due to the neutral hydrogen, ions, and electrons, Φ is the gravitational potential, and $\vec{V_i}$ and $\vec{V_H}$ are the mean drift velocities of the ions and neutral atoms. These equations show that the neutral gas is coupled to the magnetic field only through the friction represented by the term $(n_i F_{iH} + n_e F_{eH})$ between the neutral gas and the plasma.

Now in a normal interstellar cloud where $T \approx 100$ degrees, the density of the plasma is very small as compared with the neutral gas, so that the terms involving the partial pressure and gravitation of the plasma are negligible, and as a cloud contracts the tendency of the magnetic field to straighten out is counteracted by this frictional force. For initial conditions in which the cloud mass $= 10^3 M_\odot$, $H_0 = 10^{-6}$ gauss, and $n_H(0) = 10$, and in which isotropic compression has occurred, it is found that the friction balances the magnetic force if

$$\left|\vec{V_H} - \vec{V_i}\right| \approx 1.5 \times 10^3 \, n_i^{-1}. \quad (10)$$

Since $V_H < 10^5$ cm/sec, the drift of the plasma relative to the neutral gas is small provided that $n_i > 1.5 \times 10^{-2}$. Now if $n_i/n_H = 10^{-4}$, since $n_H = 2 \times 10^4$ after compression, then $n_i = 2$, and the normal proportion of plasma is too high for any sensible drift to take place during the time $(2.1 \times 10^{13}$ sec) of free fall; i. e., $V_H - V_i \approx V_H$.

However, if we consider that the cloud is dense enough for the starlight to be absorbed by dust grains at its periphery, then the amount of ionization decays because of the capture of ions and electrons by the dust grains. The time constant for the decay has been estimated by MESTEL and SPITZER to be about 1.4×10^{12} sec, i. e., it is small compared with the time of free fall. Thus under the most favorable conditions the ion density can fall to a sufficiently low value so that the magnetic field can move the plasma through the infalling neutral gas, and so the field is not compressed by the contraction.

The condition that the magnetic force just balances the friction is rapidly reached and thereafter maintained, and as the cloud contracts the magnetic field tends to straighten itself out as it drags the plasma through the gas. If the situation is stabilized at a point in which starlight producing ionization balances the decay due to collisions with dust, then the magnetic field which remains in the condensation will be rather greater than its initial value. Very little of the magnetic energy released becomes converted to kinetic energy of mass motions.

Thus the condensation of a protostar from a cloud containing a magnetic field can take place if the conditions are such that the coupling between the magnetic field and gas as a whole is weak. However, increase in the degree of ionization, i. e., heating of the gas cloud, makes the condensation much more difficult. Though this is of no interest as far as H II regions in our Galaxy are concerned, since it is not believed that stars are formed in these regions, it may be of some importance in considering the situation when the oldest stars condensed, as will be seen from the later discussion.

10. Accretion of matter

In considering the general problem of the formation of stars it is important to consider the situation in which a prestellar nucleus, a protostar, or a star is already present, and the gravitational attraction it exerts attracts and captures more mass. An early discussion of this problem was given by EDDINGTON (41), and a more efficient process was proposed later by HOYLE, LYTTLETON and BONDI (42), (43). Here it was shown that capture can occur even if there is not a direct collision between an atom and a star, if the space density of the gas is high enough. Thus if we consider two parallel streams of gas passing symmetrically on either side of a star, they will be attracted and cross each other directly behind the star. In the crossing region collisions will occur in which atoms will be left with zero outward or transverse momentum, and the star's gravity then causes them to fall into its surface. For material of sufficiently small density and temperature, i. e., with the approximation that the heat generated would be rapidly radiated away, so that pressure effects could be ignored, the rate of gain of mass is given by

$$\frac{dM}{dt} = \alpha \frac{2\pi \varrho G^2 M^2}{v^3} \qquad (11)$$

where ϱ is the density of the interstellar gas, M is the mass of the star and v its velocity relative to the interstellar gas; α is a constant of the order of unity. On the other hand, in the limit in which the velocity is zero and pressure effects are taken into account the rate becomes (44)

$$\frac{dM}{dt} = \alpha' \frac{2\pi\varrho G^2 M^2}{c^3} \tag{12}$$

where c is the velocity of sound in the gas. In the case in which both velocity and pressure effects are important, it has been supposed by BONDI that the rate is approximately given by

$$\frac{dM}{dt} = \alpha \frac{2\pi\varrho G^2 M^2}{(v^2+c^2)^{3/2}}. \tag{13}$$

Now it has been generally shown that for normal interstellar conditions, accretion is relatively unimportant. For example, let us suppose that $M = 10\,M_\odot$, $R = 6\,R_\odot$, $L = 10^{37}$ erg/sec. Then from equation (11)

$$\frac{dM}{dt} \approx 10^{55} \frac{\varrho}{v^3}.$$

Thus if $\varrho = 10^{-23}$ g/cm³ and $v = 10$ km/sec,

$$\frac{dM}{dt} \approx 10^{14} \text{ g/sec.}$$

On the other hand, the luminosity measures the rate of conversion of mass from hydrogen to helium, and thus the rate of evolution of the star is given by

$$-\frac{dM}{dt} \approx 10^{18} \text{ g/sec.}$$

Thus in these fairly normal conditions in a low-density interstellar cloud, the accretion of hydrogen would be negligible as far as the star's evolutionary rate is concerned, i. e. as far as replacing its mass is concerned. It will only become important when regions of high density are encountered by a star of very low velocity, in which case the rate of accretion will be primarily determined by equation (12). However, all of the observational arguments suggest, and the previously discussed theories concerning protostar formation demand, that stars are born in regions where the density is far greater than the mean. Thus it is important to decide: (i) whether accretion provides an alternative mechanism for producing stars which are apparently quite young, in regions of dense cloud complexes; in this case they will only be rejuvenated stars having cores which are far older; or (ii) whether accretion following initial condensation and contraction can increase the masses of protostars by a significant amount. This first problem has been studied by MCCREA (45).

(i) It is necessary to take into account the slowing down of the star in the dense medium into which it moves. The resistive force is given by (46)

$$F = -2\pi\varrho \frac{G^2 M^2}{v^2} \log_e\left(1 + \frac{s^2 v^4}{G^2 M^2}\right) = -2\pi\varrho \frac{G^2 M^2}{v^2} \beta \tag{14}$$

where s is the radius of the effective sphere of influence of the star. It has been shown by McCrea that if x_1 and t_1 are the distance and the time elapsed to slow the star down to a velocity $\ll c$, then

$$t_1 \approx \frac{v_0^3}{(4\alpha + 3\beta) \, 2\pi \varrho G^2 M_0} \tag{15}$$

$$x_1 \approx \frac{v_0^4}{(5\alpha + 4\beta) \, 2\pi \varrho G^2 M_0} \tag{16}$$

where v_0 and M_0 are the initial velocity and mass, respectively. Equations (11) and (12) now show that the major part of the accretion will take place after the star has effectively been brought to rest. The time taken for the mass to grow to a value very much greater than the initial mass is given by

$$t_2 - t_1 = \frac{c^3}{\alpha' 2\pi G^2 \varrho M_0}. \tag{17}$$

Calculations based on these expressions are shown in Tables 1 and 2.

These results show that even in dense cloud complexes where the densities are as high as $10^3/\mathrm{cm}^3$, accretion will still *not* in general be an effective mechanism for rejuvenating stars, since the initial velocities of stars entering such complexes will be high enough to carry them right through before they reach velocities so low that spherically symmetrical accretion can occur. This is based on the fact that regions of very dense gas

Table 1. *Times and distances for reducing a star to rest relative to a cloud of density 2.5×10^{-21} g/cm³ ($T = 100°$)*

		v_0	1 km/sec	2 km/sec	5 km/sec
Isothermal case ($\alpha = 2$)					
$M_0 = 2 M_\odot$		t_1 (years)	3.1×10^6	2×10^7	2.6×10^8
		x_1 (parsec)	2.4	32	990
$M_0 = 5 M_\odot$		t_1	1.5×10^6	9.3×10^6	1.1×10^8
		x_1	1.2	14	440
Adiabatic case ($\alpha = 0$)					
$M_0 = 2 M_\odot$		t_1	4.0×10^6	2.5×10^7	3.0×10^8
		x_1	3.1	38	1150
$M_0 = 5 M_\odot$		t_1	2.0×10^6	1.2×10^7	1.4×10^8
		x_1	1.5	18	520

Table 2. *Times for large mass-increment in cloud of density 2.5×10^{-21} g/cm³ and mean molecular weight 1.4*

	$T(°K)$	30	50	100
Isothermal				
$M_0 = 2 M_\odot$	$t_2 - t_1$ (years)	3.8×10^6	8.3×10^6	2.3×10^7
$M_0 = 5 M_\odot$	$t_2 - t_1$	1.5×10^6	3.3×10^6	9.4×10^6
Adiabatic				
$M_0 = 2 M_\odot$	$t_2 - t_1$	3.7×10^7	8.0×10^7	2.3×10^8
$M_0 = 5 M_\odot$	$t_2 - t_1$	1.5×10^7	3.2×10^7	9.0×10^7

rarely extend more than ~ 50 parsecs, while the peculiar velocities of stars relative to the gas in the Galaxy are predominently ~ 5 km/sec or greater. In the few cases in which $v < 1$ km/sec the mechanism will be effective. In the case of expanding associations the outward velocities are not explicable if an accretion theory in this form is assumed.

(ii) From Tables 1 and 2 it is possible to reach some conclusions regarding the importance of accretion following protostar formation by another mechanism. Thus, let us consider the formation of stars following the formation of a condensation either by the concentration of dust, or by the mechanism of ionization and compression by an O-type star. In these cases, the surrounding matter will fall into the initial condensation (spherically symmetrical accretion), and its amount of growth in a time t will be given by the entries in Table 2 for the time taken to increase the mass by a large amount in a time $t_2 - t_1$, where $t = t_2 - t_1$. For initial masses in the range $2 - 5\,M_\odot$ the times are of the order of 10^7 years. However, the rate of increase is proportional to $1/M_0$, so that for much larger masses, which are to be expected in an initial condensation, these times can become very short indeed. Furthermore, if we are trying to explain the birth of expanding associations in a cloud complex through the presence of an O-type star, spherically symmetrical accretion may still take place in the expanding compressed layer, since the material falling into the initial condensation can have had initially only a very small velocity component relative to this initial condensation, because they both form part of the outward-moving layer surrounding the O-type star.

In discussing the accretion theory we have neglected the effect of the radiation of a star on its rate of accretion, a point which has been considered by a number of authors (47), (48), (49). Qualitatively this will have the following effect. Provided that the conditions of very low relative velocity and high density are satisfied, so that initially the rate of accretion is greater than the rate of mass depletion, the mass will grow and subsequent changes in the star's internal structure will mean that its luminosity will increase. Thus its output in the ultraviolet will increase and consequently the radius of the ionized sphere surrounding it will be increased. MESTEL (47) has shown that if the accretion rate is to be large, the radius of the ionized sphere must not be more than about 10^3 stellar radii. The absorbing power of a cloud is very much increased by the high densities near the star, so that a cloud whose mean density is quite high ($\gtrsim 10^{-21}$ g/cm^3) is able to keep the ionized sphere sufficiently small for appreciable accretion to occur. When a star has accreted to a critical mass, the ionized sphere increases rapidly to the value given by STRÖMGREN and accretion effectively ceases. The effect of radiation is thus to restrict further the conditions under which accretion is a significant factor in a star's evolution. However, if accretion takes place on to a condensation which has been produced by gravitational collapse in a dense cloud, then the protostar will probably be so young that isothermal conditions will still prevail in the contraction. Thus the accretion mass will be very cool, and the effect of radiation can be neglected.

It is of interest to point out that accretion, while unimportant in the evolution of the main body of stars in the Galaxy, may, in the case of stars which are not mixing, have the effect of disguising their true chemical composition. For example, a white dwarf of mass $1\,M_\odot$, $v = 10\,\text{km/sec}$, moving in the disk of the Galaxy through gas of mean density $10^{-24}\,\text{g/cm}^3$, will accrete at a rate of 10^{11} g/sec. In a time of 5×10^9 years such a star will accrete about 1.5×10^{28} g or less than 10^{-5} of its initial mass. However, this is very much more than the normal fraction of the mass contained in the atmosphere, consequently the material accreted will gradually form a new atmosphere. Now the white dwarf may easily be composed almost entirely of helium, while the atmosphere which it has accreted will be composed predominently of hydrogen. Consequently, white dwarfs may appear from their spectra to be rich in hydrogen, whereas in fact they are hydrogen-poor. Such effects may explain some of the white dwarf types described by GREENSTEIN (50).

This concludes our discussion of the theories of star formation which have been proposed in recent years to explain the current births of stars and the observations relating to them. It will be seen from this account that there is, as yet, no adequate theory which takes into account all of the factors involved. In particular, no description of the fragmentation process which is nearly always assumed to occur after the initial gravitational collapse has taken place is yet available. Also the effects of rotation have not been discussed adequately in describing star formation in general.

Before leaving this topic and turning to the evolution of stars after thay have contracted out of the interstellar medium, it is of interest to describe a theory proposed by HOYLE (51) to explain the formation of the oldest groups of stars. Although he considers initial conditions which are not applicable to the disk of our Galaxy today, the ideas are of interest, (a) because an attempt is made to provide an explanation of the formation of the globular cluster stars which we shall discuss in considerable detail in the following chapters, (b) because some ideas concerning a possible fragmentation process are introduced, and (c) because a discussion of the contraction of a protostar up to the point at which it becomes opaque to radiation is given.

11. Protostar formation in the absence of stars and dust

HOYLE has considered an extragalactic cloud of atomic hydrogen with a density $\sim 10^{-27}$ g/cm^3, and has shown that the thermal properties of atomic hydrogen lead to a dichotomy of temperatures. It is supposed that the heat energy of the cloud has been generated from the kinetic energy of mass motions which developed when the cloud formed. Only initial mass motions with velocities ≥ 10 km/sec are considered. The two temperature ranges that this leads to are 10000—25000 degrees and $\geq 3 \times 10^5$ degrees. The condition for contraction is that

$$2K + 3(\gamma - 1)U + \Omega < 0,$$

where K is the energy of mass motions.

Now in this model the mass motions have already been dissipated in heating the hydrogen, and if we put $\gamma = 5/3$, the condition for contraction for a mass M of radius R is that approximately

$$\frac{GM^2}{R} > 3\Re MT$$

or

$$\varrho > 3 \times 10^{-21} T^3 \left(\frac{M_\odot}{M}\right)^2. \tag{18}$$

For the situation in which $\varrho = 10^{-27}$ g/cm^3, and $T = 1.5 \times 10^4$ degrees (corresponding to the situation in which the hydrogen is about 2/3 ionized) we find that the critical mass for a condensation is given by

$$M_c = 1.4 \times 10^{10} M_\odot. \tag{19}$$

No cloud less massive than this can condense to form a galaxy. If we now consider the condensation of a cloud with mass $> M_c$, we can consider two types of compression, isothermal and adiabatic. Under adiabatic conditions all of the gravitational energy is converted to thermal energy, so that if U_0 and Ω_0 are the initial values of U and Ω, $\delta U = \delta \Omega$ for each contraction step, and $U - U_0 = \Omega - \Omega_0$. For equilibrium $\Omega = 2U$, so that the adiabatic compression leads to a rise in the thermal energy until hydrostatic equilibrium is reached, a situation which obtains for stellar masses.

In an isothermal contraction in which the dimension is decreased by a factor x, the gravitational energy released is of the order of $GM^2(1-x)/R$ while the thermal energy generated by direct compression is of the order of $3M\Re T \log_e x$, and to maintain isothermal conditions this energy must be removed from the gas. In order that the gravitational energy is dissipated in this way these two quantities must be nearly equal, and this demands that the condensation must be only just unstable against contraction, and it must not contract too far.

If $GM^2/R \gg 2U$, i.e., very little of the gravitational energy is converted to heat, this gravitational energy must be dissipated into dynamical motions, and the dynamical pressures thus exerted must be almost sufficient (apart from the heat and viscous dissipation) to re-expand the cloud to its initial dimensions. Thus the condition that the gravitational potential energy greatly exceeds the thermal energy at equilibrium leads to the condition that there will be no permanent contraction of the cloud as a whole. However, within the cloud, density fluctuations will occur, which do satisfy the condition $GM^2/R \approx 2U$. These regions, therefore, can contract by factors of 2 or 3, forming sub-condensations within the cloud, and within these, further condensations can take place. A large cloud can in this way fragment very rapidly. Since at each stage the density will increase by a factor of x^3, and since the time scale of the contraction is of the order of $t_0 = (G\varrho_0)^{-1/2}$, the time required for a large number of stages is given by

$$t = t_0(1 + x^{-3/2} + x^{-3} + x^{-9/2} + \ldots) = t_0(1 - x^{3/2})^{-1}. \tag{20}$$

If $x = 3$, the total time will increase over that for the initial step by only about 20%. The important conclusion derived from this result is that the condensation of galaxies, and the condensation of the first stars within them, are inextricably mixed together, as far as the initial time scale and the early dynamical effects are concerned. With the initial conditions set by HOYLE, this time is of the order of 10^9 years for galaxies with $M > 1.4 \times 10^{10} M_\odot$.

The fragmentation process must cease when the density of the object has reached such a point that the isothermal condition fails; the opacity becomes so high that the radiation is not able to escape freely. To determine the mass of the system at this point we can proceed as follows.

The opacity of the fragment arises partly from scattering and partly from absorption, and in the present case the scattering is unimportant. Absorption becomes important when the absorption in the radiation field becomes comparable with the energy radiated by the gas. Thus the transition to the adiabatic condition implies that the black-body distribution has been built up. The fragment then becomes similar to a star, radiating at a rate L, and for the transition we require that $d\Omega/dt > L$. The gravitational energy released per fragment in the nth step of the fragmentation is given by

$$\frac{GM_0^2}{R_0 x^{3n/2}}(x-1),$$

the release occupying a time of the order of $x^{-3n/2}(G\varrho_0)^{-1/2}$, and the rate at which energy is radiated is given by $L x^{-3n/2}(G\varrho_0)^{-1/2}$. Therefore the transition comes when

$$L = \frac{G^{3/2} M_0^2 \varrho_0^{1/2}}{R_0}. \tag{21}$$

The outward flux at distance r from the center is given by

$$\frac{16 \pi a c r^2 T^3}{3 \varkappa \varrho} \frac{dT}{dr}. \tag{22}$$

If the fragments are composed of pure hydrogen $\varkappa$ becomes extremely small for temperatures below about 5000 degrees, since even in the presence of blackbody radiation the hydrogen would be predominantly neutral. Then the luminosity $L = 4\pi a c R^2 T^4$ at the surface. Thus for the plausible values of the initial mass of the cloud from which this fragmentation takes place, it is possible to calculate the final masses of the protostars if they have condensed from pure hydrogen. However, it is important to consider the effects when a small proportion of the mass is composed of non-hydrogenic material, since the opacity is considerably changed by such an admixture, because the low ionization potentials of the metals provide enough electrons at the temperature range 3000—5000 degrees for the formation of H⁻. In this case Hoyle has argued as follows. The expression for the outward flux becomes, using a value of $\varkappa$ obtained from the tables of Chandrasekhar and Münch[1] and assuming $T = 4500°$,

$$L = \frac{160 \pi a c R_0 T_0^3}{32 \, y \, \varrho_0^2 \, \mathfrak{R}} x^{-3n}, \tag{23}$$

and equating this with the value when the adiabatic condition is reached, it is found that

$$x^{3n} = \frac{160 \pi a c G^{1/2}}{27 \, y \, \mathfrak{R}^3} \frac{T_0}{\varrho_0^{3/2}} \left(\frac{GM_0}{R_0} \approx 3 \, \mathfrak{R} \, T_0 \right) \tag{24}$$

where y is the number ratio of metals to hydrogen. Thus if $\varrho = 10^{-27}$ g/cm³ and $T_0 = 10^4$ degrees, this gives

$$x^{3n/2} = 2.3 \times 10^9. \tag{25}$$

The mass of the fragments when this final stage is reached is given by $M_s = M_0/x^{3n/2}$, and if $M_0 = 3.6 \times 10^9 \, M_\odot$, $M_s \approx 1.6 \, M_\odot$. On the other hand, we have for a pure hydrogen star that

$$L = 4\pi a c T^4 R^2 = \frac{G^{3/2} M_0^2 \varrho_0^{1/2}}{R_0}. \tag{26}$$

Putting

$$R = R_0 x^{-3n/2}, \quad \text{and} \quad R_0^3 = \frac{3 M_0}{4 \pi \varrho_0},$$

and substituting, we find that

$$x^{-3n} = \frac{G^{3/2} M_0 \varrho_0^{3/2}}{3 \, a c \, T^4}. \tag{27}$$

---

[1] Chandrasekhar, S., and G. Münch: Astrophysic. J. **104**, 446 (1946).

Thus if $M_0 = 3.6 \times 10^9 M_\odot$, $\varrho_0 = 10^{-27}$ g/cm^3 and $T = 5 \times 10^3$ degrees,

$$x^{3n} = 1.10 \times 10^{20}, \quad \text{and} \quad x^{3n/2} = 1.05 \times 10^{10}.$$

Thus the masses of pure hydrogen protostars, M_s, are given by

$$M_s = \frac{3.6 \times 10^9 M_\odot}{1.05 \times 10^{10}} = 0.34 \, M_\odot. \tag{28}$$

These results are of considerable interest, but the numerical values obtained are not of especial significance. The following facts must be borne in mind. The critical masses which have been used by HOYLE have been derived for plausible values of the initial density and temperature, and have led to values of the masses of old stars which may be reasonable. However, in many cases the masses of elliptical systems are much greater than $3.6 \times 10^9 M_\odot$. For example, the average masses of galaxies in the Virgo cluster are near $10^{11} M_\odot$ and these are mainly elliptical systems, and if their stars are completely analogous to the globular cluster stars, their average masses are near $1.5 M_\odot$. To reconcile these results with those derived above, it is necessary that there be a considerable degree of fragmentation which takes place in an initial cloud of $\sim 10^{11} M_\odot$ and which does not result in the production of a cluster of masses of $\sim 10^8 M_\odot$ but more likely a series of fragments that remain bound to each other, and then the stars which condense during the further fragmentation lose their identity with respect to a single fragment. HOYLE has treated the case of more massive systems ($\sim 5 \times 10^{11} M_\odot$) in which he supposes that initially the high-temperature condition ($T \geq 3 \times 10^5$ degrees) held. In these cases he has shown that a large number of dwarf systems with masses $\sim 10^8 M_\odot$ will be formed. (In these, the later fragmentation stages are similar to those above.) He has supposed that the later dynamical evolution of these will lead to shrinkage, disintegration, and the escape of some systems. These might be identified with some clusters that have been discovered in our local group of galaxies, at least two of which have color-magnitude diagrams similar to that of a globular cluster (52), (53).

Furthermore, it is also probable that in these condensation processes, much gas will be left over, since the fragments will only condense in regions having higher than average densities. Thus this gas may condense under suitable dynamical conditions at a later epoch, and, for example, the presence of magnetic fields may also slow up this further condensation (54). Neutral atomic hydrogen has been detected through 21-cm observations in both the Coma, Hercules, and Corona Borealis clusters of galaxies (55); of these the Coma and Corona clusters are highly condensed and consist predominently of S0 systems. This might well be primeval hydrogen contributing a considerable proportion (perhaps ~ 20% in the Coma cluster) of the total mass. In some galaxies of the local group a considerable amount of gas, comparable in mass to the stars, is present. Part of this may be a remnant of the original cloud. However, in general the fraction of the mass contained in gas within galaxies and perhaps within clusters as well is an index of the distance travelled along the evolutionary path of the galaxy or the cluster as a whole.

This treatment has not taken into account the type of condensation which may occur if a fairly uniform contraction occurs with the development of dynamical motions and the turbulent dissipation of these motions. Thus the effect of the contraction may be to amplify the turbulence so that if sufficiently large density fluctuations were set up, the fragmentation and subsequent contraction would take place through them. This point demands quantitative investigation.

Another alternative is that the condensation becomes adiabatic at an earlier stage than has been suggested by HOYLE, due to the presence of molecules which absorb strongly in the infra-red. Then the condensation will be very slow and fragmentation may not occur appreciably. Other sources of opacity which have been neglected are the excitation of neutral hydrogen giving rise to the 21-cm radio radiation, and the extra ionization produced by cosmic ray particles (whose effect is hard to estimate at an early stage of galactic evolution).

These factors might lead to the condensation of stars, even those made of pure hydrogen, having masses far larger than those derived above. If all pure hydrogen stars are of very small masses, considerable problems arise when the chemical evolution of the galaxy is considered.

In this treatment the angular momentum of a protogalaxy and the magnetic field which may exist in the initial cloud have both been neglected. Both of these must be important at all stages of the fragmentation process and also in determining the final mass beyond which fragmentation does not occur.

The final stage of the hierarchy described by HOYLE may be called a protostar. If we take as an example the final mass of $1.6 M_\odot$, the central density is $\varrho_0 x^{3n}$, which for the value $x^{3n} = 5.3 \times 10^{18}$ gives a central density of 5.3×10^{-9} g/cm^3, while the central temperature is approximately 10^4 degrees. The radius of the protostar is about 5×10^{13} cm. This is very far from the conditions that exist when a star moves on to the main sequence. Some attempt at describing these intermediate conditions has been made by McVITTIE, and these will be discussed at the beginning of Chapter II.

II. Theoretical evolutionary tracks
12. Gravitational contraction of protostars on to the main sequence

So far we have described the various proposals which have been suggested for the formation of protostars. Once these have formed the stages of contraction and fragmentation will continue to take place, until the protostar becomes opaque and the adiabatic condition is fulfilled. Further contraction will then occur at a rate which is determined by the opacity of the stellar material, i. e., by the rate at which the gravitational energy can be released. It is this stage of the evolution, taking the star almost on to the main sequence, which we shall discuss here. However, before doing this it is necessary to describe briefly a paper by McVittie (56). He has discussed in mathematical terms the stages of contraction from the initial state, in which the compression in the interstellar gas allows the Jeans criterion to be fulfilled, down to stellar dimensions.

McVittie has described the non-adiabatic contraction of a gas cloud to a complete polytrope, using a method of gas dynamics with gravitation. The assumption is made that the collapse follows a kind of homology transformation in which the change of temperature is not uniform. A large number of models are possible, depending on the kind of time dependence assumed for the inward gasvelocity, and no important conclusions bearing on the most probable mode of contraction can be drawn. The conclusion that the initial temperature has to be very low for low-density contractions to proceed, and that there is therefore a dependence $T \propto \varrho^{1/3}$, is more simply derived from the simple statement of the Jeans criterion.

A wide range of initial densities and temperatures have been treated without considering the limitations imposed by the Jeans criterion. Any discussion of the contraction of a single mass to stellar dimensions is irrelevant if the initial conditions assumed are such that the critical mass is much greater than 100 $M_\odot$, since stars of masses greater than this probably do not exist. This imposes limits on the initial density considered, from 10^{-20} g/cm³ for a low-temperature limit of 30 degrees to 3×10^{-11} g/cm³ for $T \approx 10000$ degrees.

The phase of evolution after the protostar has become opaque to radiation is the so-called Kelvin-Helmholtz contraction phase. An estimate of the Kelvin-Helmholtz contraction time, t, can be made as follows (57). Within the star

$$E = -(3\gamma - 4) U = \frac{(3\gamma - 4)}{3(\gamma - 1)}, \qquad (29)$$

where E and U are the total and the internal energies. If the configuration contracts so that the potential energy is changed by $\Delta \Omega$, a fraction

$(3\gamma-4)/3(\gamma-1)$ is radiated while the internal energy gain is $\Delta U = [3(\gamma-1)]^{-1}\Delta\Omega$. Thus the luminosity is given by

$$L = -\frac{(3\gamma-4)}{3(\gamma-1)}\frac{\Delta\Omega}{\Delta t}. \tag{30}$$

Now the condensation has contracted from a condition with radius R_0 (given by the conditions of star formation) to a condition with radius R, in time t, and $R_0 \gg R$. If the configuration is a polytrope of index n,

$$\Omega = \frac{3GM^2}{5-n}\left(\frac{1}{R}-\frac{1}{R_0}\right) \approx \frac{3GM^2}{(5-n)R}$$

so that

$$\int_0^t L\,dt = -\frac{(3\gamma-4)}{(\gamma-1)(5-n)}\frac{GM^2}{R} = \bar{L}t, \tag{31}$$

where $\bar{L}$ is the mean luminosity, or

$$t = -\frac{(3\gamma-4)GM^2}{(\gamma-1)(5-n)R\bar{L}}. \tag{32}$$

Thus, for example, taking a model which is a crude approximation to the solar model, we put $n = 3$, $\gamma = 5/3$, and $\bar{L}$ equal to the observed luminosity of the sun and find that $t = 2.4 \times 10^7$ years.

However, this ignores the fact that as the star contracts its luminosity is a function of its radius, the form of this function depending on the stellar model. It has been shown by THOMAS (*58*) that one of the conditions which must be fulfilled if a gas sphere is to contract homologously is satisfied if the opacity has the KRAMERS form and $\gamma = 5/3$. Further conditions for the stability of such a contraction have been stated by him and by ROTH (*59*), who has made an exact integration. Such a model, neglecting radiation pressure, has been described by LEVÉE (*60*) who has shown that a homologously contracting gas sphere has an effective polytropic index ranging from 2.97 at the center to 3.25 at the surface, so that putting $n = 3$ is a good approximation. However, the form of the mass-luminosity relation for a model with KRAMERS opacity is

$$L \propto M^{5.5}\mu^{7.5}R^{-0.5}. \tag{33}$$

It is easily shown that the rate of shrinkage onto the main sequence, dR/dt, is proportional to LR^2 for constant mass and chemical composition, and integration of the model of LEVÉE for the sun gives

$$t = -2.52 \times 10^7 \int_{R_0/R}^{1} \frac{dR}{R^2} = 2.52 \times 10^7 \text{ years},$$

in good agreement with 2.4×10^7 years obtained above. However, if we put $L \propto R^{-1/2}$ from equation (33), we find that

$$t = -2.52 \times 10^7 \int_{R_0/R}^{1} \frac{dR}{R^{3/2}} = 5.04 \times 10^7 \text{ years.}$$

Thus the effect is to double the times obtained using the final luminosity L. When the mass, radius, and luminosity are measured in solar units and all models are supposed to be homologous, the total time for gravitational contraction is given by

$$t = 5.04 \times 10^7 \frac{M^2}{L} \text{ years} \tag{34}$$

for stars which approximate to the model with $n = 3$. Similar calculations to these have been given by SANDAGE (61).

A number of tracks of gravitational contraction have recently been computed by HENYEY, LELEVIER and LEVÉE (62), using detailed opacity tables, for a series of masses and chemical compositions. The initial configurations have all been obtained by homology transformations from the model of LEVÉE. These authors have stated that their starting method necessarily leads to the introduction of certain transients which quickly disappear, and they also remark that a considerable latitude of choice (in starting models) leads after a few time-steps to very nearly the same results. Their starting models must always satisfy the conditions of dynamical and thermodynamical stability together with the third condition of THOMAS. It would not appear to be out of the question that protostar condensations, in moving from the isothermal to the adiabatic condition, begin to contract but reach configurations which violate the stability conditions. Their further development is not known but it may be that this instability leads to fragmentation.

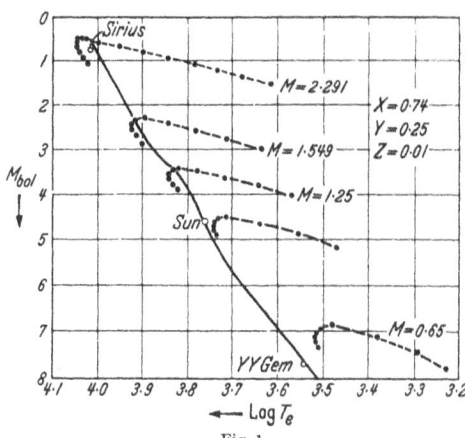

Fig. 1.
Gravitational contraction tracks for different stellar masses

Tracks given by HENYEY et al. are shown in Fig. 1. The luminosity slowly increases as the radius decreases, so that the tracks move from right to left. The approximate equation for the track is

$$\Delta M_{bol} = -2 \Delta \log T_e, \tag{35}$$

given by SANDAGE (61). The tracks reach a maximum value of M_{bol}

which approximately is the point at which thermonuclear energy release becomes incipiently significant. When this occurs the region very near to the center ceases to contract, and the rate of gravitational energy release is reduced, but the thermonuclear source is not able to make up this deficit completely. Thus the value of M_{bol} is reduced. At the point at which the thermonuclear source becomes important, convection may set in. An earlier investigation by HARRISON[1] suggested that even in the stage of purely gravitational contraction a convective core would appear, but this result is in error[2].

The time scales associated with each of the tracks in Fig. 1 are given in Table 3. The times t_1 and t_2 are, respectively, the time of maximum luminosity and the time at the end of the track (the end has been defined as the point at which changes on the original time scale become imperceptible).

Table 3. *Time scales for gravitational contraction of different masses. Times at maximum luminosity and at the end of the track are t_1 and t_2, respectively*

Mass ($M_\odot$)	t_1 (maximum luminosity)(years)	t_2 (end of track) (years)
0.65	7×10^7	1.5×10^8
1.00	1.6×10^7	3×10^7
1.25	8×10^6	1.4×10^7
1.549	4×10^6	8×10^6
2.291	1.8×10^6	3×10^6

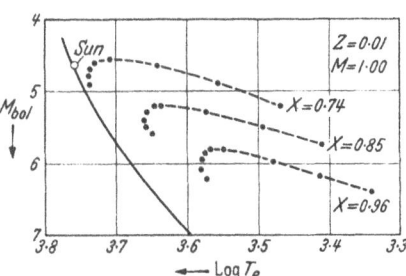

Fig. 2. Gravitational contraction tracks for different concentrations of hydrogen

Tracks for a constant mass in which the amount of hydrogen has been varied have also been computed. They show that increasing the proportion of hydrogen reduces the maximum value of M_{bol} achieved in the gravitational contraction. Such tracks are shown in Fig. 2. Similar displacements are found when variations in Z, the proportion of the heavy elements, are put in.

Work foreshadowing that of HENYEY et al. on gravitational contraction and the color-magnitude diagrams which are to be expected for young clusters was carried out by SALPETER (*63*), and we shall refer to this again when we discuss these color-magnitude diagrams in Chapter III.

Just prior to their arrival on the main sequence, the internal temperatures of the stars will have risen high enough so that they can burn the small amounts of deuterium, lithium, beryllium, and boron in their cores. Following this stage, nuclear energy generation either by the

---

[1] HARRISON, M. H.: Astrophysic. J. **102**, 216 (1945).
[2] Cf. E. J. ÖPIK: Contr. Baltic. U. No. 35 (1947).

pp-chain or the CN-cycle, depending on the stellar mass, will begin. Revised energy generation rates have recently been given (64), (65). They show that for stars above $\sim 2\,M_\odot$ the CN-cycle dominates, while below this the pp-chain is all-important. Once nuclear energy generation has begun the chemical evolution of the star is underway. The factors which govern its further evolution are the extents to which mixing and mass loss can take place. A number of possible situations can be envisaged:

(a) Complete mixing between energy-generating core and envelope, and no mass loss.

(b) Complete mixing between core and envelope with steady mass loss.

(c) No mixing between core and envelope and no mass loss.

(d) No mixing between core and envelope but steady mass loss.

In the following sections we shall discuss first the mixing problem, and then its bearing on the possible situations (a) to (d).

13. Mixing in main-sequence stars

An initially chemically homogeneous main-sequence star will begin to increase its molecular weight μ as it transmutes hydrogen to helium. If it can remain completely chemically homogenous so that μ increases throughout the whole star uniformly, and if its mass remains constant (this is situation (a) referred to above), then from the mass-luminosity-composition relation and the mass-radius-composition relation it can be shown that a star will move upward and to the left of the main sequence in the H-R diagram, accelerating because equal luminosity increments will occur in decreasing intervals as the luminosity increases. This is the type of evolution which was originally considered by GAMOW and his collaborators (66). In order that such a uniform increase can take place, it is necessary that mixing must take place at a rate which is fast compared with the rate of change of μ in the energy-generating core. There are many mechanisms by which such mixing might take place.

(α) *Convective mixing.* In a zone in which convection is important it can be supposed that mixing is both rapid and efficient. In the absence of rotation this is deduced immediately from the properties of the turbulent mixing currents. Thus it is reasonable to suppose that in stars with convective cores, the cores are chemically homogeneous, while in stars in which there is a deep outer convective zone extending inward from the stellar atmosphere, the chemical composition of the atmosphere is characteristic of the whole of this outer zone. However, the only stars which may be convective throughout are the very faintest and least massive stars on the main sequence (67), (68). It appears, therefore, that these are the only stars which could evolve upward and to the left in this manner. However, the luminosities of these stars are so small that the time taken to transmute a large proportion of their mass is very large; in the case of Kruger 60 A, a red dwarf, this is about 10^{12} years. Consequently this mode of evolution is of no practical significance.

(β) *Rotational mixing.* In the envelope of a non-rotating star in which radiative energy transport prevails, mixing is negligibly small since the diffusion effects are unimportant over timescales $\leq 10^{10}$ years *(69)*. However, in a rotating star currents will be set up in planes through the axis of rotation, the degree of mixing depending on the equatorial velocity. Investigations by SWEET and ÖPIK *(70)*, *(71)*, have shown that the estimate of the amount of mixing by EDDINGTON *(72)* was far too large. On the basis of the mixing time for a Cowling model calculated by these workers, STRÖMGREN *(73)* estimated massive stars with abnormally high rotational velocities, i. e., the Oe and Be stars, would be well mixed while even some of the normal O and B stars might also be thoroughly mixed. However, more recently MESTEL *(74)* has pursued the subject further and has concluded that the conditions for mixing are even more stringent than those derived by SWEET and ÖPIK. He has concluded that no continuous mixing between core and envelope can take place in a uniformly rotating Cowling model since the non-spherical distribution of matter set up by the rotational currents themselves tends to choke back the motion. If the rotation increases inward sufficiently the star will build up a mixing zone of definite mass. If the inner region is constrained by a magnetic field to rotate uniformly the mass of the mixing zone is very sensitive to increases of angular velocity above a certain limit. In a star in which the angular velocity decreases outward, however, the mass of the mixing zone is likely to increase slowly with increasing angular momentum. Thus in starting from the Cowling model in calculations of stellar evolution, it is usually supposed that rotational mixing does not take place. However, MESTEL has also shown that the mass of the mixing zone is determined by a parameter which varies as the star's dimensions change. The efficiency of mixing increases as the star contracts and a star which is unable to mix in its initial state with a convective core may possess a mixing zone after contraction into a shell source state.

14. Evolution of stars down the main sequence

An idea basic to some theories of stellar evolution is that proposed by FESSENKOV *(75)* and developed by MASSEVICH, PARENAGO, SOROKIN, KRAT and others. This is that stars condense on to the main sequence in certain mass ranges and then evolve along it. The idea appears to have been developed in order to overcome a difficulty associated with the theoretical mass-luminosity-composition law for homogeneous stars, and the observed mass-luminosity relation. It has been pointed out that the very strong dependence of luminosity on μ means that, for example, a star which condenses with $\mu = 0.53$ ($X = 0.90$, $Y = 0.09$, $Z = 0.004$), as compared with a star with $\mu = 0.62$ ($X = 0.70$, $Y = 0.28$, $Z = 0.02$), will have a luminosity a factor of about 3 smaller. Thus there should be some scatter in the mass-luminosity relation if the stars remain throughly mixed, because as stars age, μ will steadily increase. Alternatively, there should also be scatter if there is a spread in chemical composition in gas clouds of the same age out of which stars condense. If the observed mass-luminosity relation is very tight, so that such scatter is precluded by the observations, then it has been argued that the solution is that stars eject mass, thus becoming less massive. It is then postulated that they eject mass at a rate proportional to their luminosity in order that they remain on the mass-luminosity relation. As will be discussed in Chapter III, the formation of stars in two discrete mass ranges could then populate the main sequence.

A detailed discussion based on the general arguments has been given by MASSEVICH (76). She took standard models burning on the CN-cycle, with a convective core and radiative envelope (the basic models were those of SCHWARZSCHILD[1] and HARRISON[2]). It was assumed that throughout the evolution the star remained thoroughly mixed. Three sets of evolutionary tracks along the main sequence were computed — each was derived from a slightly different initial model, but the three sets of tracks differed very little — and they were compared with suitable main sequence observations. The agreement between theory and observation was fairly good. For each track a limiting mass was obtained, essentially from the method of calculation, above which the calculations had no validity. The limits were $5\,M_\odot$, $8\,M_\odot$ and $15\,M_\odot$, respectively. Clearly, as the stars move along such an evolutionary track, the hydrogen/helium ratio decreases, but the heavy element composition remains constant. At the time that these calculations were carried out, it was thought that such a standard model would adequately represent the sun. Thus MASSEVICH concluded from these arguments that the sun had originally reached the main sequence as a star of between 5 and $8\,M_\odot$. The computed changes in mass and luminosity with time were plotted; the initial decrease in mass was large, but slowed down as the star decreased in luminosity, so that in the last 3×10^9 years it was practically unchanged. Its maximum lifetime was estimated to lie between 8.7 and 10×10^9 years. Over the range of applicability of the calculations, the relation

[1] SCHWARZSCHILD, M.: Astrophysic. J. **104**, 203 (1946).
[2] HARRISON, M. H.: Astrophysic. J. **108**, 310 (1948).

between dM/dt and L was found to be approximately linear, as had initially been postulated by FESSENKOV.

The rate of mass loss was considered further by MASSEVICH (77), assuming that the stars between O 8 and G 4 are constrained to follow the empirical mass-luminosity relation. She found that a star of mass $14.2\,M_\odot$ with an initial hydrogen content $X = 0.88$ must lose mass at an average rate of $0.0064\,M_\odot$ per 10^6 years until its mass is $2.53\,M_\odot$. The computed loss for a solar mass is $0.07\,M_\odot$ per 10^9 years. If the rate of mass loss is written in the form

$$\frac{dM}{dt} = -\frac{k}{3}L,$$

this relation, together with a mass-luminosity relation of the form

$$L = M^4 \times \text{const.},$$

leads to a theoretical luminosity function which is in agreement with the observed luminosity function for the solar-neighborhood stars.

More recently MASSEVICH (78) has considered various evolutionary tracks for completely mixed models with mass loss, for a range of stellar masses, for different assumed values of Z, and for a variety of assumed opacity laws. She has also considered ranges of models in which the mass has been kept constant.

At this point it is necessary to make some general criticisms of the postulates upon which evolution of stars down the main sequence is based, and upon which some models have been constructed.

The rate of mass loss can be calculated from the models and the observed mass-luminosity relation, once the form of the relation has been assumed. The basic justification for putting

$$\frac{dM}{dt} = -\frac{k}{3}L$$

appears to be that it is possible in this way to populate the main sequence starting from only one or two discrete masses for condensing stars. If it is assumed that there is a whole spectrum of masses in the process of star formation, then this postulate is no longer necessary. There would appear to be no theoretical justification for assuming such a law, and no observational evidence that mass ejection is proportional to luminosity is available. The energy required to eject mass from the surface of the star must come from nuclear sources, and it can be shown that in this case we would not expect in general that a linear relation between dM/dt and L would hold.

A more serious difficulty associated with the models just discussed is that they all depend on the assumption of complete mixing. As we described earlier, work carried out by SWEET, ÖPIK, and MESTEL has led conclusively to the result that mixing in radiative zones is never of importance in the time scales which are of interest.

Evolution of stars in which both mass loss is assumed, and in which chemical inhomogenities are developed, i. e., mixing does not occur, will be described later.

15. Stellar models with inhomogeneities in chemical composition

The modern theory of the evolution of stars off the main sequence, worked out in recent years by SCHWARZSCHILD, HOYLE and their many collaborators, depends on the development in an originally homogeneous star of a core having a molecular weight, μ_c, different from that of the envelope, μ_e. Thus in this theory one of the basic postulates on which Gamow's early theory is based (that of complete mixing) (66) has been discarded. These investigations take us right up to 1956. However, it is worth recalling that in the later sections of a paper written in 1943 GAMOW (79) anticipated in a qualitative discussion some of the developments that have come since.

During the last twenty years work on models with a chemical discontinuity between core and envelope has been carried out by a large number of workers in order primarily to elucidate the problem of the structure of the red giants (80). From the evolutionary standpoint it is important to consider sequences of models which represent the transition from the main sequence models to the red giants. Key investigations were therefore those of SCHOENBERG and CHANDRASEKHAR (81) and HARRISON (82). They investigated the equilibrium configurations of stellar models having (i) convective cores in which the energy was generated and radiative envelopes with different molecular weights, and (ii) isothermal cores and radiative envelopes with different molecular weights. They used KRAMERS opacity and neglected degeneracy. The major conclusions were as follows.

As the ratio of the molecular weights μ_c/μ_e increases with time, the radius and mass of the convective core both decrease. At some epoch the energy generation in the core ceases, the convection stops, and the core becomes isothermal, so that the series of models with isothermal cores must be used. The calculations then showed that for each assumed value of μ_c/μ_e there is a maximum mass of the isothermal core beyond which no fit can be made with the radiative envelope to make an equilibrium model. This critical mass is a function of μ_c/μ_e. It has its largest value for a homogeneous star but diminishes to a lower limit, determined by these authors to be 0.1 of the mass of the star, when $\mu_c/\mu_e = 2$. This is known as the SCHOENBERG-CHANDRASEKHAR limit. When a star has reached this limit its only nuclear energy production is confined to a thin shell between the isothermal core and the radiative envelope.

Following the work of SCHOENBERG and CHANDRASEKHAR, who did not take into account possible degeneracy in their models with isothermal cores, GAMOW and KELLER (83) considered isothermal cores with partial degeneracy and concluded that under certain conditions such models could give radii large enough to fit the red giants. The result was contested by HARRISON who computed a number of partially degenerate models that did not have large radii. More recently HAYASHI (84) and SCHWARZSCHILD, RABINOWIZC, and HÄRM (85) found that under special conditions models with partially degenerate cores could give rise to models fitting well the radii and luminosities of red giants. However, they concluded that the conditions, which demand that the fraction of the mass in the exhausted core lie between very narrow limits (for smaller fractions the star lies near the main sequence, while for larger fractions it appears probable that no equilibrium configurations exist) are so stringent that partially degenerate cores are only a contributing factor, and not the main cause, for the large radii of red giants.

An important step forward was made by SANDAGE and SCHWARZSCHILD (86), who considered the quasi-equilibrium states through which an unmixed model passes after reaching the Schoenberg-Chandrasekhar limit. Their starting point is a model with an isothermal core which has reached the limiting mass, a shell source burning on the CN-cycle, and a radiative envelope. They then consider a series of models in which it is supposed that the hydrogen-poor core is contracting, so that an additional energy source is present in the core; this core is now in radiative equilibrium. It is found that as the cores contract the envelopes of these models greatly expand, while the shell source remains exceedingly thin, about 90% of the energy generation arising in a thickness of only 0.4% of the radius. As the core contracts only part of the gravitational energy is released as radiation flux, the remainder going to increase the internal energy of the core, so that the central temperature increases as the radius increases. This sequence of models suggests that stars originally on the main sequence will, after developing inhomogeneities in the cores, evolve in the following way. The core will grow until the Schoenberg-Chandrasekhar limit is reached. SANDAGE and SCHWARZSCHILD point out that the fraction, q, of the total mass at which this occurs is about 12% for a hydrogen content, X, of 0.6. During the evolution from the chemically homogeneous main-sequence model the luminosity will increase slightly, becoming about 1 magnitude brighter, the radius will increase by about 70% from the main-sequence model, while the effective temperature remains nearly constant. Thus throughout this evolution the star will remain near the main sequence. However, the contraction of the core and the subsequent envelope expansion mean that the star will then move rapidly to the right of the main sequence in the H-R diagram.

The results of the theory are applied in Chapter III, when we consider the observations of color-magnitude diagrams and luminosity functions of star clusters. The limiting mass of 12% can be used in deducing the ages of clusters from the break-off point on the main sequence. Thus the age is given by

$$t = 0.007 \, c^2 \, \frac{XqM}{L} \tag{36}$$

where $Xq = 0.07$. This is approximately independent of the assumed hydrogen content of the envelope, X, since for varying X the jump in the molecular weight at the core-envelope boundary varies. It has been shown by HARRISON (82) that q varies in such a way that the product Xq is nearly constant.

This formula is only approximate, however, since it does not take into account a gradual brightening of the star as it goes to the Schoenberg-Chandrasekhar limit. In general the formula for the age can be written

$$t = 6.2 \times 10^{18} \, XM \int_0^q \frac{dq}{L} \tag{37}$$

and integrated from the starting point of the star on the main sequence. This relation has been used extensively to date a number of clusters, as will be discussed later. Uncertainties in the ages due to other factors will also be described in Chapter III.

The Sandage-Schwarzschild models suggest that the stars move extremely rapidly to the right after leaving the main sequence. They do not indicate that the envelope expansion ceases over the range that they have taken, and this is in disaccord with the observations of globular clusters. Neither do they indicate an increase in luminosity, i. e., a track both upward and to the right in the H-R diagram, which is demanded to explain the red giants in globular clusters. Thus these authors proposed that a new physical process must occur which they had not taken into account, and in this connection they speculated both about the onset of helium-burning and about the possible effects of mixing. However, to elucidate the further evolution we must consider the work of HOYLE and SCHWARZSCHILD. Before doing this we shall consider briefly an independent approach to the early stages of evolution, using the mass-loss theory which has been developed in the USSR.

Evolutionary tracks for homogenous and non-homogenous models, both with constant mass and assuming mass loss, have been constructed by SOROKIN and MASSEVICH (87). They consider three stages in the life of a single star. In its initial state it is a massive main-sequence star and it evolves down the main sequence by losing mass, remaining completely mixed throughout. Such evolution has been described in

Section 14, where objections to it have also been discussed. As the star reaches the region of the sun in the mass-luminosity diagram, the mass ejection becomes extremely small and the star enters a stage of "quiet evolution". In this stage, mixing apparently ceases and the star remains on the main sequence for a period estimated for $1\,M_\odot$ to be $6\text{—}8 \times 10^9$ years. The onset of this stage of quiet evolution is related in this theory to the break in the main sequence for the solar-neighborhood stars. The third stage of evolution apparently begins when a limit, similar to the Schoenberg-Chandrasekhar limit, is approached. Several possibilities are investigated. In one it is supposed that as the energy sources in the core are exhausted, or as the convection recedes, catastrophic contraction takes place. This must presumably lead to violent ejection and reorganization of the star's structure. In a second model it is supposed that the contraction takes place non-catastrophically. The luminosity increases, mixing takes place, probably some mass is ejected, and the star begins its quiet evolutionary phase again. However, now the value of X/Y will have decreased. A diagram showing these stages of evolution is reproduced in Fig. 3.

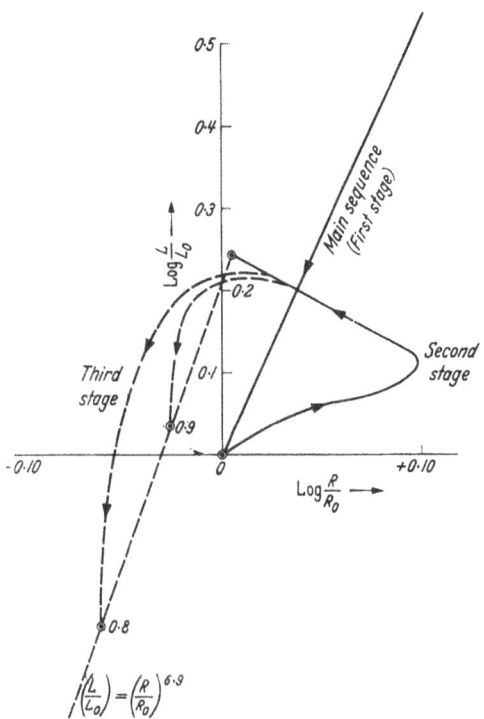

Fig. 3. Evolutionary track for a star, according to SOROKIN and MASSEVICH (87)

Further work along these lines was carried out by MASSEVICH (88), (89), (78), using non-homogeneous models with convective cores. She computed the times taken for stars of constant mass to reach limiting values of μ_c/μ_e, and values comparable with those of SANDAGE and SCHWARZSCHILD were obtained. Similar calculations were also carried out for stars which were also losing mass. In this case the time scales were found to be increased.

These authors do not appear to have discussed the interpretation of the color-magnitude diagrams for globular clusters (a discussion by

MASSEVICH of galactic clusters on the basis of the concept of mass loss is given in Chapter III). It would seem that the turn-off point which is the characteristic dating point for these clusters on the basis of the arguments of SCHWARZSCHILD, SANDAGE, and others, would be reached in this case, after all the stars originally above this point in the initial main sequence had lost mass and effectively ceased moving down the main sequence, and had then burned out their cores and moved to the right.

16. Evolutionary tracks of HOYLE and SCHWARZSCHILD

HOYLE and SCHWARZSCHILD (90) have attempted to construct evolving models which will explain the observed features of the color-magnitude diagrams of globular clusters (Fig. 4). Although they have termed their investigation a preliminary reconnaissance of the problem, it is at the same time the most ambitious attempt at understanding details of stellar evolution so far made. The major features are as follows. The tracks have been computed for stellar masses of 1.1 and 1.2 $M_\odot$; these correspond approximately to the masses at the top of the truncated main sequence in the observed color-magnitude diagrams. Of course the evolutionary track of a single star is not identical with the line occupied by the stars in a given cluster, since this latter is the locus for stars of a range of masses at a given time, while the former is the locus for stars of a unique mass at different times. The difference between these two is small, and the reason is essentially that the time taken for stars to evolve after they have left the main sequence is short as compared with the time that the stars spend on the main sequence; consequently only a narrow mass range is demanded to populate the whole observed H-R diagram.

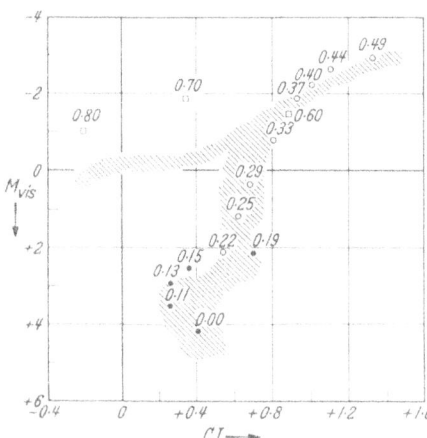

Fig. 4. Evolving stellar models of HOYLE and SCHWARZSCHILD (90), compared with observed color-magnitude diagram of globular clusters

In the initial stage, HOYLE and SCHWARZSCHILD have constructed models ($M = 1.1 M_\odot$) with isothermal partially degenerate cores, a thin energy-producing shell using the CN-cycle, and a radiative envelope; it is supposed that the core consists of essentially pure helium while the radiative envelope has a composition consisting predominantly of hydrogen. They have used the CN-cycle throughout, but the run of values of the shell temperature shows that initially the energy generation process must be due to the pp-chain, though the temperature soon increases so that the CN-cycle dominates. In this treatment they used Z (carbon and nitrogen) $= 0.005$, which they considered to be typical of the compositions of extreme population II stars. The sequence of models is shown in Fig. 4, where they are superposed on the mean diagram for M 3 and M 92. Models with this particular structure are represented by the closed circles; they are given up to a mass of the exhausted core $q = 0.19$. The next model obtained under those assumptions for $q = 0.22$ is not shown on this diagram, but it lay very far to the right of the other computed points, a result in agreement with the earlier work of SANDAGE and SCHWARZSCHILD (86). HOYLE and SCHWARZSCHILD have attributed the small discrepancies between theory and observation in the range up to $q = 0.19$ to the approximations that have been made in this region. For example, since the star starts off on the pp-chain, the material is not sharply divided in composition at the core boundary; this does not occur until the CN-cycle dominates. Also, they have divided the outer envelope into two zones, the outer one in which the Kramers opacity alone is operating, and an inner one in which the electron scattering alone is at work. Since, ideally, contributions from both sources of opacity should be taken into account throughout the envelope, the result of this has been that they have slightly exaggerated the luminosities. Slight changes in luminosity can also be obtained if the assumed composition of the envelope (90% hydrogen and 10% helium) were changed.

The failure of these models beyond $q = 0.19$ was attributed by them to their simplification of the boundary condition at the surface, i. e., that the density and temperature tend to zero at the surface. In fact the condition which must be fulfilled at the surface is that

$$T \to T_{eff}, \varrho \to \frac{\mu H G M}{k T R^2} \text{ as } r \to R,$$

i. e., T and ϱ approach the atmospheric values as the surface is reached. It has been argued that the envelope solutions which correspond to the idealized mathematical conditions at the surface differ negligibly in the interior of the star from the solutions that satisfy the boundary. However, in these models HOYLE and SCHWARZSCHILD have shown that this is only true for their models up to $q = 0.19$. Beyond this point the models fail badly in this respect. Stated in more physical terms, the models for q greater than 0.19 give a density which falls too soon as the surface is approached, i. e., when ϱ reaches the photospheric value, $T = T_1 > T_{eff}$. Now in a radiative zone the dependence of ϱ on T is of the form $\varrho \propto T^{3.25}$ whereas models taking into account the true surface condition require a dependence of ϱ on T which is smaller than this. More than this, in the models here studied there is a very thin convective zone immediately below the photosphere in which hydrogen is in a critical stage of ionization, and in this zone ϱ varies with T more rapidly than in a radiative zone, so that the discrepancy is even worse.

To overcome these difficulties it is proposed that the convection zone lying below the photosphere extends into the interior to a considerable depth. Below the hydrogen ionization level the temperature-dependence of the density is $\varrho \propto T^{3/2}$, a much slower decline than that in the radiative zone. Thus the star adjusts its photospheric density to the required value by adopting a structure with a sizeable outer convective zone, the parameter of adjustment being the depth of this zone.

Models have then been computed by elaborate techniques taking into account the effect of an outer convective zone on the structure of the star, and in this way it has been possible to fit the observed sequence in the globular clusters along the subgiant branch and up towards the giant branch. As the stars move along this sequence the calculations show that the temperature in the hydrogen-burning shell as it moves outward remains fairly constant at about 20×10^6 degrees. This situation holds until the luminosity of the star is about $100 \, L_\odot$. At this stage the core begins to evolve in such a way that the density in the shell tends to decrease. This effect, together with the necessity of increasing the rate of energy generation, means that the temperature must rise to 40 to 50×10^6 degrees as the star reaches the giant branch.

This rise in temperature does not in itself affect the isothermal condition in the core. Departure from the isothermal condition only occurs when gravitational energy is released. The rate of this release parallels the luminosity. Since increased energy generation means that the helium core grows faster, towards the top of the giant branch (as distinct from much lower down as was supposed by SANDAGE and SCHWARZSCHILD) it is found that the rate of energy generation by the gravitational process is sufficient to maintain a temperature gradient between the center and the periphery of the exhausted core. Still, however, the release of gravitational energy is a negligible source of energy for the star as a whole.

The increase of temperature in the core has little effect on the star's structure as long as the inner parts of the core remain degenerate. However, for a large enough rise of the central temperature the helium-burning process

$$3 \, \alpha \to C^{12}$$

will commence. For this to occur demands temperatures in excess of 10^8 degrees. When helium-burning begins in a degenerate zone we have the peculiar situation that there is no natural balancing process between energy generation and energy outflow as is the case for normal stellar matter. If energy production is inadequate to balance the outward flux, the shrinkage that occurs is accompanied by cooling and by a further falling behind of the energy generation. If energy generation exceeds the outward flux, the resulting expansion is accompanied by further heating which leads to more energy being generated. This situation is inherently unstable, and it can only be stabilized when the temperature has risen to such an extent that the material becomes non-degenerate.

In the Hoyle-Schwarzschild models it is supposed that it is at this point, where the increased energy generation leads to the creation of a non-degenerate core, that the stars cease to ascend the giant branch. Thus the tip of the giant branch is reached when helium-burning has begun.

This, then, has traced the evolution of stars in globular clusters up to the top of the giant branch. The next problem is to consider how the stars evolve after helium-burning has commenced, and to explain why they move off and populate the horizontal branch and evolve into the RR Lyrae star region and beyond. HOYLE and SCHWARZSCHILD have only given a tentative discussion of the way in which the later stages of evolution may take place and these stages demand much more investigation. They have considered some models which now contain a non-degenerate core in which helium-burning is taking place; this core must contain an inner convective zone because of the strong dependence of helium-burning on temperature. Outside the helium-burning core there is a hydrogen-burning shell source, while the envelope is supposed to be radiative throughout. This latter postulate means that the true surface condition has again been neglected; the authors argue that if their models turn out to be to the right of the giant branch then their neglect of the surface condition is inappropriate, and only the calculated luminosity is of significance, while if they are to the left of the giant sequence all is well and good. In fact, as can be seen from Fig. 4, the model for which $q = 0.60$ lies on the giant sequence but considerably lower than the tip, indicating that the assumption of a non-degenerate structure by the core forces a star to retrace its steps down the giant branch. The models for $q = 0.70$ and $q = 0.80$ lie considerably off the horizontal branch, though they do show a trend towards rapid evolution to the left. However, when $q = 0.60$ the energy generation in the core by helium-burning is comparable with that in the shell by hydrogen-burning, but the energy generation through the conversion of 1 g of helium into carbon is only about 6×10^{17} ergs. Thus q cannot increase from 0.60 to 0.70 before all of the helium is consumed. So the later models using a helium-burning core and a hydrogen-burning shell are ruled out on energetic grounds. In fact, the helium in the center is rapidly exhausted and a helium-burning shell source is then present. The core will now consist of carbon. Further heating of the core could lead to generation of energy by reactions of the type $C^{12}(\alpha, \gamma) O^{16}$, $O^{16}(\alpha, \gamma) Ne^{20}$ etc. However, no models as complicated as these have yet been considered.

The models of HOYLE and SCHWARZSCHILD are thus able to explain the evolution of stars as far as the giant branch in globular clusters, the chemical compositions of the models used being characteristic of these extreme population II systems. A plausible explanation is also given of the termination of the giant branch. The models are only applicable over a narrow mass range, near $1.1-1.2 \, M_\odot$, and any extrapolation of the evolutionary sequences very far outside this range will probably fail.

To conclude the discussion of the theoretical evolutionary tracks in globular clusters, it is necessary to mention the recent work of HASELGROVE and HOYLE (91) and the ambitious program of future work outlined by HOYLE. They have initiated a program of computation of evolutionary tracks using electronic computing machines, and have first of all used this method to rediscuss the problem of the age of stars in globular clusters with special application to M 3. They have computed a series of models using the EDSAC 1, starting with main-sequence models with $M = 1.25 \, M_\odot$, $X = 0.9309$, $Y = 0.0666$ and Z (carbon and nitrogen) $= 0.0025$, parameters which are reasonable for population II stars. No mixing between energy-producing core and envelope is assumed. The models develop a core with $q = 0.08$ which eventually is burned out and gives rise to an isothermal core and a shell source. (Their results depend on the old rate of the CN-cycle, which is now superseded, and the assumed abundance of C and N.) Evolution off the main sequence is then followed, the correct surface boundary condition being used, and the models move on to the giant branch. Fitting of the track with the observed main sequence in M 3 gives them an age for this system of 6.5×10^9 years. HOYLE has recently outlined a program of work which will take the stars right to the tip of the giant branch and may allow him to consider the evolution onto the horizontal branch. The problem has been programmed for an IBM 704 machine.

An interesting problem, particularly when considering the chemical evolution of stars, is that of the evolutionary path which will be followed by stars that have condensed out of pure hydrogen, so that the CN-cycle cannot operate at any stage until helium-burning has begun. In a recent paper HAYASHI (92) has claimed that pure hydrogen stars burning on the pp-chain in a shell can also explain the location of the giants in the globular-cluster H-R diagrams.

As a digression in their discussion of the evolution of globular cluster stars, Hoyle and Schwarzschild have considered the evolution of the giants of population I by carrying out similar calculations in which the chemical composition has been supposed different, i. e., they have increased the metal content (which was taken as very small for the globular cluster stars) to the normal value for population I stars. Although there are many uncertainties in such calculations, particularly since the characteristics of convective transport in the envelopes are largely unknown, they have obtained models which to some extent diverge from and tend to lie 2 to 3 magnitudes lower than, but roughly parallel to, the giant sequence in population II. This is in reasonable agreement with observation. Thus it appears that just this difference in composition is sufficient to explain the difference in luminosity between the giants of the two populations. Only in the photospheric regions does the metal content play a critical role. It greatly affects the extent of the envelope, thus changing the radius considerably for a given luminosity, but hardly affects the deep interior at all.

17. Evolution of massive stars off the main sequence

In considering the development of a chemical inhomogeneity between core and envelope and its role in the evolution of a star off the main sequence, the detailed way in which this inhomogeneity develops has so far been ignored. Essentially it has been supposed that it develops discontinuously. However, there are two possibilities when a star initially having a well-mixed convective core and a radiative envelope begins to evolve. The mass of the convective core may grow, or it may shrink. If it grows, then a discontinuity will be set up at the boundary, and the star will be sharply divided into two zones of different compositions. If it shrinks, convection will die out in the outer regions of the core, and since this material will have been processed to a lesser extent than material in the inner core, a continuous gradation of chemical composition will be set up. Thus at a given time the star will be broken up into three zones: a central convective core with a uniform composition characteristic of its life so far, an intermediate radiative zone, inert from the point of view of nuclear energy generation but with a variable chemical composition reflecting the retreat of the convective core, and an outer radiative zone with a composition characteristic of the original material out of which the star condensed.

This intermediate zone was neglected in the calculations of Schoenberg and Chandrasekhar (81), Harrison (82), and Sandage and Schwarzschild (86), for the evolution of stars off the main sequence. However, it has been taken into account in the work of Tayler (93) and of Kushwaha (94), and since it is apparent that for massive stars, where electron-scattering opacity is dominant, this intermediate zone is of importance, these models are the best available at present for investigating the way in which massive stars begin to evolve. For stars of small mass, Kramers opacity is more important and the extent of the intermediate zone will decrease. For example, in the original models of SCHOENBERG and CHANDRASEKHAR the initial mass of the convective core was 14.5% of the total mass, while the mass limit for the isothermal core was 10%.

Both TAYLER and KUSHWAHA have supposed that the CN-cycle is operating. In his first paper TAYLER assumed electron-scattering opacity throughout while in the second he used only Kramers opacity. KUSHWAHA used a formula in which both opacity terms were added together, and this is probably the best approximation so far used. The major difference between these models and those in which the transition zone can be neglected (i. e., those with $M \approx 1 - 1.2\,M_\odot$) is as follows. In models such as those of SANDAGE and SCHWARZSCHILD, the star reaches the limiting core mass before gravitational contraction of the core and expansion of the envelope moves it rapidly to the right in the M_{bol}-log T_e plane, having brightened by about a magnitude before abruptly leaving the vicinity of the main sequence. But massive stars with considerable intermediate zones will move almost immediately to the right off the main sequence and will only slowly increase in luminosity while doing this. Thus, for example, TAYLER has carried one of his models through until nearly 20% of the mass has been converted, but the isothermal core condition has still not been reached.

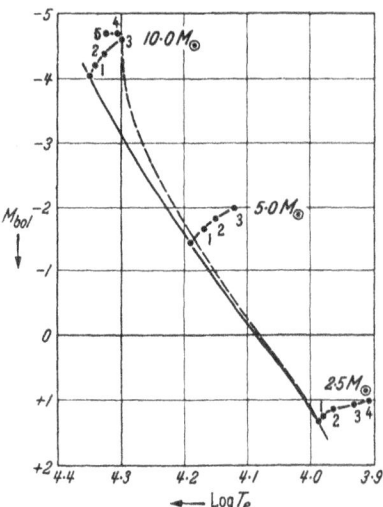

Fig. 5. Evolutionary tracks for massive stars, according to KUSHWAHA (94)

KUSHWAHA has considered masses of 2.5, 5, and 10 $M_\odot$. He has also attempted to take into account the fact that after a sufficient time has elapsed the decrease of hydrogen in the core as well as the general shrinkage of the core means that energy generation in the intermediate zone begins to be felt again. The theoretical evolutionary tracks for his models, which are quite similar to those of TAYLER, are shown in Fig. 5. The dotted line joins the points corresponding to the same age for the three stars. The starting models for the three stars were shown by him to lie quite close to the observed main sequence in this mass range. The dotted line should be compared with the observed H-R diagrams for such clusters as the Hyades and the Pleiades which are discussed in Chapter III.

For the most massive star KUSHWAHA found that in the last inhomogeneous models the effective temperature reaches a minimum and then begins to increase again, thus forming an elbow in the evolutionary track which begins to move back towards the main sequence. Such an effect was also found for a less massive star by HENYEY, LELEVIER and LEVÉE (95). It is believed to be caused by the sharp reduction in the hydrogen content in the energy-generating core at this epoch which necessitates a strong temperature increase and thus a shrinkage of the star.

None of the tracks have been computed to the point where energy production in the core ceases and the isothermal condition is reached. Beyond this point it must be supposed that the release of gravitational energy and the development of a shell energy source involving both the intermediate zone and part of the unburned envelope, i. e., a thick shell source, will begin. The central temperature in the later stages of the most massive models of KUSHWAHA is already about 40×10^6 degrees (a rise of about 10×10^6 degrees over the initial models), so that after further contraction helium-burning will not be long delayed.

It is generally believed that the next important stage of evolution of the massive stars is into the red supergiant region of the H-R diagram. Construction of models of such stars would appear to be the next stage of the theoretical development.

III. Associations and clusters

In our Galaxy, several different types of star groups have been identified and studied. These include binary and multiple stars, stellar associations, galactic clusters, moving clusters, globular clusters, and star clouds. With the exception of star clouds, which are large aggregates consisting of whole regions of spiral arms containing a star density higher than the average, and which often contain galactic clusters, strong arguments suggest that each of these star groups had a common origin, and thus the stars have the same ages; i. e., they are coeval.

Apart from binary and multiple stars which have been discussed to a limited extent in Chapter I, we shall now describe each of these groups in turn. The first groups to be described are the associations, since we shall discuss the groups in an age sequence starting with the youngest.

18. Definitions of O- and T-Associations

That the O and early B stars are found in associations was shown by AMBARTZUMIAN (96), (97), (98). He defined such associations as follows. They are groups containing ~ 100 stars of types B0 or earlier, with diameters of 30—200 parsecs. They are often centered upon a cluster containing O stars, or a multiple star of the Trapezium type. In them the space density of the early-type stars, although much greater than the mean space density of such stars, is *less* by an order or half-order of magnitude than the mean space density of stars of all types in the solar neighborhood.

AMBARTZUMIAN showed that such associations are unstable groups, having a positive total energy, and must be in the process of disintegration through the general effect of the rest of the Galaxy. Hence, since the lifetimes of such groups cannot exceed a few million years, they must have formed recently. This work provided the first definite indications that star formation had occurred very recently in the Galaxy, and was presumably occurring at the present time. It thus resolved the paradox of the high-luminosity O and B stars, whose maximum ages before they must burn all their nuclear fuel at their present rate of energy output (a few times 10^7 years) were inconsistent with the known age of the earth and therefore of the sun, if all stars were supposed to have originated at the same time.

Another kind of association, the T-association, was also considered by AMBARTZUMIAN *(96)*, *(97)*, *(98)*. He put forward the theory that these also are groups of young stars, recently formed, and in the process of a disintegration that will feed their members into the general galactic field. The T-associations are groups of T Tauri stars, such as the group in the Taurus dark clouds discovered by JOY *(99)*. T Tauri stars are irregular variables associated with dark or bright nebulosity, in whose spectra emission lines appear and may vary in strength. Their absorption-line spectral types are in the range F 8 — M 2. As more dense interstellar regions are surveyed for faint stars, so more T-associations are found. Sometimes O- and T-associations are found together, as in the Orion nebula *(100)*, *(101)*, *(102)*, *(103)*, *(104)*.

That there is a genetic relation between T Tauri stars and dense nebulae, rather than that the T Tauri stars are simply stars passing through nebulae and interacting with them, is demonstrated by the fact that their space density in, say, the central regions of the Taurus clouds, is about 5—15 times the space density of stars in the same absolute magnitude interval near the sun *(27)*, *(104)*.

19. Properties of T Tauri Stars

Further evidence that the T Tauri stars are young objects is as follows. HERBIG *(24)* and PARENAGO *(103)* showed that they lie above the main sequence, and this has been beautifully confirmed by the accurate photometry of the very young clusters NGC 2264 and NCG 6530 *(105)*, *(106)*, *(107)*. The evolutionary interpretation of these and other young clusters is described later in this chapter. Young stars which are still in the process of gravitational contraction and have not reached the main sequence lie above and to the right of it. It is sufficient to point out here that their location is that to be expected from the prediction of SALPETER (*63*) and the calculated tracks of HENYEY et al. (*62*).

An important observation by JOY (*99*) and SANFORD[1] is that the spectra indicate that the upper layers of the T Tauri atmospheres where the emission lines originate are *rising*, relative to the underlying layers which give the absorption spectrum. At first sight this is surprising, since velocities indicating the *infall* of material into a contracting star might be expected. It is possible, however, that the observed velocities are those of shock fronts set up by disturbances arising after the infall of material. However, it must be admitted that at present there is no direct evidence for the infall of material. A T Tauri star may thus be nearly isolated from the interstellar medium and contracting in a self-consistent way, with the mass gain through accretion small or negligible.

The T Tauri stars might be expected to give information on such parameters of star formation as the mass distribution function and the amount of rotation. The absorption lines of the stars bright enough to be observed with moderate dispersion are wide and diffuse; if this is due to rotation, then the stars must be rotating faster than main-sequence stars of the same spectral types. The gravitational contraction tracks are from right to left and slightly upwards in the H-R diagram, so that stars now of types F 8—K will finally become main-sequence stars of types A—G. This would argue that the masses of the T Tauri stars are $\sim 2.5\ M_\odot$. HERBIG (*108*) has pointed out that the rotations now observed will, if angular momentum is conserved, result in the correct order of magnitude for rotations of main-sequence stars. The rotations in NGC 2264 (*105*) set in at type F 8 and steadily increase with advancing spectral type. This question of the angular momentum of contracting stars may be intimately connected with the explanation of the H-R diagrams of young clusters.

20. Expansion of O-Associations

Observational confirmation that the O-associations are unstable groups comes from studies of the motions of stars in associations, which show that they are indeed expanding (*109*), (*111*), (*112*), (*113*). If the measured expansions are assumed to have occurred at a constant velocity, then by tracing back the paths until minimum distances between the stars are reached, ages of 1.3×10^6 years, 4.2×10^6 years, 4.5×10^6 years, and $\sim 50 \times 10^6$ years are derived for the ζ Per, Lacerta,

---
[1] SANFORD, R. F.: Publ. Astr. Soc. Pacific **59**, 134 (1947).

II Cephei, and Cassiopeia-Taurus associations, respectively. These ages indicate how recently such star groups have been born.

During the process of star formation, it is apparently possible for stars sometimes to acquire very high velocities which remove them rapidly from their place of origin. BLAAUW and MORGAN (113) showed that the stars AE Aur (O9.5 V) and μ Col (B0 V) have velocities that are almost equal in magnitude and opposite in direction, and very high for early-type stars (128 and 127 km/sec, respectively). When the paths are traced back in space and time, they meet in a region very close to the Trapezium stars in the Orion nebula, 2.6×10^6 years ago. Other high-velocity early-type stars have been traced back to associations where they probably originated (112); we note particularly 53 Ari, which apparently came from the Orion association 4.8×10^6 years ago.

The "expansion ages" of associations and individual stars are at first sight in good agreement with the maximum possible ages on the theory that these stars spend a time on the main sequence given by the dating procedure described in Chapter II. A similar age is obtained under rather different assumptions by MASSEVICH (114). However, MÜNCH has pointed out some discrepancies. Firstly (115), there are four high-luminosity B 1 stars, very similar spectroscopically to ζ Per, but at such large distances from the galactic plane that the times elapsed since they could have been in the plane are about $3—4 \times 10^7$ years. These times are greater than the maximum ages given by equation (38), and an order of magnitude greater than the expansion age of ζ Per, which is so similar spectroscopically to these stars. This discrepancy can be avoided if it is supposed that the stars originated in high galactic latitudes (114). Alternatively the variation in velocity with distance from the plane may be greatly different from that computed. Both of these possibilities seem unlikely.

The second discrepancy is in the age of 53 Ari (116), the high-velocity star which presumably originated in the Orion association 4.8×10^6 years ago. This is a β CMa variable star, as are also 12 Lac and 16 Lac which belong to the Lacerta association. The ages of these stars may be an order of magnitude greater than that obtained from the expansion hypothesis. However, this may indicate only that the evolutionary arguments which have been proposed for the β CMa variables are in error. It should also be borne in mind that there may be a spread in the times of formation of stars in an association, as has been pointed out by MASSEVICH (114).

AMBARTZUMIAN has shown that the Trapezium-type multiple stars so often found in O-associations are unstable and must be disintegrating. SHARPLESS (117) pointed out that the observed systems might formerly have been much closer together. At that epoch a system consisting of one O5, one O7, and ten B5 stars, too close to be resolved, would be indistinguishable in spectrum and absolute magnitude from a single O6 star.

21. Expansion of T-Associations

Knowledge of proper motions and radial velocities of T Tauri stars, which may eventually make possible a study of systematic motions in T-associations, are still lacking.

22. Rate of star formation in O- and T-Associations

The postulate can be made that the O- and T-associations are the seats of all star formation in the Galaxy at the present time. Sometimes both types of association are found together, as in the Orion Nebula. There is some evidence suggesting that there is a real break in the main sequence of solar-neighborhood stars at or near spectral type G. The arguments come from two directions:

(i) the break in the mass-luminosity relation which is well established for solar-neighborhood stars with masses slightly below that of the sun;

(ii) different distributions of space velocity for stars above and below type G, which will be discussed later.

As a consequence of this break it has been proposed by PARENAGO and MASSEVICH ($80a$) that the upper and lower parts of the main sequence have been populated by different mechanisms and from different starting points; i. e., the O-associations will populate the upper part and the T-associations the lower part. Whether there is a fundamental difference between modes of star formation in O- and T-associations, or whether they merely represent different parts of a continuous mass spectrum, it is instructive to look at estimates that have been made for the numbers of stars produced in associations.

AMBARTZUMIAN (118) has estimated that the total number of O-associations currently existing in the Galaxy is about 10^4. Taking an average age for associations as 3×10^7 years, then the total number which have existed in a time 5×10^9 years is $\sim 10^6$. If each association contains 10^2—10^3 stars, this gives 10^8—10^9 for the total number of stars which have been produced by O-associations.

HERBIG (108) has estimated that the minimum present number of T Tauri stars in the Galaxy is 5×10^5. Taking what he regarded as a conservatively large estimate of 5×10^6 years for the time spent by a T Tauri star in association with the nebula in which it is formed, i. e., in the T Tauri stage, this means that 5×10^8 stars will have been produced in this way in 5×10^9 years. If the galactic distribution of the T Tauri stars were uniform, then the above rate would give about

30 stars within 10.5 parsecs of the sun. Thus the estimated production rate by this process is galactically significant.

It has been suggested that the late-type dwarf stars with emission lines found in the solar neighborhood, which are not associated with nebulosity, may represent a later stage in the life of low-mass T Tauri stars which have still not reached the main sequence (*108*), (*119*). For low-mass stars the contraction time scale is 10—100 times longer than the estimated time for the association to disperse. The dwarf Me stars tend to lie above the main sequence (*120*), and their velocity dispersion is only about half that of normal dwarf M stars. This would be reasonable if dMe stars had formed in and escaped from nearby nebulosity and clouds, while the dM stars, being older, represent a less flat galactic subsystem. If this suggestion is correct, then the energy source for the emission lines must persist for a time after the star has left the nebulous region. However, the dMe stars in the solar neighborhood may be unconnected with T Tauri stars and may have another origin altogether.

AMBARTZUMIAN (*118*) has supposed that the O-associations feed stars into the flat subsystem of stars, while T-associations put stars into intermediate subsystems. There may not, however, be a clear-cut distinction. For example, some O and B stars, ejected from associations with high velocities, may eventually reach high galactic latitudes. Knowledge of these aspects of the problem is still rudimentary.

The overall estimates obtained by AMBARTZUMIAN and HERBIG for the total numbers of stars formed in associations have been determined on the supposition that these rates have remained constant over the last 5×10^9 years. The results then show that the associations have contributed a large number, but nothing like the total number, of stars now present in the disk of our Galaxy. However, the number of stars which can be formed is ultimately determined by the amount of gas and dust available, and there is some evidence suggesting that the present amount of gas is only $\sim 1\%$ of its initial value. Thus, whatever the detailed process of star formation is, it must be a decreasing function of the age of the Galaxy. We do not know what the most important modes of star formation were in the early life of our Galaxy, though we might suspect that processes other than associations dominated in this epoch. It seems clear, however, that it is unreasonable to extrapolate the formation of stars in associations over the whole life of the Galaxy. It is more reasonable to conclude that star formation in associations is the dominant mode of star formation at the present epoch, that it will continue to dominate, though it will decrease in magnitude as more of the gas is condensed into stars, but that it may not have had great significance in the early history of the Galaxy.

23. Early observational work and interpretations of galactic clusters

Galactic clusters grade indefinitely into multiple stars in one direction and small star clouds in the other. Some may even be accidental groupings, though this fraction will be very small. Moving clusters are simply the nearer galactic clusters in which it has been found possible to measure radial or transverse motions. Early work (prior to 1930) on clusters has been well summarized by SHAPLEY (*121*) and TRUMPLER (*122*). The characteristic by which all clusters have been first detected is that the star density reaches a value significantly greater than that of the stars in the surrounding field. This then immediately points up a fundamental difference between clusters and O-associations. As has already been pointed out, in stellar O-associations the local space density of association members is *less* than the space density of non-members, although for the particular classes of stars in the association, it is greater than for the corresponding classes of non-members. It is of interest, however, that often stellar associations are centered on galactic clusters. The other major difference between clusters and associations is in their dynamical states. The stars in an association form an expanding system with a total positive energy, while a galactic cluster is a loosely bound system with negative total energy which can only be disrupted by outside influences.

We shall refer here only to the description of galactic clusters by TRUMPLER (*122*), and the general deductions which he made from his observations, as follows. Clusters containing O and B main-sequence stars (i. e., massive stars) have very few or no red or yellow giant stars (the few being generally supergiants). Clusters whose main sequences do not extend upwards as far as types O and B most frequently contain an appreciable number of stars in the giant branch.

It was clear from these early observations that the H-R diagrams of individual clusters tended to have a smaller scatter in their main sequences than did the solar-neighborhood field stars. This feature, together with an apparent dispersion when the main sequences of several clusters were fitted together, led to an interpretation in terms of the well-known relation between luminosity, surface temperature, and hydrogen content. KUIPER (*123*) concluded that stars in each cluster had a common origin and therefore equal hydrogen content, but that the hydrogen content differed from one cluster to another. For the Hyades cluster, from four binaries whose masses were known, the low value of the hydrogen content $X = 0.13$ was found. Later, PARENAGO and MASSEVICH (*124*) derived a value of the heavy-element content $Z = 0.12$ for the Hyades. The discussion to follow will show that both of these values are unacceptable today.

24. Modern observational work and interpretations; dating of clusters

The accumulation of accurate photometric observations and parallaxes for field stars and clusters in recent years has shown that below $M_v = +5$ main sequences of clusters and field stars, if plotted together

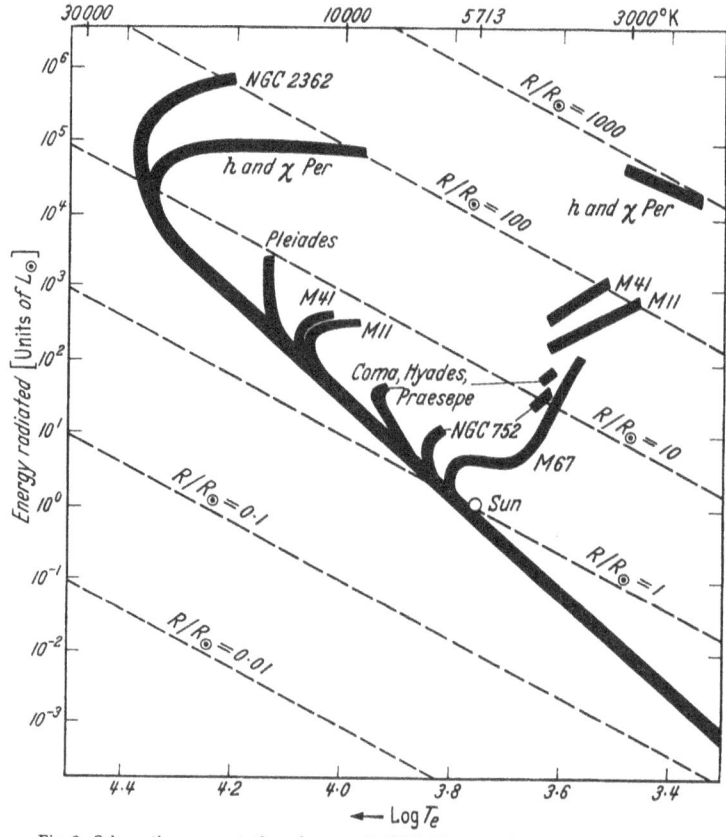

Fig. 6. Schematic representation of composite H-R Diagram of various galactic clusters

in an H-R diagram, actually agree well and form a narrow sequence (120), (125). It can therefore be postulated now that the hydrogen content in unevolved field stars and cluster stars in the solar neighborhood is approximately the same. The small differences in the heavy element content Z, which may be expected if continuous enrichment through element synthesis in stars is going on, will not materially affect X, since the presently accepted value of Z is only a small fraction of X. Spectroscopic observations show that there is little variation in the hydrogen content of stars relative to the heavy elements in different galactic clusters.

The dispersion in the upper end of the main sequences of clusters, shown by Kuiper's composite H-R diagram, has, however, been substantiated by modern observations and can now be interpreted in terms of the theory of the early stages of stellar evolution described in Chapter II. Assuming that no mixing takes place between the core and the envelope, a star begins to move above and to the right of its original place on the main sequence as soon as it has converted an appreciable amount of the hydrogen in its core to helium. This takes place on a shorter time scale for the more luminous stars. Computed evolutionary tracks are shown in (86), (93); see also Fig. 5. Composite H-R diagrams of clusters are shown in (126), (127); see also Fig. 6 which gives a composite diagram in the M_{bol}, log T_e plane and is taken from (128). The resemblance between the computed tracks and the upper ends of the cluster main sequences is to be noted.

The abrupt end of all the observed main sequences is explained by the fact that, after a star reaches the Schoenberg-Chandrasekhar limit, its structure changes rapidly. Since the time taken to reach this limit depends on the luminosity and mass, this provides a method of finding the age of stars that are just about to leave the main sequence (73) and hence a method of dating clusters (86). A convenient formula for the time, τ, at which a star of mass M and luminosity L reaches the Schoenberg-Chandrasekhar limit has been obtained from the Sandage-Schwarzschild models and is given by (128)

$$\tau = 1.10 \times 10^{10} \frac{L_\odot M}{L M_\odot} \text{ years}. \tag{38}$$

For the more massive stars this method should be modified according to the evolutionary theory of TAYLER and KUSHWAHA described in Chapter II.

Tables 4 and 5 give the ages of a number of associations and galactic clusters as determined by various workers. MICZAIKA (129) used the formula given by STRÖMGREN (73) applied to the brightest stars in each cluster. LOHMANN (130) used Tayler's theory (93) to compute sequences of stars at different ages, starting from an "age zero" main sequence. VON HOERNER (131) used the position of the left-hand edge of the point of greatest curvature at the top of the main sequence. He used Tayler's theory and a mean, not an "age zero", main sequence. SANDAGE (128) used equation (38) and fitted the observed cluster sequences to an age zero main sequence. A review article by ARP (132) should be consulted for the technical details concerning the derivation of an age zero main sequence. Two ways of determining such a sequence have been used. Observed main sequences for different clusters have been corrected theoretically for the effects of the early stages of evolution by JOHNSON and HILTNER (141). Alternatively, an empirical main sequence has been obtained by fitting together the observed main sequences of various clusters, using in each case that portion more than three magnitudes below the observed turn-off, by JOHNSON and SANDAGE (149) and SANDAGE (133).

The ages derived by MASSEVICH (114), (134), (135), listed in Tables 4 and 5, were obtained by using a different theory from the other workers, allowing mass loss to occur in unmixed stars as was described in Chapter II.

It will be noted that there is quite a range in the various age determinations for any one cluster. For the older clusters, the most accurate dating procedure is probably that of SANDAGE. It is certainly necessary to use, as he did, an age zero main sequence which is supposed to be completely unevolved, for comparison with the individual clusters. Fitting the lower main sequence to this is probably less ambiguous than

Table 4. *Ages of associations*

Name	Age (years)	
	v. HOERNER (131)	MASSEVICH (114)
γ Cyg	4×10^5	
II (P) Cyg	4×10^5	
I (h and χ) Per	4×10^5	$3 - 4 \times 10^5$
II Cep	7×10^5	
Orion	3×10^6	
I Gem	3×10^6	
Sco - Cen	$<4 \times 10^6$	
II (ζ) Per	5.5×10^6	
Lacerta	6.8×10^6	

Table 5. *Ages of clusters*

Name	Age (years)				
	MICZAIKA (129)	LOHMANN (130)	VON HOERNER (131)	SANDAGE (128)	MASSEVICH (114, 134, 135)
h and χ Per	$>1 \times 10^6$		4.4×10^6	1×10^6	2×10^6
NGC 663			6.8×10^6		
NCG 2362		2×10^7	6.8×10^6	1×10^6	
IC 4665		6×10^7			
NGC 2264			1.5×10^6		
			($<15 \times 10^6$)		
NGC 457			1.5×10^7		
Perseus moving cluster			2.0×10^7		
Pleiades	4×10^7	1.0×10^8	8.0×10^7	2×10^7	10^7
M 34			1.1×10^8		
NGC 7243			1.7×10^8		
M 41			1.7×10^8	6×10^7	
NGC 2516			1.7×10^8		
M 11		2.9×10^8	2.0×10^8	6×10^7	
M 39		3.0×10^8	2.3×10^8		
UMa stream	5×10^8		3.0×10^8		
Hyades	5×10^8	1.1×10^9	8.7×10^8	4×10^8	3×10^8
Praesepe	5×10^8	1.5×10^9	3.0×10^8	4×10^8	
Coma			5.9×10^8	3×10^8	
NGC 752		2.7×10^9	2.3×10^9	1×10^9	
M 67		3.8×10^9	4.6×10^9	5×10^9	

using the point of greatest curvature at the upper end of the main sequence, as was done by VON HOERNER. For the younger clusters, Sandage's assumption of homologous models probably leads to error, and the theory for more massive stars should be used, the clusters being compared with an unevolved main sequence. At the upper end of the main sequence it is more difficult to derive such an unevolved sequence, and modifications in the presently accepted one may yet be made. The validity of Massevich's values depends on whether or not mass loss occurs in the way proposed.

The present-day H-R diagram of those cluster stars which are not on the main sequence represents the locus of points through which stars of different masses will pass. Since the lifetime of an evolving star is short, relative to the time spent on the main sequence, all evolving stars in the "snapshot" of a cluster at the present time will have come from only a small part of the main sequence, i. e., a small mass range. The sequences of stars not on the main sequence will thus very nearly represent the evolutionary track of a star having a mass slightly greater than that corresponding to the upper limit of the main sequence in that cluster. The number of stars in any part of the H-R diagram (off the main sequence) will be proportional to the time spent in that part of the evolutionary path.

We now consider in more detail the H-R diagram of a few galactic clusters, and conclusions that can be drawn from them.

25. The very young Orion Nebula cluster

578 stars down to a limiting apparent magnitude of $15^{m}\!.2$ have been found by PARENAGO to belong to a cluster in the region of the Orion association *(103)*, *(136)*. Fig. 7 shows the color-magnitude diagram reproduced from PARENAGO's work; it is remarkable for the following features. The main sequence extends only between types O and A5, and there is a definite break between this and the large scattered group of stars below. The latter lie above the normal position of the main sequence, and may be called subgiants. They include a great number of variables, presumably of T Tauri type.

The measures for the fainter stars were supplemented by measures from published prints of the field, which have low accuracy. However, it does not seem possible that any such systematic effect could have caused the break between the main sequence at A5 and the group of subgiants, since this break occurs at about $M_{pv} = 9.5$. A photoelectric investigation by JOHNSON *(137)* confirms the abrupt end of the main sequence at the same place as found by PARENAGO. Although JOHNSON measured few of the stars fainter than this, and gives no data from which it may be determined whether they are cluster members, he confirms their general location in the diagram. His measures also give a suggestion of a break between the position of these stars and the end of the main sequence.

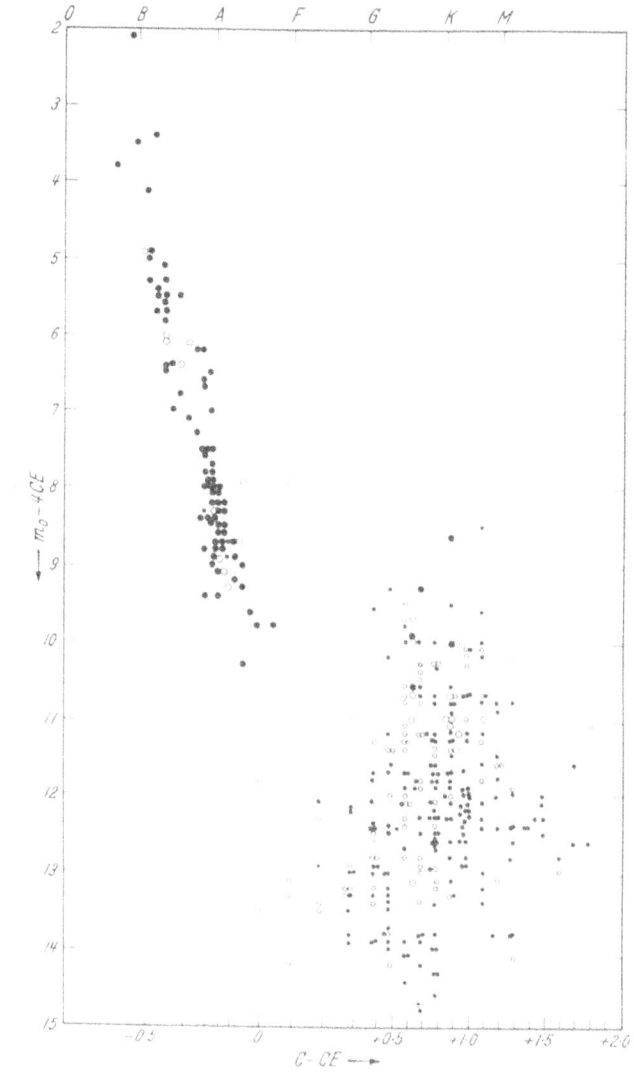

Fig. 7. Color-magnitude diagram of the Orion Nebula cluster, by PARENAGO (*103*)

26. The very young clusters NGC 2264 and NGC 6530

A photometric and spectrographic study of NGC 2264 by WALKER (*105*) has shown it to resemble the Orion Nebula cluster in that it has a normal main sequence extending only from $M_v = -5$ (type O7) to $+1.5$

(type A0), below which there is a group of stars lying above the normal main sequence. This cluster differs from the Orion Nebula cluster in that there is no break between the main sequence proper and the group of subgiants. The latter are spread into a broad band parallel to the main sequence and about two magnitudes above it. Their spectra agree with the photometry in indicating luminosities that place them above the main sequence.

In addition, the majority of these stars fainter than $M_v = +3.5$ are T Tauri variables having hydrogen emission lines. WALKER concluded from the bright blue stars in this cluster that it is very young, so young in fact that the less massive stars may still be in the process of condensation, and may not yet have reached the main sequence. He gives 3×10^6 years as the time taken by an A0 star to reach the main sequence by gravitational contraction from a starting configuration with surface temperature 4000°. The most luminous star in NGC 2264 is the O7 star S Mon, and WALKER gives an upper limit to its age as about 5×10^6 years. The agreement here is good; however, when the contraction time of the *less* luminous stars is considered, a problem arises, which is discussed in the following section.

The color-magnitude diagram of NGC 6530 by WALKER (*106*) is very similar to NGC 2264, in that it has a normal main sequence extending only from O5 to about A0. Below this the stars lie above the main sequence, and in this region there are numerous T Tauri stars. The age of this cluster is concluded to be about 3×10^6 years, like NGC 2264.

27. Possible explanations of the H-R Diagrams of the Orion Nebula cluster, NGC 2264 and NGC 6530

These three very young clusters are the only ones which at the time of writing have any detailed H-R diagrams constructed for them. Thus we can attempt to discuss the evolution of such clusters by using this first material. Two major features demand explanation. The first is the break in the main sequence in the Orion Nebula cluster, observed by both PARENAGO and JOHNSON, and shown in Fig. 7. In the other two clusters there does not appear to be a break. Second, in all three clusters an explanation is needed for the lower-luminosity stars which lie to the right of the main sequence. A number of different explanations can be considered.

(i) PARENAGO concluded that the H-R diagram of the Orion Nebula cluster supports his previous hypothesis, that there is a real break in the main sequence near solar type and that the upper part is populated by stars formed in O-associations and the lower part by stars formed in T-associations. The cluster is very young because of the O-type stars in it, and it was suggested that stars are probably being formed in this region at the present time. The theory of evolution down the main sequence through mass loss by completely mixed stars was applied, and an age of $\sim 10^7$ years was derived from this theory as the time taken for an O-type star to reach A5.

(ii) The second explanation is based on the assumption that all the stars in each of these clusters are coeval, and stars of all masses according to the mass-spectrum mentioned in Chapter I began to contract towards

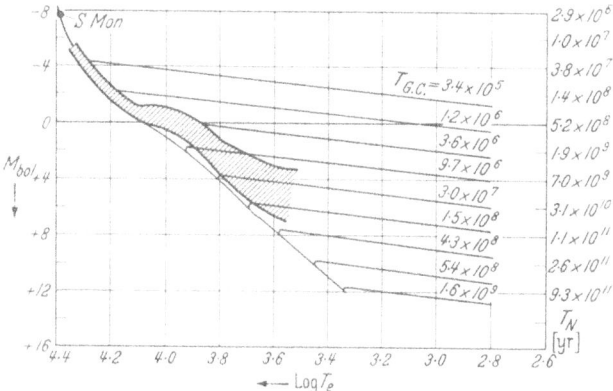

Fig. 8. Computed gravitational contraction tracks and times ($T_{G.C.}$) for stars of various M_{bol}. Times to reach S-C Limit (T_N) are also shown. Shaded area represents observations of NGC 2264

the main sequence. The difficulty in explaining the diagram on the basis that the stars in the upper part have had time to reach the main sequence while those in the lower part are still gravitationally contracting has been discussed by SANDAGE [61], and is illustrated by his diagram reproduced in Fig. 8. Here the color-magnitude diagram of NGC 2264 has been plotted in the coordinates of M_{bol} and $\log T_e$. Computed contraction tracks for stars of various masses and their time scales are also shown. The ages derived by WALKER from the upper end of the main sequence of NGC 2264 and from the contraction time of an A0 star are both 3×10^6 years, but if this is the age of the cluster, then below A0 one would expect a band of stars to be spread out more or less parallel to the contraction tracks and not, as is actually the case, parallel to the main sequence. In NGC 6530 these stars do populate a region which is funnel-shaped, widening towards the fainter stars, with its upper boundary nearly horizontal, like the contraction tracks. The fainter stars which lie near the main sequence, however, constitute just the same problem in this cluster as in NGC 2264. On the simple theory, there does not appear to have been time for the coolest stars to have contracted to their observed position, which would take about 100 times as long as for the A0 stars to contract.

A way out of this difficulty may be found if it is supposed that fragmentation occurs. Let us consider the fragmentation of large contracting masses of gas. According to Hoyle's theory, the last fragmentation of a cloud of about 2—3 $M_\odot$ will give fragments of radius $\sim 10^{14}$ cm. Of course the initial conditions are very different here from those assumed by HOYLE. This radius is considerably larger than the radius at the end of the tracks in Fig. 8, and in this theory the fragments would only just at this point become star-like, with the adiabatic condition superseding the isothermal. The fragments would still have to undergo the long contraction times indicated in Fig. 8. The locus of points from which stars would have reached their present positions in 3×10^6 years extends over to the right of Fig. 8 for present-day types A 0 and earlier, but rapidly comes in close to their present positions for the less massive stars.

The number of stars later than type A 0 is about 20 times the number of those earlier than A 0. If there were a mechanism for the break-up of large contracting clouds that would result in stellar fragments with surface temperatures in the observed range, this could provide the answer. Stars with masses much in excess of 50 $M_\odot$ are not observed; the complete break-up of only a few contracting gas masses of about 100 $M_\odot$ or greater could provide the numbers of low-mass stars, but it is not clear that such fragments could rapidly become stellar, as would be necessary to explain the observations.

A possible mechanism of break-up, not considered by HOYLE, is rotation. It has recently been pointed out by PRENDERGAST *(138)* that the break in the main sequence observed in the Orion Nebula cluster at A 5 is fairly near to that at which there is a sharp drop in the rotational velocities of main-sequence solar-neighborhood stars, i. e., at F 5. This break, if measured in angular momentum units, is very sharp indeed. To explain the break it may be supposed that below a certain value of total angular momentum a protostar is able to contract without breaking up into separate condensations, while above this value fragmentation must take place. The appearance or non-appearance of a break in the observed number of stars would then be related to the ratios of the masses into which the protostars fragment and the original mass distribution. This hypothesis would suggest that all of the low-mass stars lying to the right of the main sequence were once members of binary or multiple systems and that most of them have been disrupted by tidal forces. A study of binary systems in these clusters is highly desirable. Also studies of the rotations of the stars in the Orion Nebula cluster would determine whether the break is coincident with a jump in rotational velocities for these cluster stars alone and not just close to, but not necessarily coincident with, the corresponding jump in the rotations of the general field stars. It would also be expected that the upper part of the main sequence of the Orion Nebula cluster would be populated by stars that have maximum values of rotation compatible with their masses.

(iii) A third possible explanation for the position of the fainter stars in the H-R diagram might be forthcoming through the effect of dust grains. According to WHIPPLE and SPITZER, the presence of dust speeds up the contraction by a large factor over the time for pure gravitational contraction of gas alone. The extent to which this could be important in low-mass condensations after they had become adiabatic with surface temperatures greater than a few hundred degrees is not clear. More work needs to be done in considering this possibility.

(iv) Finally, it is possible that the observed distribution of stars in the H-R diagram is due to a spread in starting-times for condensations of different masses. The presence of numerous small condensations would then have to be assumed to occur before and perhaps facilitate the start of a few larger ones. However, OORT and SPITZER and BIERMANN and SCHLÜTER have suggested the opposite; the presence of a massive O or B star might, through the jet action of ionization upon gas in its vicinity, cause local density increases and lead to the condensation of smaller stars in large numbers. The sequence of events needed to explain these observations might be obtained if the ideas of KRAT and UREY are correct, i. e., small solid condensations are present in large numbers and are gravitationally attracted, leading to the formation of stellar masses.

28. Young clusters

(α) h and χ Persei

This double cluster contains very luminous B-type stars with absolute magnitudes as high as $M_v = -7$. BIDELMAN (*139*) showed, as had previously been hinted[1], that it is surrounded by a corona of M-type supergiants at the same distance as the cluster and therefore presumably physically connected with it, indicating a genetic relation between bright blue stars and M-type supergiants. Recently photometric and spectroscopic observations have been made on the brightest blue stars (*140*), (*141*), and on the red supergiants (*142*). In the H-R diagram the high-temperature stars form a band extending from the main sequence, upwards and to the right to spectral type A 5 Ia, where they come to an abrupt end. There is then a very wide Hertzsprung gap, and then the sequence of M-type supergiants.

MASSEVICH (*114*) discussed evolution of the double cluster and the association around it, considering only that part of the H-R diagram to the left of the Hertzsprung gap, and using her theory of evolution

---
[1] ADAMS, W. S., A. H. JOY and M. L. HUMASON: Astrophysic. J. **64**, 225 (1926).

with mass loss but without mixing (78). For the cluster, with its well-defined main sequence curving to the right at its upper end, she derives an age of 2×10^6 years. In the association a well-defined sequence is present at about the same position, indicating the same age, but in addition there are stars on the main sequence above this turn-off point, whose ages must be a factor of about 10 less than the age of the cluster. The measures are compared with a computed initial age zero main sequence, which agrees well with the initial main sequence of JOHNSON and HILTNER (141) but extends upwards further.

Also included in her H-R diagrams are Tayler's (93) computed evolutionary tracks for a star of constant mass 10 $M_\odot$, and a track computed by MASSEVICH under the same assumptions for a constant mass of 20 $M_\odot$; these take stars off the main sequence into the B supergiant and giant region in times of 3×10^6 to 10^7 years. She concludes that both the cluster and the association were formed at about the same time (10^6 to 10^7 years ago), but that the process of star formation has continued in the outer association for a considerable time after it ceased in the nucleus. This is taken as possible support for the Oort-Spitzer-Biermann-Schlüter hypothesis concerning star formation; the cluster formed first, and ionization from the hot stars in it has facilitated star formation in the region around it.

MASSEVICH postulates that the A-type supergiants, of which there are more in the association than in the cluster, are stars in the process of gravitational contraction and have not yet reached the main sequence. However, she does not discuss the M-type supergiants in the association, and the existence of these together with a wide Hertzsprung gap makes it more probable, in our opinion, that the A-type supergiants are evolving away from the main sequence, prior to a rapid structural readjustment that will take them into the M-supergiant region.

Finally, the question of the classical cepheids in the vicinity of h and χ Persei, to which BIDELMAN drew attention, should be mentioned. One of these, SZ Cas, is probably a member of the cluster. If so, it is not as luminous as the B and M supergiants. The evolutionary stage reached by the cepheids is a problem that may have more light thrown upon it by some investigations under way at present of galactic clusters containing cepheids.

(β) Pleiades

The turn-off point in the Pleiades comes at a lower luminosity than that in h and χ Persei; the cluster is therefore somewhat older. There are no red giants in this cluster. MITCHELL and JOHNSON (143) have found what they believe to be a break in the main sequence near $M_v = +6.5$ and have suggested that this marks the place below which stars are still gravitationally contracting on to the main sequence. The break is very much smaller than that in the Orion Nebula cluster. It lies about 9 magnitudes below the brightest stars observed in the Pleiades. Tayler's theory was used to derive an age of 1.5×10^8 years for this cluster, in good agreement with that obtained by LOHMANN (130). If the break in the lower main sequence is real and is connected in some way with the gravitational contraction, then a similar age is derived from the time for contraction of stars at this luminosity.

29. Clusters of intermediate age

The observation on which the ages of the intermediate clusters in Table 5 are based are mainly by BECKER, COX, JOHNSON, SANDAGE, STOCK, WEAVER, and their associates; a full list of references is given by VON HOERNER (131). A recent one may be added: the H-R diagram of NGC 1664 has been obtained by LARSSON-LEANDER (144) and an age of 1×10^9 years assigned.

Of the well-observed clusters in this group, M 41, M 11, Hyades, Praesepe, and NGC 1664 each have a Hertzsprung gap and then a sequence of red giants. The luminosities of the giants are approximately the same as the luminosity of the stars at the turn-off point in the main sequence.

If white dwarfs represent the final stage of stellar evolution, then it is to be expected that in all clusters of intermediate and old age in which they are observationally accessible, white dwarfs will be found. There is evidence for their presence in the Hyades and probably in Praesepe and the Coma cluster (145), (146).

30. The old clusters NGC 752 and M 67

The H-R diagrams of NGC 752 and M 67 are unusual for galactic clusters. The observations of NGC 752 (147) have so far shown no main-sequence stars above $M_v = +2.5$ or below $M_v = +4.5$. There is a Hertzsprung gap, and a few red giants. The upper limit to the main sequence is real, but there is a possibility that if photometry were carried to fainter stars, and if cluster membership could be checked, a lower extension of the main sequence might be found. The characteristic curve at the upper end, and the absence of main-sequence stars brighter than $+2.5$, show that this is a relatively old cluster. ROMAN (148) has found that the spectra show weak metallic lines, indicating a slightly lower than normal metal content. According to the idea that the interstellar medium is continuously enriched by element synthesis in stars, a correlation of heavy-element content with age might be expected. ROMAN has computed models, and derived a minimum age of 1.5×10^9 years for the cluster, in reasonable agreement with the ages given in Table 5.

M 67 is a rich cluster in rather high galactic latitude. The main sequence contains practically no stars brighter than $M_v = +3.5$ *(149)*; it curves round into a subgiant sequence running approximately horizontally and then upwards into the red giant region. There is no Hertzsprung gap. The absence of a well-populated main sequence above $+3.5$, implying that all the brighter stars originally on the upper main sequence have passed through their evolution and have come to the end of their lives, shows that this is an old cluster. In fact, it is the oldest galactic cluster so far studied (see Table 5). The spectra, however, do not show weak-lined characteristics. BAADE has reported, from inspection of his plates, that there may be a large number of white dwarfs in M 67 *(127)*. Since these are stars at the ends of their lives, there would be expected to be many in such a rich, old cluster as M 67. Finally, there appears to be a thinly-populated sequence of stars which looks rather similar to the horizontal branch in globular clusters and may be the analogue in old galactic clusters of what is a characteristic and well-defined feature in globular clusters. (Cf. the later discussion in this chapter.)

31. General conclusions from H-R Diagrams of galactic clusters

From the H-R diagrams just described, we may make the following empirical deductions concerning the course of stellar evolution.

(i) If clusters are very young, the lower-mass stars will not yet have reached the main sequence.

(ii) As clusters age, the brighter stars evolve off the main sequence, so that the upper extent of this can be used to indicate the age of clusters. VON HOERNER *(131)* noted that there should be a range of only 9.5 in bolometric magnitude on the main sequence in any one cluster, the upper end being limited by evolution off the main sequence, and the lower end by the time taken for gravitationally contracting stars to reach the main sequence. This conclusion appears to be confirmed by the recent observations of the Pleiades *(143)*.

(iii) The Hertzsprung gap is wedge-shaped. It is widest in those clusters whose main sequences extend the furthest upwards, i. e., the youngest clusters, and disappears in the oldest (M 67). The existence of such a gap between the main sequence and the giant or supergiant region shows that stars above a certain mass near $1.5 \, M_\odot$ move very rapidly in their evolutionary paths after reaching the Schoenberg-Chandrasekhar limit. That evolution here is rapid, in the absence of degeneracy in the core, was shown by SANDAGE and SCHWARZSCHILD *(86)*. REDDISH *(150)* suggested that whether or not there is a Hertzsprung gap may depend on whether or not the evolving stars are massive enough to have convective cores. This has also been discussed by SANDAGE *(128)*.

(iv) After passing through the Hertzsprung gap, stars enter the giant or supergiant region. Evidence from most of the clusters does not indicate whether a star becomes stabilized at the left or right end of the giant sequences, but the diagram for M 67, where there is a continuous sequence of stars, strongly suggests that a star moves from left to right along the red-giant sequence.

(v) The luminosities of red giants are apparently related to the masses of the main-sequence stars from which they evolved. The most massive ones which were of types O and B become supergiants; less massive stars, as in M 41, become giants of luminosity class II; still less massive ones become normal (luminosity class III) giants with little dispersion in luminosity over a large mass range. In other words, after leaving the main-sequence region, stars of masses greater than about $2 M_\odot$ move approximately horizontally in the H-R diagram. This was already indicated by the position of giants in the mass-luminosity plane. The place of the five giants of types G 5 III to K 5 III in the very young cluster NGC 2264 in this scheme is not clear. WALKER pointed out that they may be extreme cases of gravitationally contracting stars that have not reached the main sequence, or that possibly they may not be cluster members. Alternatively, they may be stars that have undergone considerable mass loss and have reached a giant configuration appropriate to lower-mass stars.

(vi) In every cluster the red giant branch comes to an abrupt end; these observations suggest either that a rapid change then takes place which removes them completely from the giant branch or that stars move back along the giant branch, presumably because of further structural changes. We saw in Chapter II that globular clusters provide some evidence for the subsequent history of population II stars, but of the galactic clusters so far observed, only M 67 contains enough stars to provide any clues to the later, apparently rapid, evolution of population I stars. In that cluster there is a sparse distribution of stars in a "horizontal branch", and it is possible that this is the track of population I stars of masses near $1.5 M_\odot$, when they are nearing the end of their lives.

32. Empirical deductions about evolution of field stars in the solar neighborhood

(α) *The dissolution of galactic clusters*

Because of the effect of differential galactic rotation and shear, and also because the times of relaxation for all but the richest galactic clusters are small compared to 10^9 years[1,2], we should expect that there

---
[1] AMBARTZUMIAN, V. A.: Ann. Leningrad State Univ., No. 22, p. 19 (1938).

[2] CHANDRASEKHAR, S.: Principles of stellar dynamics. University of Chicago Press, 1942.

will be a steady loss of stars from clusters into the general field. The looser, poorer clusters dissipate faster than rich clusters. For example, MARKARIAN[1] has discussed the dissipation of clusters containing O-type stars.

OORT (*151*) has considered a problem posed by counts of the numbers of galactic clusters with main sequences extending up to various ranges of spectral type. On the basis of the dating procedure just described, time intervals, ΔT, between the ages corresponding to the various main-sequence limits were derived. Assuming a uniform rate of formation of clusters during the last 5×10^9 years, the number of clusters in each range should be proportional to ΔT, but the actually observed numbers of clusters depart from this by factors of more than 100 for the oldest ones, in the sense that there are too few old clusters. The data were taken from the list of TRUMPLER (*122*). Yet the predicted times for complete dissolution of clusters are very much larger than the relaxation times, and are in general $\sim 3 \times 10^9$ years.

At least three arguments which might explain this discrepancy can be put forward. Firstly, all the preceding arguments leading to the dating procedure may be erroneous. This appears highly unlikely. Secondly, clusters may, because of other factors, such as tidal effects due to interstellar matter and the acceleration of the rate of dissipation as the cluster is depleted, evaporate at a faster rate than that calculated. Thirdly, cluster formation may not have occurred at a uniform rate during the last 5×10^9 years. Finally, there may be an effect of observational selection in the available material.

The discrepancy is large, and is already apparent for clusters of ages 10^8 years. Further, if 3×10^9 years is the average time for total dissolution of a cluster, an estimate for the rate of star production in the general field by this means yields a smaller number than that estimated previously through the dissolution of associations, simply because of the very much longer lifetimes for clusters.

SPITZER (*152*) has studied the disruption of galactic clusters by gravitational encounters with interstellar cloud complexes. He has shown that the effect is a full order of magnitude greater than the disruption resulting from passing field stars alone. Although conclusive results are not possible at present, this work strongly suggests that such an effect can account for the discrepancy pointed out by OORT. According to Spitzer's preliminary calculations, the disruption time for an extended cluster like the Hyades will be as little as 6×10^7 years, while that for the Pleiades probably does not exceed 10^9 years and may be considerably less. Only a relatively dense cluster, such as M 67, can apparently survive disruption for a period as long as 5×10^9 years.

---

[1] MARKARIAN, B. E.: Soobs. Bjurakan Obs. 5 (1950).

However, at an epoch earlier than a few times 10^9 years ago, the possibility that the rate of formation of clusters may have been different from the rate in the more recent past should not be forgotten. When more gas and dust were present, the conditions for star formation and for the stability of clusters may have been different, and what we see now is the tail end of a decaying activity. This problem is connected with possible evolutionary stages of galaxies discussed later.

33. Color-Magnitude diagrams of globular clusters; early observations, difference from solar-neighborhood stars

Globular clusters are, as their name suggests, highly condensed clusters containing $\sim 10^4$—10^6 stars. They are distributed in a spherical halo about the center of our Galaxy. Early work on their color-magnitude arrays by SHAPLEY and others [see for example (*121*)] revealed the fact that stars in globular clusters occupy a different area of the H-R diagram from that populated by galactic cluster and solar-neighborhood stars. We are concerned here with the differences between H-R or color-magnitude diagrams of galactic and globular clusters, which have an important bearing on the ages, origin and evolution of these clusters.

Owing to the great distances of globular clusters, SHAPLEY's studies did not extend to fainter absolute magnitudes than about $+1$, and it never became clear what was the relationship, if any, between the globular and galactic clusters. However, SHAPLEY's studies showed that the absence of blue high-luminosity stars in the former, and the upward-inclined yellow and red giant branches extending to stars of similar color to those found in the galactic clusters and the solar neighborhood, but about three magnitudes brighter, were characteristic. It was these features which provided BAADE (*153*) with the clue leading to his resolution of stars in the disk of M 31 and its companions, and his demonstration of the existence of two stellar populations, which will be discussed in Chapter IV. The globular clusters belong to the extreme population II.

The observations of SHAPLEY and others revealed also the existence of a bifurcation of the characteristic color-magnitude diagram of globular clusters into a horizontal branch near $M_v = 0$, and a nearly vertical band of steadily increasing star density extending to fainter magnitudes and abruptly cut by the limiting magnitude to which the observations reached. Attempts to extrapolate the observations[1] to link them with the main sequence in the solar neigborhood were hampered by the fact that the measures extended only to $m_v = +17$.

---

[1] JOHNSON, H. L., and M. SCHWARZSCHILD: Astrophysic. J. **113**, 630 (1951).

34. Modern observations and interpretation: Dating of globular clusters

The tie-in between the two different kinds of color-magnitude diagrams was not provided until accurate photometric data extending to absolute magnitudes approximately as faint as that of the sun were obtained (154), (155). It was then apparent that the red giant and subgiant branches in these diagrams lead into main sequences which are at approximately the same place as the population I main sequence, but which extend upward only to $M_v = +3.5$. The break-off point at this magnitude was taken to mean that all stars brighter than this had evolved off the main sequence. It was used by SANDAGE (155) to date the typical globular cluster M 3 by the method described earlier in this chapter. An age of 5×10^9 years was found; the theory of HASELGROVE and HOYLE gives an age of 6.5×10^9 years (Chapter II), which is probably more accurate.

The fact that the subgiant and giant branches of globular clusters form a continuous band of stars joining the main sequence, as in the old galactic cluster M 67, implies that the evolving stars, after leaving the main sequence, move continuously towards higher luminosities and lower surface temperatures, until they reach a limiting magnitude and color beyond which their evolutionary paths must have an abrupt discontinuity. The theoretical tracks of HOYLE and SCHWARZSCHILD (Chapter II) have succeeded in explaining these observed subgiant and giant branches. The time scale for stars in this stage is much shorter than that on the main sequence. Therefore, the color-magnitude diagram, which is actually the locus of points reached by stars in a small mass range, is very nearly represented by the actual computed tracks, as shown by Fig. 4, where the theory is compared with observations in M 3 and M 92.

The horizontal branch is a very characteristic feature of globular cluster diagrams, and presumably represents the evolutionary history of stars in this mass range after they reach the top of the giant branch. The theory of stellar evolution has not, at the time of writing, been carried further than the giant stage. The relation of the subsequent evolution to the horizontal branch of globular cluster diagrams is therefore not clear at present, nor is the evolutionary significance of the sharply defined region on the horizontal branch where the RR Lyrae variable stars occur. Some discussion of this and later stages will be given at the end of this Chapter.

35. Effect of chemical composition; fitting of main sequences of globular and galactic clusters

The first accurate color-magnitude diagrams of globular clusters preceded the work on M 67, and the comparison between them is interesting. The main sequences of M 67 and M 3 both break off at about the same point, implying that they have similar ages (*149*). The fitting together of these diagrams may be revised somewhat in the near future, and Table 5 shows that LOHMANN (*130*) derived a smaller age for M 67 than did VON HOERNER (*131*) and SANDAGE (*128*). However, the evolving stars probably have nearly the same masses in the two clusters. Of the different parameters which determine the state of stars, namely age, mass, and composition, only the chemical composition is left as the possible cause of the difference.

Although no spectrophotometric determinations of chemical compositions of globular cluster stars have yet been made, work on spectral classification of individual red giants (*156*), (*157*), (*158*) and of the integrated light from clusters (*159*) has shown that the spectra differ from those of stars in the solar neighborhood. This difference can be explained by the globular cluster stars having a considerably lower abundance of the heavier elements (calcium and the iron group in particular), relative to hydrogen. The theoretical evolutionary tracks of HOYLE and SCHWARZSCHILD, which agree so well with the observations of M 3 and M 92, were computed for a considerably lower abundance of the heavy elements than the accepted solar-system value. HOYLE and SCHWARZSCHILD showed that a higher abundance of these elements causes the giant branch of the track to lie at lower absolute magnitudes, and this is just what is observed in the comparison between M 67 and M 3.

There are some theoretical reasons for supposing that a continuous gradation with age of the heavy-element content in stars might be found, in the sense that the youngest stars would have the largest amounts of these. The globular clusters are certainly old stellar systems, as would be demanded by this hypothesis, but M 67 is also old and yet its chemical composition is apparently similar to that of younger stars in the solar neighborhood.

It is not possible at present to determine the true age difference between M 67 and any of the globular clusters, because of uncertainty in locating the true position of the main sequences of globular clusters. This requires photometry at the limits of the best equipment available today. At the moment, fitting of globular clusters is usually done by assuming that the RR Lyrae stars on the horizontal branch always have $M_v = 0.0$. However, this may not be true. For the same reason, the exact nature of the small difference in color-magnitude diagrams of different globular clusters cannot be located at present. There are certainly small differences, but it is not known at present whether the diagrams should be fitted together by means of the RR Lyrae variables or the main sequences, or possibly even the giant branch.

It is not entirely clear what the theoretical difference between the main sequences in the M_{bol}, $\log T_e$ plane should be, as a function of the heavy-element content, Z. Even if there is no appreciable difference in these coordinates for the range covered in the comparison, there may still be a slight shift in the color-magnitude diagrams, owing to different amounts of the blanketing effect in spectra with strong and weak lines. This effect may cause extreme population II stars to be spuriously called subdwarfs. Thus there is no theoretical reason at present for expecting the globular cluster main sequences to lie as far below the solar-neighborhood main sequence as indicated by the preliminary photometry by BAUM (160) in M 13. A first check of this (161) indicates that the effect is not as large as found by BAUM.

36. Horizontal branch stars in clusters

The existence of a well-populated sequence of stars forming the horizontal branch in the color-magnitude diagrams of globular clusters suggests that this represents the evolutionary track, after leaving the giant branch, for stars with masses near $1.2\,M_\odot$. Counts of numbers of stars in the horizontal branch suggest that the time spent here by a star is comparable with the time spent in the giant branch but that it is much shorter than the time spent on the main sequence. In all the clusters for which accurate measures of colors and magnitudes are available, the giant branch bifurcates; one part is the subgiant branch leading up from the main sequence, while the other leads down, through a sparsely populated region, into the horizontal branch, which is at or near $M_v = 0$. There is a gap between the giant and horizontal branches in, for example, M 13 (162).

In most globular clusters there is a short, sharply defined section of the horizontal branch which is filled exclusively with RR Lyrae variable stars. Quite wide variations exist in the relative numbers of stars on the blue and red sides of this "gap" where the variables occur. As SANDAGE (155) and ARP (162) showed, the number of variables in a cluster is strongly correlated with the relative populations of the blue and red sides of the variable star region, in the sense that clusters with the greatest number of variables (e. g., M 3) have almost equal stellar density on either side of this region, while M 13, with only a few variables, has no stars on the red side of the variable region. There is also a difference in the shape of the horizontal branch from cluster to cluster; for example, it is far from horizontal in M 13 and M 10, where the bluer stars are progressively fainter (164), (162). There is variation also in the faint magnitude limit to which the blue end reaches. Assuming that the horizontal-branch stars represent a later evolutionary stage than the red giants, it is not clear whether a star, after leaving the tip of the red-giant branch, travels back down it, into the left-hand fork, along the horizontal branch from right to left, and down into the faint blue dwarf region, or whether it jumps from the red giant tip to the left end of the horizontal branch, and travels along this from left to right. That a star does travel along the horizontal branch is suggested by the continuity of star counts through the variable star gap in those clusters with an appreciable number of variables, but clusters like M 10 do not necessarily suggest this. The onset of helium-burning or some other mechanism may lead to a structural adjustment which moves the star back towards the horizontal branch. Alternatively, it is possible that the star may first oscillate up and down the giant branch.

The "gap" where the cluster-type variables occur is a very sharply defined region of the H-R diagram (165). The extreme sharpness of the boundaries of this region was confirmed in M 3 and M 92 by ROBERTS and SANDAGE (166) and by WALKER (167). ROBERTS and SANDAGE found an excellent correlation between mean color index and period for the variables in M 3. BELSERENE (168) examined the periods of the variables in M 3 and found evidence for both abrupt and gradual changes, but both increases and decreases were found, with no systematic effect. The observations by ROBERTS and SANDAGE in most cases confirmed these apparent changes, but still no systematic effect was apparent. Hence, although a method is available here in principle for determining the direction of evolution, it has not led so far to a solution of the problem.

A third possibility must be borne in mind. Stars, after leaving the red giant branch, might undergo such structural changes as to move them very rapidly in the H-R diagram, but they might reach a stable configuration whose locus is defined by the horizontal branch. The position which they would take up on this branch might be defined by the amount of mass-loss experienced in the red giant stage, the amount of mixing, and the way in which mixing set in. Those stars finding themselves in a limited range of surface temperature might be unstable against pulsation (regular pulsation seems likely to be a phenomenon depending on the outer layers of a star and not on conditions within the core). After a period of structural stability, the stars might again become unstable and move rapidly off the horizontal branch, probably towards a white dwarf stage characterised by exhaustion of nuclear fuel and star death.

Although so far the existence of stars on a horizontal branch has been established only in globular clusters, there may be a region in old galactic clusters (population I) which is analogous to this. There is an indication of this in the color-magnitude diagram of M 67, where the evolving stars have masses near the evolving stars in globular clusters.

This sequence is much more sparsely populated than the corresponding region in globular clusters, and, if real, must therefore represent a more rapid evolution. It is not known yet whether this feature is a definite characteristic of old clusters like M 67, whether it is characteristic of both high- and low-mass stars in population I, or whether it is unique in M 67.

37. Luminosity functions of clusters

The luminosity function $\varphi(M)$ for any assembly of stars is defined by

$$dN = \varphi(M)\, dM,$$

where dN is the number of stars per cubic parsec with absolute magnitude lying between M and $M + dM$. In a group of stars of different ages, such as the stars in the solar neighborhood, the observed luminosity function depends upon both the distribution of masses with which the stars formed (the original mass function), and the removal of all but the younger members of the bright end of the function through the effects of stellar evolution. In coeval groups, however, only the original mass function is important, since the main sequence is always effectively undepleted, up to some limiting magnitude above which all stars have been removed by the effects of evolution.

SALPETER (169) showed that a change of slope in the observed solar-neighborhood luminosity function near $M_v = +4$ could be accounted for by partial depletion of the brighter stars by evolution, and derived an "initial" luminosity function, $\psi(M)$. From this he obtained an initial mass function, $\xi(\mathfrak{M})$, defined by

$$dN = \xi(\mathfrak{M})\, d(\log \mathfrak{M})\, \frac{dt}{T}$$

where dN is the number of stars in the mass range $d\mathfrak{M}$ created in the time interval dt per cubic parsec, and T is the age of the oldest stars observed. The rate of star formation was assumed to be constant during time T. The form of the mass function which he derived is

$$\xi(\mathfrak{M}) = 0.03 \left(\frac{\mathfrak{M}}{\mathfrak{M}_\odot}\right)^{-1.35}$$

in the range 0.4 to 10 solar masses.

Comparison between $\varphi(M)$, $\psi(M)$, and $\xi(\mathfrak{M})$ in the solar neighborhood and in various clusters leads to the following results. Whereas the general luminosity function, $\varphi(M)$, drops considerably below the values in galactic clusters at the bright end of the main sequence, the "initial" luminosity function, $\psi(M)$, shows good agreement down to about $M_v = +6$, and consequently the initial mass function is also approximately the same in the solar neighborhood and in those galactic clusters which have been studied (105), (106), (127), (170), (171). The effect of loss of low-mass stars through evaporation is apparent, however, particularly in the old galactic cluster M 67 *(172)*. Luminosity functions in globular clusters agree with the solar-neighborhood $\psi(M)$ over the available range of overlap *(173)*, *(127)*, *(174)*.

The luminosity functions of stars off the main sequence in M 67 and M 3 have been used by SANDAGE *(175)* to derive semi-empirical evolutionary tracks, since stars in a given stretch of the subgiant or giant branches must have come from a stretch on the main sequence containing an equivalent number of stars. By this means the times to travel various segments of the subgiant and giant branches have been obtained.

IV. Properties of stellar populations; problems of star formation and evolution on the galactic scale

38. Stellar populations

The concept of different stellar populations has repeatedly been used in differentiating between galactic and globular clusters, and indeed this concept is so familiar today that it permeates all discussion of stellar evolution. Before discussing stellar evolution on the galactic scale, however, we shall briefly recall Baade's work and describe the situation as it appears at present.

In 1944 BAADE *(153)* succeeded in resolving the stars in the central region of the Andromeda Nebula and in its two elliptical companions, M 32 and NGC 205, by using red-sensitive plates and a fairly narrow-band red filter. He thus showed that these regions were populated by stars whose brightest members were red, high-luminosity giants. This population was different from that characterising the solar neighborhood and spiral arms of galaxies, where the brightest members were high-luminosity O- and B-type stars. It was apparently similar to the population of stars in globular clusters, where also the brightest members were red giants.

Thus BAADE introduced the concept of two kinds of stars: population I was that found in spiral arms of galaxies, galactic clusters, and the solar neighborhood. Besides the O and B high-luminosity stars, the classical cepheids, M supergiants, and the carbon and S stars were among the typical members of population I, and the presence of dust was characteristic. Stars in the nuclei, halo regions, and between the arms of spiral galaxies, throughout elliptical galaxies, and in globular clusters were called population II. The most characteristic members of population II were the RR Lyrae variables. Dust was always absent in population II regions. A few members of population II might be found in the solar neighborhood, but were characterized by high space velocities relative to the sun, since they had come into the sun's vicinity from other parts of the galaxy with galactic orbits different from the sun's. These results were foreshadowed many years earlier by the work of OORT (176).

Later work, and a combination of studies of galactic structure, stellar distribution, and stellar motions showed that although populations I and II provided convenient general divisions, they themselves could be subdivided and actually there was a continuous range of populations. These conclusions are based on the work of many astronomers, notably OORT (177), PARENAGO (178), and VON WEIZSÄCKER (19). The following is a schematic subdivision of the whole range in our Galaxy into convenient classifications, which was discussed at the Vatican Conference on stellar populations (179):

Extreme population II (halo, globular clusters),
Intermediate population II
Disk population
Intermediate (older) population I
Extreme population I.

The physical parameters which lead to this division are age and chemical composition. How these are correlated with the parameters of location and velocity in the Galaxy will be further discussed in the remainder of this chapter.

39. Spatial and velocity distribution of different kinds of stars

The spatial distribution of stars in the Galaxy may be described by sets of ellipsoids or subsystems with varying degrees of flattening toward the equatorial plane of the Galaxy. If we consider the Galaxy in its initial stages of condensation into stars, and if this began soon after or at the same time as the Galaxy became differentiated from the rest of the Universe, then the first stars to condense would presumably have a roughly spherical distribution (like the system of globular clusters and halo stars). Stars, once formed, would remain "frozen" into the configuration possessed by the Galaxy at the time of their condensation, while the remaining matter would lose turbulent energy and begin to contract towards the equatorial plane defined by the rotation of the Galaxy. Thus stars forming at successive epochs would to a large extent delineate the shape possessed by the material out of which they formed, i. e., the shape of the Galaxy at the time of their formation; cf. (16).

Evidently the correlation of position of stars in the H-R diagram with the ellipsoids or subsystems describing their spatial distribution is a useful clue in studying relative ages of and evolutionary links between various kinds of stars.

Stellar velocity distributions are strongly correlated with spatial distributions. Stars in highly flattened subsystems, like the young stars of high temperature and luminosity, have small velocities in the direction perpendicular to the galactic plane (the z-direction), and their total velocities are close to the circular velocity given by Keplerian motion at the corresponding distance from the galactic center. Stars in intermediate subsystems have larger z-velocities in order to maintain such a distribution against gravitational attraction by the main body of the Galaxy; their galactic orbits thus have an appreciable inclination to the galactic plane. Stars forming a spherical subsystem (globular clusters, halo stars) have the largest z-velocities.

Components of stellar peculiar motions in the galactic plane are correlated in magnitude with the z-component, but here an additional factor must be considered besides the peculiar motion of the star at the time of its formation. SPITZER and SCHWARZSCHILD (180) showed that the largest interstellar cloud complexes (of masses $10^6 M_\odot$) could accelerate stars and cause an increase in velocity with age. Since the interstellar clouds lie close to the galactic plane, such an effect would be mainly apparent in the components of velocity in the plane, and would also be larger for stars that spend a larger proportion of their time in the plane. Whether or not this effect is important, the end result is the same: older stars, on the average, have larger peculiar motions, and stars of similar ages and past histories should have similar velocities.

The ordering of stars in different parts of the H-R diagram according to spatial distribution and velocities has been studied for many years by many workers. This work has been described in the book by TRUMPLER and WEAVER (181). Reference may also be made to PAYNE-GAPOSCHKIN (182). As an example, we reproduce in Table 6 the correlation between velocity and spectral type which has been obtained by PARENAGO (178). Here σ_1, σ_2, and σ_3 are the components of the velocity ellipsoid, where σ_3 is in the direction which makes a fairly small angle with the direction of the galactic center. PARENAGO has also grouped the stars into subsystems according to spatial distribution as follows:

Very flat: interstellar gas, B stars, classical cepheids, supergiants, galactic clusters.

Rather flat: A, gA, gF, gG, gM, R and N stars.

Intermediate: dG—dM stars, red variables, long-period variables except for those with periods 150—200 days, subgiants, white dwarfs, planetary nebulae.

Spherical: long-period variables with periods 150—200 days, subdwarfs, high-velocity giants, globular clusters, RR lyrae variables.

This grouping clearly shows that the youngest systems have a very flat distribution, while the oldest are spherical. Those of intermediate age lie between. The correlation between velocity dispersion and shape of subsystem is also clear, and the correlation of both with age is in agreement with our previous arguments.

However, the progression of velocities and shape of subsystem as the main sequence is descended is not yet fully explained. There is an approach towards equipartition of kinetic energies between stars of different masses, but complete equipartition is not attained (*183*), (*184*), (*185*), (*186*). Several factors may contribute to this. In the first place, it is conceivable that during star formation some kind of quasi-equipartition is set up in a condensing agglomerate, so that the stars of lowest mass might be formed with the largest peculiar motions. Secondly stars fainter than $M_v = +4$ are on the average older than stars brighter than this, since the oldest brighter stars will come to the end of their lives in times less than 6×10^9 years while the oldest fainter ones will still be in existence. Thus either the accelerating mechanism due to the interstellar clouds or the effects of decaying hydromagnetic turbulence and equatorial contraction of the Galaxy will result in larger velocities for the lower part of the main sequence; the latter would not necessarily give a continuous increase of velocities with decreasing mass, and indeed a levelling off is found. Thirdly PARENAGO (*178*), (*187*), has interpreted the change in velocity dispersions as being due to a real break in the main sequence between types G4 and G7, stars in the two parts being thought to have different origins and evolutionary histories.

Table 6. *Components of the velocity ellipsoid for stars of different spectral types, after* PARENAGO (*178*)

Sp. Type	σ_1 km/sec	σ_2 km/sec	σ_3 km/sec
B0-B5	4.8	7.6	9.7
B 7-A 2	5.5	10.5	16.2
A 3-A 8	7.9	9.3	19.1
A 9-dF 1	9.5	12.8	23.9
dF 2- dF 4	11.7	17.0	26.8
dF 5-dF 7	16.7	21.4	31.8
dF 8-dG 2	22.6	27.5	46.0
dG 3-dG 7	27.3	29.6	49.7
dG 8-dK 2	22.9	30.5	52.0
dK 3-dK 6	22.6	29.2	50.6
dM	22.5	25.6	45.7
Cepheid Variables	5.4	8.6	12.5
gA-gF	10.3	14.2	26.8
Subgiants	23.7	27.1	42.5
gG 0-gG 8	14.6	17.9	25.6
gG 9-gK 1	15.7	20.5	30.5
gK 2-gK 5	17.3	20.5	30.6
gM	16.3	22.5	31.2
Supergiants	7.7	10.2	12.7
High-velocity giants	100	60	50
Red variables	18.7	23.1	37.8

A link with the work on luminosity functions can be seen. There is a slight deficiency of faint stars in some relatively young galactic clusters. This is in agreement with the velocity dispersions of stars at different points on the main sequence, and again the question is raised as to whether there is a real difference between conditions of star formation at an earlier epoch, when the Galaxy contained more interstellar matter, and recent times, when star formation is confined to the galactic plane and the amount of interstellar matter is a smaller fraction of the total mass of the Galaxy. In a discussion by VAN WIJK (*188*) of the place of

Die Entstehung von Sternen

origin of the high-velocity stars in the solar neighborhood, the possible effect of local density and amount of turbulence on the luminosity of the resulting stars is mentioned.

The correlation between the chemical composition of stars and their population characteristics has been mentioned previously. Extreme population II has a lower abundance of the heavier elements, relative to hydrogen, and there is in general a gradation of heavier-element content running through the various population types. This is reflected in the fairly good correlation between velocity and those spectral characteristics that depend on the heavier element content (189), (190).

Vyssotsky (191) has shown that stars of spectral types dG, gK, and dM can be divided into two groups, A and B, which are distinguished by spectroscopic criteria. The dG stars of population B show strengthening of CH and of the hydrogen lines and weakening of the metallic lines. The gK stars of population B show strengthening of CH and Ca I λ 4227.

Table 7. *Velocity dispersions in Vyssotsky's population groups A and B*

Spectral Type	Dispersion (km/sec)	
	Population A	Population B
A	11 ± 0.3	....
dG	17 ± 1	24 ± 2
gK	16 ± 2	27 ± 2
dM	18 ± 1	>30

The dM stars of population A are the dMe stars, with hydrogen or Ca II emission. The two groups A and B have different velocity dispersions as may be seen from Table 7 (191). According to the velocities, population A is related to population I, flatter subsystems, and the spiral arm population; while population B is related to the less flat, intermediate subsystems. The very noticeable division at type dM confirms earlier work by Delhaye (192) and suggests that the dMe stars are relatively young; the dMe stars with hydrogen emission as well as Ca II have a velocity dispersion of 15 km/sec, an even lower value than that in Table 7. The possible connection between these stars and T Tauri stars was discussed in Chapter III.

Detailed work on the spatial and velocity distributions of particular classes of stars should continue to give information on the genetic relationships between these classes. For example, the gK stars are most closely correlated with the F main-sequence stars[1], and Sandage (133) pointed out that most of the K0—K2 normal giants in the solar neighborhood should have evolved from the main sequence at type F, although a certain proportion should have come from type A.

The distribution of the M giants shows they apparently belong to an intermediate subsystem and do not delineate spiral structure. They show a strong concentration towards the galactic center, which is more

---

[1] Vyssotsky, A. N.: Astron. J. **56**, 62 (1951).

pronounced for the late type M stars than for the earlier *(193), (194), (195)*. Thus the majority of the M giants are probably fairly old, and have therefore evolved from stars that are not very massive and lie on the main sequence perhaps in the late A—F range. The discovery of ordinary M giants in galactic clusters is awaited with great interest; their spatial distribution suggests that they might be found in clusters like M 67. So far, they have been found in one cluster only, NGC 6940 *(195a)*.

40. Spiral structure in the Galaxy

O-associations, H II regions, individual B stars, and neutral atomic hydrogen all delineate spiral structure when their distributions are plotted spatially *(196), (197), (198)*. In recent years the 21-cm observations have proved to give an extremely powerful technique. The 21-cm observations suggest that the neutral atomic hydrogen in our Galaxy comprises only ~ 1% of the total mass *(199)* and is very closely confined to the spiral arms. Since it seems unlikely that a comparable amount of interstellar matter is tied up in the form of molecular hydrogen and dust grains, the total mass of interstellar matter is therefore a small proportion of the mass of the Galaxy.

It has been made clear in the preceding sections that young stars may be expected to be confined to regions where interstellar gas and dust are located. This, with the exceptions noted in Section 20, is in general found to be the case. Therefore, we may conclude that star formation at the present time is confined to the spiral arms of our Galaxy *(200)*, i. e., to a region containing only a small proportion of the total mass. Apart from the O and B stars, certain other groups of stars which belong to flat subsystems, and which are therefore relatively young, might be expected to delineate spiral structure. These are the M supergiants, the S and some carbon stars, and the classical cepheids.

The early M-type supergiants are in fact associated with H II regions and O and B stars, confirming the genetic relationship shown by the observations in h and χ Persei *(193)*. The S stars are also spiral arm objects, associated with O and B stars, and therefore probably represent a later evolutionary stage of massive stars *(201), (202)*. Carbon stars show clustering tendencies; there is apparently a separation between classes R and N *(193)*, the N stars showing a tendency to spiral structure.

In the northern hemisphere, association between carbon stars and O and B stars has not been found, but such an association has been found in the southern hemisphere *(194)*. The distribution of the classical cepheids does not in general delineate spiral structure, except for some indications in the southern hemisphere *(203)*. Except for one probable member of h and χ Persei, the classical cepheids are not associated with OB stars *(118)*, but are apparently correlated with stars of late B type. Some classical cepheids have been shown to be members of galactic clusters *(204)*, *(205)*.

41. Stellar evolution in external galaxies: General remarks

To determine whether stars in members of the local group and in other nearby galaxies have similar mass ranges, population characteristics, chemical compositions, and evolutionary tracks to those in our Galaxy, colors, magnitudes, and spectra of individual stars would ideally be required. However, such data are limited to stars of high luminosity, and less precise data can give much information. Integrated spectra studied by MORGAN and MAYALL *(206)*, following the pioneer work of HUMASON and MAYALL, are a very useful tool. Integrated colors have been studied by many workers of whom we will mention only HOLMBERG *(207)* and STEBBINS and WHITFORD *(208)*. These colors show relatively little dispersion among elliptical nebulae, but there is a considerable range in the colors of spiral galaxies, indicating a range in the relative proportions of the stellar populations that are contributing to the light.

A search for the population indicators such as classical cepheids, OB stars, H II regions, and, in the nearer galaxies, RR Lyrae variables, can also be made. Another useful indicator of the stellar population of a galaxy, which may therefore throw light on conditions for star formation and stellar evolution, is the mass-to-light ratio *(209)*, *(210)*, *(211)*. This depends upon difficult observations with large telescopes in order to derive masses from the rotational velocities of galaxies, or else upon certain assumptions when the velocity dispersions in groups of nebulae are used.

Finally, 21-cm observations to yield the total mass and distribution of neutral atomic hydrogen may lead to tentative conclusions about the relative rates of present-day star formation in different galaxies. Some results of observations and their bearing on evolution in other galaxies are given in the succeeding paragraphs.

42. M 31

M 31 is thought to be similar to our Galaxy in form, content, and chemical composition. Both are Sb systems *(212)*, *(213)*. In M 31, high-luminosity blue stars are confined to the spiral arms, where the interstellar dust and H II regions are also located, and thus the usual correlation between young stars and interstellar matter is observed. The amount of dust in the spiral arms decreases with increasing distance from the center *(200)*. An investigation of the relative ages of the young stars in the inner and outer arms would be interesting. A study of the novae in M 31 (*214*) showed that their distribution is intermediate between a highly flattened disk population and a nearly spherical globular cluster population.

The 21-cm observations made at Leiden suggest that the fraction of the mass which is neutral hydrogen is only a few percent (*215*), as in our Galaxy, although the results of HEESCHEN (*216*) lead to more neutral hydrogen than do the Leiden results, and we await clarification of this discrepancy. However, we conclude that probably the rate of star formation at present in M 31 is similar to that in our Galaxy and is also confined to the spiral arms.

Spectra of individual stars and integrated spectra do not indicate any difference in chemical composition between M 31 and our Galaxy. MORGAN and MAYALL (*206*) have constructed an H-R diagram that would reproduce the observed spectrum of the nucleus; it has a giant branch running from G 8 to M, a Hertzsprung gap, and a main sequence containing F 8 to G 5 stars. Fainter main sequence stars could be present in the same proportions as in the solar neighborhood without contributing to the integrated spectra. The disk of M 31 has probably the same population as the nucleus. It is noteworthy that spectra indicating a deficiency of metals and carbon, oxygen, and nitrogen are not found: thus most of the light in M 31 comes from stars that are apparently "old population I" or disk population, and this is believed to be the case in our Galaxy also.

This conclusion is also reached from a study of the count-to-brightness ratio in M 31, i. e., the number of resolved stars brighter than a definite limit, divided by the total surface brightness (*217*). This is clearly quite sensitive to the stellar population, and gives information about a similar range of the luminosity function to that given by integrated spectra. The value for the main body (disk) of M 31 agrees well with that computed for the solar neighborhood by means of van Rhijn's luminosity function and differs from that computed for the globular cluster M 3. BAUM and SCHWARZSCHILD (*217*) have therefore concluded that the main contribution to the light from the main body of M 31 is from old population I stars similar to those in our Galaxy. The color of M 31, $+1.0$ (*208*), is also consistent with this.

The most recent study of the mass-to-light ratio measured in solar units, f, in M 31 (*211*) yields the value of 16, and the data are consistent with this being constant over the nucleus and disk. The mass-to-light ratio gives information about the luminosity function far down the main sequence, where the stars contribute to the mass but not to the light. Using a value of $f = 4$ in the solar neighborhood, SCHWARZSCHILD computed that 77% of the mass of M 31 is due to population II stars while 92% of the light is from old population I stars. For a mass-to-light ratio of 2.5 in the solar neighborhood, as given by GLIESE (*120*), the corresponding figures are 86% and 91%, respectively.

From these different methods of analysis, it seems probable that the evolutionary history of M 31 and its stars is very similar to that in our Galaxy.

43. M 33

Data for M 33, which is of type Sc, are not as extensive as for M 31. Red supergiants are found in the spiral arms in association with blue high-luminosity stars, in agreement with our picture of the evolution of these stars in our Galaxy. They are being studied at present by HUMASON and SANDAGE. Preliminary 21-cm observations (*218*) suggest that neutral atomic hydrogen contributes about one fifth of the mass of M 33. However, preliminary Leiden observations (*219*) do not confirm this and suggest a smaller fraction.

The color of M 33 is bluer than that of M 31; $C = 0.63$ (*208*). The integrated spectra also reveal an entirely different distribution of stellar population in the H-R diagram (*206*). Although data for constructing an H-R diagram that would reproduce the spectrum are not yet available, the percentage of A- and F-type stars must be far higher in the main body of M 33 than in that of M 31.

The mass-to-light ratio, f, also indicates a difference in stellar population. Its value is 4 (*211*), similar to that in the solar neighborhood. Probably the mass and light are largely contributed by intermediate population I stars.

These facts all suggest that the present-day rate of star formation relative to the total mass is larger in M 33 than in M 31, so that a larger proportion of brighter main-sequence stars is present.

44. The Magellanic clouds

Data on the Magellanic Clouds have been taken from a review article by BUSCOMBE, GASCOIGNE, and DE VAUCOULEURS (*220*), where full references are given. The Clouds were formerly classified as irregular, but primitive structure suggestive of barred spirals has been found, particularly in the Large Cloud. Many OB stars, H II regions, and bright red variables indicate that population I is well represented in both Clouds, while globular clusters containing RR Lyrae variables show that population II is also present, although it probably makes up a much smaller proportion than in M 31 and our Galaxy. Relatively few novae (intermediate population II) have been observed. Many galactic clusters are present in both clouds. Neutral hydrogen is present in both Clouds in large quantities (227), and the fraction of the total masses made up by this is probably of the order of a half in each case.

On the other hand, dust is observed in the Large Cloud but not in the Small Cloud; its absence in the latter case while neutral hydrogen and OB stars are present is surprising. Possibly the lower density of the Small Cloud has an effect on the formation of dust. Another possibility (not suggested in the review article quoted), which cannot be ruled out at present, is that the chemical composition of the Clouds, particularly the Small Cloud, is different from our Galaxy in having a smaller abundance of the elements heavier than helium.

The mass-to-light ratio, f, is 1.5 in the Small Cloud and 0.7 in the Large Cloud, and this low value suggests a higher proportion of extreme population I stars than is present in the solar neigborhood. The blue colors of both Clouds are also consistent with this; the color index on the (P, V) scale of Eggen is $+0.18$ in the Large Cloud with no change between the central core and the outer regions, while in the Small Cloud it increases from -0.03 in the core to $+0.45$ in the outer parts. This latter result, which should be regarded as preliminary, is surprising since it indicates a higher proportion of bright blue stars in the central regions of the Small Cloud.

Luminosity functions for the brightest stars, as far as the data are available, seem to be similar to that for the bright stars in the solar neighborhood, but preliminary color-magnitude arrays (particularly in the Small Cloud) differ from anything found in our Galaxy in the large proportion of Ib supergiants of types F—K, which form a maximum in a band extending from color index -0.3 to $+1.5$.

The preponderance of population I and the high proportion of neutral hydrogen suggest that the present-day rate of star formation in the Clouds may be larger than in M 33, and very much larger than in M 31 and our Galaxy, although the absence of dust in the Small Cloud poses a problem. If there are evolutionary sequences in galaxies, then we might suggest that the Clouds are relatively young, and a chemical composition having a lower proportion of heavier elements would then not be surprising. The young and middle-aged stars in the Clouds might thus resemble the oldest stars in the Galaxy as regards chemical composition. Work is being carried out at present by ARP on color-magnitude diagrams in the Small Cloud, and should throw light on the evolutionary tracks of stars in clusters and in the general field. The latter may be, as in the solar neighborhood, a composite of evolutionary tracks for stars in a range of masses and ages. It will be interesting to know whether the apparent preponderance of F—K supergiants among the hight-luminosity stars is real, and if so, from what part of the main sequence these stars might have come.

45. Elliptical galaxies in the local group

Resolution of the brightest stars in M 32, NGC 185, NGC 205, and the dwarf systems in Sculptor, Fornax, Draco, and Leo I and II shows that they are bright red giants as in the nucleus of M 31. The integrated colors, where available, are close to the mean value of $+0.86$ for elliptical galaxies as a group, with the exception of NGC 205, which has a color of about $+0.7$. Integrated spectra are not included in the lists by MORGAN and MAYALL (206), but other ellipticals (not in the local group) have spectra which indicate that most of the light comes from K giants. Again NGC 205 is an exception: probably F-type stars contribute largely to its light.

M 32, NGC 205, NGC 147, and NGC 185 were searched for novae in the same survey as that of M 31 (214), and none were found. The absence of dust in elliptical galaxies has often been commented upon since BAADE pointed out that this is correlated with systems being composed of population II stars. Once more, NGC 205 is an exception, since some dust clouds are present, and BAADE (212) has pointed out that near these there are some blue stars. This is a striking confirmation of the correlation between bright blue stars and the presence of dust.

Although dust is absent from most elliptical galaxies, interstellar gas may be present, as is shown by the frequent observation of the [OII] emission lines at $\lambda 3727$ in their spectra (222). By arguing by analogy with globular clusters, where the detailed color-magnitude diagrams and theoretical and empirical evolutionary tracks enable the rate of star deaths and consequent ejection of gas to be estimated (127), it has been shown that the observed proportion of gas may be accounted for by ejection from stars; if this gas has been processed in the interiors of stars, the proportion of hydrogen to elements heavier than hydrogen may be much less than normal. We may at least be certain of one fact, the present-day rate of star formation in elliptical galaxies must be very much less than that in the Sb systems M 31 and our Galaxy.

The mass-to-light ratio in M 32 (211) is $f = 200$, where the mass is derived upon the assumption that an asymmetry in velocities and form in M 31 is due to the gravitational pull of M 32. It is remarkable that this ratio is so much higher than that in the solar neighborhood and in the galaxies already discussed. Support for this high value is given by similarly high values found for other elliptical galaxies which are not members of the local group. Firstly, a direct determination in the nearby system NGC 3115 leads to a value of $f = 100$. Secondly, study of the orbital motions of pairs of galaxies (224) gives an average value of $f = 300$; while this is uncertain, more weight can be given to the factor of 6 by which f in average ellipticals exceeds f in average spiral galaxies.

Thirdly, an uncertain and remarkably high value of $f = 800$ has been determined in the Coma cluster of galaxies (211). However, this depends on a number of uncertain factors: the small velocity sample available for calculating the kinetic energy may not be representative; there may be many more faint galaxies in the cluster than has been thought.

There have been suggested (211) three possible explanations for the high value of f in ellipticals, all of which imply different conditions for star formation than those in our Galaxy and in M 31. Firstly, the luminosity function might be similar in form to that in the solar neighborhood but shifted about 5 magnitudes fainter. The average stellar mass would then be about a quarter the average near the sun. Secondly, the luminosity function might have a different form from that near the sun, with a much larger proportion of faint dwarfs. Thirdly, the excess mass might be contributed by white dwarfs, which might be much more numerous than near the sun. This would imply a much higher rate of star deaths in earlier times, and consequently a larger number of massive stars. The original luminosity function might have been similar in shape to that in the solar neighborhood but shifted 7 magnitudes brighter. The average stellar mass would have been 6 times larger.

The high mass-to-light ratio in elliptical galaxies points to a difference in stellar content between them and globular clusters, where the ratio is only of order 1. The colors are different also; the average color index of globular clusters is $+0.50$ as compared with $+0.86$ for elliptical galaxies. The spectral energy distribution curve in M 32 is also different from the average in globular clusters (225), and the integrated spectra so far observed indicate a possible difference in chemical composition (159), (206) in that the elliptical nebulae do not indicate a lowered abundance of the metals. And yet both have been classified as "pure population II"; the brightest stars in both are similar, and the count-to-brightness ratio in one elliptical, NGC 205, is very similar to that in the globular cluster M 3 (217). A preliminary color-magnitude diagram for the Draco system (226) looks very similar to that for globular clusters, and RR Lyrae variables have been found in that system. Possibly the stellar content of dwarf ellipticals like the Draco system is not the same as that in larger systems like M 32 (227).

A theoretical luminosity function for M 32 has been constructed so as to be consistent with the observed spectral energy distribution and to yield a mass-to-light ratio of the right order of magnitude (228). Fainter than absolute magnitude $+6$ this theoretical function rises much more steeply than the function in the solar neighborhood, corresponding to the second possible explanation given above (211); the proportion of faint dwarfs which contribute to the mass but not to the light is very much larger. Thus we see once again a hint that under different physical conditions, the mass distribution function in star formation may change and yield different luminosity functions.

46. Intergalactic star clusters in the local group

The National Geographic Society-Palomar Observatory Sky Survey plates have revealed the existence of objects which may be called intergalactic star clusters[1, 2]. These objects resemble poor globular clusters like NGC 4147, but they are at very great distances which place them right outside our Galaxy, in intergalactic space. It is not known at present whether they originated in the Galaxy and escaped from it, whether they have escaped from some other galaxy in the local group, or whether they had an independent origin. Radial velocity measures might enable the first possibility to be tested.

A study of the stellar population of these clusters, by means of color-magnitude diagrams, luminosity functions, and a search for variable stars will be particularly interesting in considering their possible origin. Some theoretical discussion concerning their possible origin has been given in Chapter I.

A preliminary observational study of one cluster (the "11^h cluster") has been made from the Sky Survey prints, and a color-magnitude diagram derived (52). A more detailed investigation of this cluster, and a preliminary study of another, the 10^h cluster, have been made from 200-inch Palomar plates (53). By assuming that the horizontal branch is at $M_v = 0.0$, their distance moduli are found to be 20.5 and 20.6 magnitudes, respectively (125 kiloparsecs). The luminosity function of the 11^h cluster, when scaled up by a factor of 14, agrees well with that for the globular cluster M 3, down to the limit of observations at $M_v = +1$. From this the integrated apparent magnitude is found to be $+14.3$. Some striking differences are found, however, between this cluster and M 3. Its diameter is about twice as large, and there is little or no central concentration. The horizontal branch is well populated but very short; it extends only from the red giant branch to the place where the RR Lyrae variable star location usually starts in normal globular clusters. There are no RR Lyrae variables. In the 10^h cluster, however, the horizontal branch appears to be more normal and there is one RR Lyrae variable. In the 11^h cluster there are two red variables, with periods probably in the range 100—200 days, and these are among the brightest stars in the cluster.

Another of the clusters, from a preliminary uncalibrated study, has a typical population II giant branch, leading into the beginning

---

[1] ABELL, G. O.: Publ. Astr. Soc. Pacific **67**, 258 (1955).
[2] WILSON, A. G.: Publ. Astr. Soc. Pacific **67**, 27 (1955).

of the main sequence, but no horizontal branch at all. More studies are clearly desirable. In spite of the general similarity to poor globular clusters in the Galaxy, there seems to be a difference in the evolutionary tracks of the stars which may be important.

47. More distant galaxies

Studies in more distant galaxies can be made by means of integrated spectra, colors (including spectral energy distribution curves), and, in some cases, the mass-to-light ratio. These may be correlated with the classification sequence according to galactic form. As has already been mentioned, the elliptical and S0 galaxies show fairly little dispersion in color about the mean value of $+0.86$, while spiral galaxies range from quite blue colors for irregular, magellanic, Sd, and Sc systems to colors similar to those of ellipticals for the Sa and Sb systems, and within each subdivision there is a considerable dispersion.

The spectra show a similar correlation (*206*), (*229*). Among those so far classified, the A-systems comprise irregular and normal and barred spirals of subdivision c; the AF systems are all of Sc type; the F-systems contain Sb and Sc; the FG systems are mostly Sb; the K-systems contain subdivisions a and b of both normal and barred spirals, ellipticals, S0 systems, and one irregular system.

So far, no evidence of differences in chemical composition have been detected, but only large differences are likely to show on low-dispersion spectra. There are, however, a few cases of lack of correlation between colors and spectra where the possibility of this being due to a remarkably low abundance of elements heavier than helium should not *a priori* be ruled out. NGC 3034 and NGC 3077, in the M 81 group, have early-type spectra (showing principally the Balmer series of hydrogen) and reddish colors (about $+0.8$) (*222*). Although these galaxies are associated with M 81, they need not necessarily have the same composition. If the chemical elements have been synthesized in stars, then if these galaxies actually condensed into stars at a later time than M 81 they would represent an earlier stage in the chemical evolution of galaxies. This, however, is very speculative. Possibly the colors are due to the presence of dust.

Most of the colors and spectra of galaxies outside the local group can be explained in the following way. The ellipticals and lenticulars contain mostly intermediate (not extreme) population II stars or old disk population I corresponding to subsystems of intermediate flattening in our Galaxy, together with some extreme population II. The spirals contain a mixture of young population I (flat subsystems) and old disk population I, together with some population II. The old population is contained in the nucleus and main body while the young population is located in the spiral arms. On going from subdivision c to a, i. e., from systems with small nuclei to those with large nuclei, the proportion of the old population increases.

The high mass-to-light ratio in ellipticals has already been mentioned, together with its implication that there is a different luminosity function for stars in these systems. At present the problems of stellar content, involving stellar formation and evolution, in different kinds of galaxies pose a large number of questions which still remain to be answered. How much extreme population II (similar to globular clusters in our Galaxy) is present in any class of galaxy? The resolution of the brightest stars in the nucleus of M 31, M 32, and their globular clusters at the same absolute magnitude certainly suggests a fair proportion in these systems. In what respect do the globular clusters of spiral and elliptical systems differ from or resemble the globular clusters of our Galaxy or various parts of the galaxies themselves? Is there a difference in population between dwarf and giant ellipticals? Is there a range of chemical compositions and, if so, is it correlated with structural classification?

A final point concerns the colors of the most distant elliptical nebulae so far observed. The apparent excess reddening (Stebbins-Whitford effect) which was thought to be present implied that there was a real difference in stellar content that was correlated with distance, and which was therefore interpreted as an aging effect (*225*), (*230*). The relatively short time scale spanned by the farthest measures did not, however, suggest that great changes in stellar content would be apparent, and it now appears probable that this excess reddening is not present (*231*).

48. Are there evolutionary sequences of galaxies?

As was discussed in Chapter I, it is generally supposed that a protogalaxy at the time of its formation as a separate unit is wholly composed of gas which contains some density fluctuations, and a certain amount of angular momentum and magnetic field. Stars begin to condense in this system, and since the rate of star formation must be closely connected with the amount of gas present, it is logical to assume that this rate is a function which decreases as the galaxy ages. If massive stars are formed at an early stage they will go through their life-history rapidly. At the end they must eject matter back into the galaxy in the form of gas, but a white dwarf remnant may be left which will play no further part in the interchange of matter between stars and the interstellar medium. Also, the lowmass stars which will continuously be formed as part of the natural mass-distribution will go through their life-histories so slowly that they will not have evolved appreciably in current estimates for the age of our sample of the universe. Thus, as time goes on, an increasing proportion of the mass of a galaxy will become tied up in the form of inert white dwarfs and low-mass red dwarfs.

It is natural to ask whether galaxies of different types form an evolutionary sequence or whether the structure of a galaxy is determined by some initial parameter, such as angular momentum or initial magnetic field. With regard to Hubble's original "tuning fork" classification diagram, he specifically remarked that the nomenclature "early" for the subdivision a and "late" for the subdivision c was not to be taken to have temporal implications, but that the sequence of classification was purely empirical. An early suggestion (232) was that a galaxy, shrinking under its gravitational attraction and possessing initial angular momentum, would tend to progress through configurations like those given in the sequence $E0 \to E7 \to Sa \to Sb \to Sc$. Individual galaxies might stop at any point in the sequence from want of angular momentum. However, the evidence on stellar populations today tends to argue against this.

That the sequence might be evolutionary in the opposite direction, $Irr \to Sc \to Sb \to Sa \to S0 \to E$, was suggested by SHAPLEY (213) because, in the light of present knowledge of stellar populations in different systems, there is a steadily decreasing proportion of young stars and interstellar matter, on the average, and an increasing proportion of old stars, in going this way along the sequence; see also (233). Once a galaxy has come to be formed principally of small stars (elliptical system), one cannot see how it could ever again contain much gas and become "young" again, with the formation of massive stars. On the other hand, a young (spiral) system, with much gas and many massive stars, could become old, but whether it would, when old, look like an elliptical galaxy is another matter.

Theoretical arguments based upon the decay of initial turbulence, flattening towards the equatorial plane, and the transfer outwards of angular momentum and its eventual loss from the system also suggest that evolution would be in the sense suggested by SHAPLEY. They have been given by VON WEIZSÄCKER and his associates (19), (234) and references given therein, (235). According to this theory, the time scale for a galaxy will depend on its mass, since the time scale for flattening connected with the decay of the original turbulence may be of the same order as one rotational period, while the time scale for loss of angular momentum (with the consequent growth of the nucleus of a galaxy) might be 10 or 20 times the rotational period. Thus galaxies born at the same instant of time might have different genetic "ages", the smaller ones becoming "old" more rapidly.

A critical discussion of this possible evolutionary sequence, with particular reference to the colors of galaxies, has stressed the large range in the colors of irregular and spiral galaxies of all subdivisions (236). Galaxies of the whole range of structural types may have the same color and hence, by implication, the same stellar content. If the same stellar content in turn implies the same age, then there is not a simple one-to-one correlation between structural form and age. In certain peculiar galaxies with intense ultraviolet continuum radiation, the blue color index may be due to extremely hot stars, more luminous than in our Galaxy, and presumably very massive. A recent investigation of one or two such galaxies, originally described by HARO (236), has been made by MÜNCH (237). This has shown that the galaxies probably have small masses and their nuclear regions are apparently excited by these highly luminous, hot stars.

HARO indicated three possible interpretations of his comparison between colors and structural classification. First, as one extreme possibility, Hubble's sequence might represent a simple and direct evolutionary progression, from irregular through spiral to elliptical systems. Second, at the opposite extreme, the initial conditions of formation might determine the structural form of a galaxy, and within that fixed structure it might undergo its specific evolutionary process, involving aging of the stars and reddening of the color. Third, a combination of the first two possibilities might represent the truth.

OORT (238) has given a qualitative discussion, arguing that evolution through the dichotomy between spiral and elliptical forms is determined largely by the initial angular momentum of the protogalaxy.

The initial conditions of angular momentum, magnetic field, turbulent energy, and size of density fluctuations may govern the initial mass-function of the condensing stars and the speed of their formation. For example, perhaps spiral arms form in regions of high magnetic field; the persistence of bars in barred spirals despite rotation might suggest the presence of strong magnetic fields which inhibit differential rotation. If for some reason star formation had been partially inhibited in one galaxy as compared to another, e. g. by magnetic field conditions, then such a galaxy would retain more of its mass in the form of interstellar matter. Again, if the initial mass-functions vary in different galaxies, then this will clearly have an effect on the apparent age of a system as judged from its present-day stellar content.

Finally, the evolution of galaxies inside and outside dense clusters can be expected to be different because of the effect of collisions in the latter. These will sweep out gas and dust from the systems in collision, and the high rate of occurrence of S0 systems in dense clusters such as the Coma and the Corona clusters may be the result of such collisions

(*239*). Although the rate of collisions is considerably less than that originally suggested by Spitzer and Baade (*239*) because of the considerable increase in the distance scale, it still seems possible that galaxies which pass near to the center of such a cluster will undergo a number of collisions in making a single transit across the central region.

Evolution in galaxies is a speculative subject at present, and needs more study from many points of attack, both theoretical and observational. One important approach, currently being undertaken, is that of studying the stellar content of galaxies as deduced from their integrated spectra in relation to their structural features (*206*). Further, the revision of the Hubble classification system by Sandage (*240*), the classification system devised by de Vaucouleurs (*241*), and the classification according to the nuclear regions by Morgan (*242*) may all be expected to contribute to our understanding of galactic evolution.

V. Conclusion

In this paper we have attempted to outline the present status of the problems relating to the formation and evolution of stars, and the properties of the stars in clusters and galaxies. It is of some interest to describe briefly some of the major problems which appear at the present to be of primary importance to solve.

In considering theoretically the condensations of stars out of the interstellar medium, various mechanisms for producing an initial compression have been proposed. Those which demand that dust is an important constituent are in some difficulty because little is known about the physical properties of dust grains, and hence about the reflecting and absorbing properties. Consequently further investigation of the physics of dust grains is closely connected with the problems of star formation.

If it is believed that the density fluctuation from which the first condensation occurs is closely connected with the hydromagnetic turbulence in the interstellar medium, we must appeal to the rudimentary theory of this turbulence. A complication which has arisen since the cosmogony in which turbulence played the major role was developed by von Weizsäcker is the gradual realization that it is necessary to develop a theory of compressible turbulence to represent the conditions in the interstellar medium. This is a matter of extreme mathematical difficulty, but it again bears directly on the mechanisms of star formation.

In the situation in which it is supposed that the process of star formation begins by the compression in the surrounding medium caused by radiation and heating of a highly luminous O or B star, a basic problem is as follows. If it is hoped to explain the formation of associations in this way, a separate mechanism must be invoked to account for the formation of the initial star or group of stars which then lead to the formation of all of the others. The ideas of OORT and SPITZER and BIERMANN and SCHLÜTER have always presupposed the existence of these first young stars. These are only some of the problems which come to mind in discussing the first condensations.

All of the critical masses which have been obtained for these first protostars are very large, as are the masses estimated for such objects as the globules observed in regions where star formation is going on. They are often large enough to be masses of protoclusters. Thus in general it is not these first condensations which contract on to the main sequence to become stars. In order for stars to be formed, either subcondensations must form within the massive one, or else it must lose the major portion of its mass or fragment. The appearance of the H-R diagrams of young clusters suggests that such processes must occur. It is possible that the rotation of the initial condensation may determine the way in which the fragmentation takes place. In this case the process is intimately related to be problem of the loss of angular momentum. Alternatively, it may be that in the process of gravitational contraction, massive protostars pass into non-equilibrium configurations. The form of the mass function for stars in clusters and also in the general field may provide a powerful tool in understanding the processes which occur after the first condensation but before the stars finally arrive on the main sequence. The processes mentioned here, and probably others, all remain to be investigated. These are problems of primary importance if we are to understand the formation of clusters and associations.

In understanding the evolution of single stars we must rely almost entirely on theoretical advances, because of the time scales involved. Considerable progress has been made in recent years in explaining the evolution of stars in a limited mass range as far as the giant branch of the H-R diagrams. However, for stars on the horizontal branch no adequate theoretical explanation is yet available. Thus the way in which stars in globular clusters evolve into the RR Lyrae star region and beyond remains so far unknown.

For massive stars little is known of their evolution after they have left the main sequence.

The mechanism of mass loss through the life history of a star also presents a puzzle. It appears that there are serious objections to the proposal made by a number of astronomers from the USSR that stars on the main sequence lose mass at a rate proportional to their luminosity. On the other hand, that mass loss is an important factor cannot be doubted if it is accepted that the final stage in a star's evolution is the white dwarf stage. For the existence of the Chandrasekhar limit for a completely degenerate configuration guarantees that all stars which originally had masses in excess of about 1.4 $M_\odot$ will lose mass at some stage. Observationally, mass loss is known to occur in some stars in the supergiant stage, probably in Wolf-Rayet stars, and in the catastrophic nova and supernova outbursts. Also in close binary systems mass exchange and probably mass loss from the system takes place. However, in all of these cases it is difficult to estimate the current rate of mass loss, and impossible to determine the average rate or the total loss of matter over the star's life history. On the theoretical side attempts are being made to take mass loss into account in constructing evolutionary tracks using electronic computing machines.

Another important problem when we are considering the H-R diagrams of clusters of different ages is to determine exactly the place where the Hertzsprung gap first becomes apparent, and to explain the origin of this gap. A further observational problem which is being attacked at the present time is to determine whether the main sequences at the faint ends of globular clusters such as M 3 and galactic clusters such as M 67 are coincident.

The investigation of star formation and evolution on the galactic scale, i. e. the stellar populations of galaxies, is only in its infancy at the present time. In the years ahead, work with large telescopes on the spectra of galaxies, determination of the mass-to-light ratios, and estimates of the amounts of gas by 21-cm techniques, can all be expected to give valuable information about the formation and evolution of stars in external galaxies.

References

1. LINDBLAD, B.: Monthly Notices roy. Astr. Soc. **95**, 20 (1934); Nature (Lond.) **135**, 133 (1935).
2. HAAR, D. TER: Astrophysic. J. **100**, 288 (1944).
3. KRAMERS, H. A., and D. TER HAAR: Bull. Astr. Netherl. **10**, 137 (1944).
4. SPITZER, L.: Astrophysic. J. **93**, 369 (1941); **94**, 232 (1941).
5. — Centennial Symposia, Harvard College Observatory, 1946, p. 87.
6. WHIPPLE, F. L.: Astrophysic. J. **104**, 1 (1946).
7. OORT, J. H.: Bull. Astr. Netherl. **12**, 177 (1954).
8. BIERMANN, L., and A. SCHLÜTER: Z. Naturforsch. **9a**, 463 (1954); Gas Dynamics of Cosmic Clouds. Amsterdam: North Holland Publishing Co. 1955, Ch. 27.
9. OORT, J. H., and L. SPITZER: Astrophysic. J. **121**, 6 (1955); Gas Dynamics of Cosmic Clouds. Amsterdam: North Holland Publishing Co. 1955, Ch. 28.
10. SAVEDOFF, M. P.: Astrophysic. J. **124**, 533 (1956); T. K. MENON, unpublished.
11. ÖPIK, E. J.: Irish Astr. J. **2**, 219 (1953).
12. EBERT, R.: Z. Astrophysik **37**, 217 (1955).
13. BONNOR, W. B.: Monthly Notices roy. Astr. Soc. **116**, 351 (1956).
14. MCCREA, W. H.: Monthly Notices roy. Astr. Soc. **117**, 562 (1957).

15. Krat, V. A.: Isv. Glavnoi Astr. Obs. Pulkova **19**, 2, No. 149 (1952); **18**, 4, No. 145 (1952); Problems of Cosmology, Moscow Academy of Sciences, Vol. 1, p. 34, 1952.
16. Urey, H. C.: Astrophysic. J. **124**, 623 (1956).
17. Su-Shu Huang: Publ. Astr. Soc. Pacific **69**, 427 (1957).
18. Struve, O.: Stellar Evolution. Princeton University Press 1950, Ch. 3.
19. Weizsäcker, C. F. von: Astrophysic. J. **114**, 165 (1951).
20. Bok, B. J., and E. Reilly: Astrophysic. J. **105**, 255 (1947).
21. Bok, B. J.: Centennial Symposia, Harvard College Observatory, 1946, p. 53.
22. Bok, B. J.: Astronomical J. **61**, 309 (1956).
23. Lilley, A. E.: Astrophysic. J. **121**, 559 (1955).
24. Herbig, G. H.: Astrophysic. J. **113**, 697 (1951); J. roy. Astr. Soc. Canada **46**, 222 (1952).
25. Haro, G.: Astrophysic. J. **115**, 572 (1952); **117**, 73 (1953).
26. Ambartzumian, V. A.: Soobs. Bjurakan Obs., No. 13 (1954).
27. Herbig, G. H.: Symposium on Non-Stable Stars, I. A. U. Monograph No. 3, Cambridge: at the University Press, 1957, p. 3.
28. Fessenkov, V. G., and D. L. Roshkovsky: Astron. J. USSR **28** (1951); **29**, 382, 397 (1952); **30**, 3 (1953); **31**, 3 (1954); Trans. I. A. U. **8**, 702 (1954).
29. Struve, O.: Sky and Telescope **13**, 181 (1954).
30. Oort, J. H.: Gas Dynamics of Cosmic Clouds. p. 246. Amsterdam: North Holland Publishing Co., 1955.
31. Weizsäcker, C. F. von: Z. Astrophysik **24**, 181 (1947).
32. Haar, D. ter: Astrophysic. J. **110**, 321 (1949).
33. Alfvén, H.: Ark. Mat. Astron. och Fysik **28A**, No. 6 (1942).
34. Lüst, R., and A. Schlüter: Z. Astrophysik **38**, 190 (1955).
35. Struve, O.: Stellar Evolution. Princeton University Press, 1950, Ch. 4.
36. Kuiper, G. P.: Pub. Astr. Soc. Pacific **47**, 121 (1935).
37. — Publ. Astr. Soc. Pacific **67**, 387 (1955).
38. Lyttleton, R. A.: The stability of rotating liquid masses. Ch. 10. Cambridge: at the University Press 1953.
39. Su-Shu Huang, and O. Struve: Ann. d'Astrophysique **17**, 85 (1954).
40. Mestel, L., and L. Spitzer: Monthly Notices roy. Astr. Soc. **116**, 503 (1956).
41. Eddington, A. S.: Internal Constitution of the Stars, p. 39. Cambridge: at the University Press 1926.
42. Hoyle, F., and R. A. Lyttleton: Proc. Cambridge Phil. Soc. **35**, 405, 592 (1939); **36**, 325, 424 (1940); Monthly Notices roy. Astr. Soc. **101**, 227 (1941).
43. Bondi, H., and F. Hoyle: Monthly Notices roy. Astr. Soc. **104**, 273 (1944).
44. — Monthly Notices roy. Astr. Soc. **112**, 195 (1952).
45. McCrea, W. H.: Monthly Notices roy. Astr. Soc. **113**, 162 (1953).
46. Dodd, K. N., and W. H. McCrea: Monthly Notices roy. Astr. Soc. **112**, 205 (1952).
47. Mestel, L.: Monthly Notices roy. Astr. Soc. **114**, 437 (1954).
48. Schatzman, E.: Gas Dynamics of Cosmic Clouds, p. 193. Amsterdam: North Holland Publishing Co. 1955.
49. Gurzadian, G. A.: Astron. J. USSR **26**, 104 (1949).
50. Greenstein, J. L.: Handbuch der Physik. Vol. 50, 1957.
51. Weizsäcker, C. F. von: Astrophysic. J. **114**, 165 (1951).
52. Bergh, S. van den: Publ. Astr. Soc. Pacific **68**, 449 (1956).
53. Burbidge, E. M., and A. R. Sandage: Astrophysic. J. **127**, 527 (1958).
54. Burbidge, G. R.: Private communication.

55. HEESCHEN, D. S.: Astrophysic J. **124**, 660 (1956); Publ. Astr. Soc. Pacific **69**, 350 (1957).
56. MCVITTIE, G. C.: Astronomical J. **61**, 451 (1956).
57. CHANDRASEKHAR, S.: An Introduction to the Theory of Stellar Structure p. 453. University of Chicago Press 1939.
58. THOMAS, L. H.: Monthly Notices roy. Astr. Soc. **91**, 122, 619 (1931).
59. ROTH, H.: Physic. Rev. **39**, 525 (1932).
60. LEVÉE, R. D.: Astrophysic. J. **117**, 200 (1953).
61. SANDAGE, A. R.: Proc. Vatican Conference on Stellar Populations, 1957, p. 149.
62. HENYEY, L. G., R. LELEVIER and R. D. LEVÉE: Publ. Astr. Soc. Pacific **67**, 154 (1955).
63. SALPETER, E. E.: Mem. Soc. roy. Sci. Liège **14**, 116 (1954).
64. BURBIDGE, E. M., G. R. BURBIDGE, W. A. FOWLER, and F. HOYLE: Rev. mod. Physics **29**, 547 (1957).
65. FOWLER, W. A.: Astrophysic. J. **127**, 551 (1958) and private communication.,
66. GAMOW, G.: Physic. Rev. **53**, 595, 908 (1938);
    — and E. TELLER: Physic. Rev. **53**, 608 (1938);
    — Physic. Rev. **55**, 718, 796 (1939);
    — and E. TELLER: Physic. Rev. **55**, 791 (1939);
    — Nature (Lond.) **144**, 575, 620 (1939).
67. OSTERBROCK, D. E.: Astrophysic. J. **118**, 529 (1953).
68. LIMBER, D. N.: Astrophysic. J. **127**, 363, 387 (1958).
69. CHAPMAN, S.: Monthly Notices roy. Astr. Soc. **82**, 292 (1922); A. S. EDDINGTON: Internal Constitution of the Stars, p. 277. Cambridge: at the University Press 1926.
70. SWEET, P. A.: Monthly Notices roy. Astr. Soc. **110**, 548 (1950).
71. ÖPIK, E. J.: Monthly Notices roy. Astr. Soc. **111**, 78 (1951).
72. EDDINGTON, A. S.: Monthly Notices roy. Astr. Soc. **90**, 54 (1929).
73. STRÖMGREN, B.: Astronomical J. **57**, 65 (1952).
74. MESTEL, L.: Monthly Notices roy. Astr. Soc. **113**, 716 (1953).
75. FESSENKOV, V. G.: Astron. J. USSR **26**, 67 (1949); Trans. I.A.U. **8**, 702 (1952).
76. MASSEVICH, A. G.: Astron. J. USSR **26**, 207 (1949).
77. — Astron. J. USSR **28**, 36 (1951).
78. — Soobs. Astron. Inst. Sternberg, No. 99, 3 (1956).
79. GAMOW, G.: Physic. Rev. **65**, 20 (1944).
80. ÖPIK, E. J.: Publ. Obs. Tartu **30**, Nos. 3 and 4 (1938); **31**, No. 1 (1943); Armagh Obs. Contr. No. 2 (1949); No. 3 (1951);
    GAMOW, G.: Astrophysic J. **87**, 206 (1938);
    CRITCHFIELD, C. L., and G. GAMOW: Astrophysic. J. **89**, 244 (1939);
    CHANDRASEKHAR, S., and L. R. HENRICH: Astrophysic. J. **94**, 525 (1941);
    SCHOENBERG, M., and S. CHANDRASEKHAR: Astrophysic. J. **96**, 161 (1942);
    HOYLE, F., and R. A. LYTTLETON: Monthly Notices roy. Astr. Soc. **102**, 218 (1942); **109**, 614 (1949);
    HARRISON, M. H.: Astrophysic. J. **100**, 343 (1944); **103**, 192 (1946); **105**, 322 (1947);
    GAMOW, G., and G. KELLER: Rev. mod. Physics **17**, 125 (1945);
    REIZ, A.: Ann. d'Astrophysique **10**, 301 (1947);
    LEDOUX, P.: Astrophysic. J. **105**, 305 (1947); Ann. d'Astrophysique **11**, 174 (1948);
    MASSEVICH, A. G.: Astron. J. USSR **25**, 168 (1948); Soobs. Astron. Inst. Sternberg, No. 30, 30 (1949);

LI HEN, and M. SCHWARZSCHILD: Monthly Notices roy. Astr. Soc. **109**, 631 (1949);
BONDI, C. M.: Monthly Notices roy. Astr. Soc. **110**, 275 (1950);
— and H. BONDI: Monthly Notices roy. Astr. Soc. **110**, 287 (1950); **111**, 397 (1951);
GARDINER, J. G.: Monthly Notices roy. Astr. Soc. **111**, 102 (1951);
MASSEVICH, A. G., V. P. MATVEYEVA, and L. N. TOOLENKOVA: Astron. J. USSR **28**, 432 (1951);
OKE, J. B., and M. SCHWARZSCHILD: Astrophysic. J. **116**, 3 (1952);
PARENAGO, P. P., and A. G. MASSEVICH: Astron. J. USSR **27**, 137 (1950);
SOOSHKINA, E. I.: Astron. J. USSR **30**, 180 (1953);
MASSEVICH, A. G.: Trudy Astron. Inst. Sternberg **22**, 21 (1953);
RESNIKOV, A. O.: Astron. J. USSR **31**, 60 (1954); **33**, 151 (1956).
80a. PARENAGO, P. P., and A. G. MASSEVICH: Astron. J. USSR **27**, 137 (1950).
81. SCHOENBERG, M., and S. CHANDRASEKHAR: Astrophysic. J. **96**, 161 (1942).
82. HARRISON, M. H.: Astrophysic. J. **100**, 343 (1944); **103**, 192 (1946); **105**, 322 (1947).
83. GAMOW, G., and G. KELLER: Rev. mod. Physics **17**, 125 (1945).
84. HAYASHI, C.: Physic. Rev. **75**, 1619 (1949).
85. SCHWARZSCHILD, M., I. RABINOWICZ, and R. HÄRM: Astrophysic. J. **118**, 326 (1953).
86. SANDAGE, A. R. and M. SCHWARZSCHILD: Astrophysic. J. **116**, 463 (1952).
87. SOROKIN, V. S., and A. G. MASSEVICH: Astron. J. USSR **28**, 21 (1951).
88. MASSEVICH, A. G.: Astron. J. USSR **30**, 508 (1953).
89. — Mem. roy. Soc. Sci. Liège **14**, 170 (1954).
90. HOYLE, F., and M. SCHWARZSCHILD: Astrophysic. J. Suppl. **2**, 1 (1955).
91. HASELGROVE, C. B., and F. HOYLE: Monthly Notices roy. Astr. Soc. **116**, 515, 527 (1956).
92. HAYASHI, C.: Progr. Theor. Phys. **17**, 737 (1957).
93. TAYLER, R. J.: Astrophysic. J. **120**, 332 (1954); Monthly Notices roy. Astr. Soc. **116**, 25 (1956).
94. KUSHWAHA, R. S.: Astrophysic. J. **125**, 242 (1957).
95. HENYEY, L. G., R. LELEVIER and R. D. LEVÉE: Publ. Astr. Soc. Pacific **67**, 341 (1955).
96. AMBARTZUMIAN, V. A.: Astrophysics and Stellar Evolution, ed Aremenian Acad. Sci. USSR 1947.
97. — Astron. J. USSR **26**, 3 (1949).
98. — Trans. I.A.U. **8**, 665 (1952).
99. JOY, A. H.: Astrophysic. J. **102**, 168 (1945); **110**, 424 (1949).
100. PARENAGO, P. P.: Variable Stars **7**, 169 (1950).
101. HARO, G.: Astronomical J. **55**, 72 (1950).
102. HERBIG, G. H.: Astrophysic. J. **111**, 15 (1950).
103. PARENAGO, P. P.: Astronomical J. USSR **30**, 249 (1953).
104. HARO, G., B. IRIARTE, and E. CHAVIRA: Bol. Obs. Tonantzintla y Tacubaya, No. 8 (1953).
105. WALKER, M. F.: Astrophysic. J. Suppl. **2**, 365 (1956).
106. — Astrophysic. J. **125**, 636 (1957).
107. HERBIG, G. H.: Astrophysic. J. **119**, 483 (1954).
108. — Proc. Vatican Conference on Stellar Populations, 1957, p. 127.
109. BLAAUW, A.: Bull. Astr. Netherl. No. 433 (1952).
110. — and W. W. MORGAN: Astrophysic. J. **119**, 625 (1954).
111. MARKARIAN, B. E.: Soobs. Bjurakan Obs. No. 11 (1953).

112. BLAAUW, A.: Astrophysic. J. **123**, 408 (1956).
113. — and W. W. MORGAN: Astrophysic. J. **119**, 625 (1954).
114. MASSEVICH, A. G.: Astron. J. USSR **34**, 176 (1957).
115. MÜNCH, G.: Publ. Astr. Soc. Pacific **68**, 351 (1956).
116. — and E. FLATHER: Publ. Astr. Soc. Pacific **69**, 142 (1957).
117. SHARPLESS, S.: Astrophysic. J. **119**, 334 (1954).
118. AMBARTZUMIAN, V. A.: Isvestia Acad. Sci. USSR, Phys. Ser. **14**, 15 (1950).
119. HARO, G.: Symposium on Non-Stable Stars, I.A.U. Monograph No. 3, p. 26. Cambridge: at the University Press 1957.
120. GLIESE, W.: Z. Astrophysik **39**, 1 (1956).
121. SHAPLEY, H.: Star Clusters. New York: McGraw-Hill 1930.
122. TRUMPLER, R. J.: Publ. Astr. Soc. Pacific **37**, 307 (1925); Lick Obs. Bull. No. 420 (1930).
123. KUIPER, G. P.: Harvard Bull. No. 903 (1936); Astrophysic. J. **86**, 176 (1937).
124. PARENAGO, P. P., and A. G. MASSEVICH: Astron. J. USSR **28**, 466 (1951).
125. JOHNSON, H. L., and C. F. KNUCKLES: Astrophysic. J. **122**, 209 (1955).
126. — Astrophysic. J. **120**, 325 (1954).
127. SANDAGE, A. R.: Astrophysic. J. **125**, 422 (1957).
128. — Proc. Vatican Conference on Stellar Populations, 1957, p. 41.
129. MICZAIKA, G. R.: Mem. roy. Soc. Sci. Liège **14**, 275 (1954).
130. LOHMANN, W.: Z. Astrophysik **42**, 114 (1957).
131. HOERNER, S. VON: Z. Astrophysik **42**, 273 (1957).
132. ARP, H. C.: Handbuch der Physik. Vol. **51** (1958).
133. SANDAGE, A. R.: Astrophysic. J. **125**, 435 (1957).
134. MASSEVICH, A. G.: Astron. J. USSR **32**, 412 (1955).
135. — Astron. J. USSR **33**, 576 (1956).
136. PARENAGO, P. P.: Trudy Astron. Inst. Sternberg 25 (1954).
137. JOHNSON, H. L.: Astrophysic. J. **126**, 134 (1957).
138. PRENDERGAST, K. H.: Private communication.
139. BIDELMAN, W. P.: Astrophysic. J. **98**, 61 (1943); **105**, 492 (1947).
140. JOHNSON, H. L., and W. W. MORGAN: Astrophysic. J. **122**, 429 (1955).
141. — and W. A. HILTNER: Astrophysic. J. **123**, 267 (1956).
142. BLANCO, V. M.: Astrophysic. J. **122**, 434 (1955).
143. MITCHELL, R. I., and H. L. JOHNSON: Astrophysic. J. **125**, 414 (1957).
144. LARSSON-LEANDER, G.: Stockholm Obs. Ann. **20**, No. 2 (1957).
145. HUMASON, M. L., and F. ZWICKY: Astrophysic. J. **105**, 85 (1947).
146. LUYTEN, W. J.: Astronomical J. **58**, 75 (1953); **59**, 224 (1954); Harvard Announcement Cards, Nos. 1202 (1953); 1244 (1954).
147. JOHNSON, H. L.: Astrophysic. J. **117**, 357 (1953).
148. ROMAN, N. G.: Astrophysic. J. **121**, 454 (1955).
149. JOHNSON, H. L., and A. R. SANDAGE: Astrophysic. J. **121**, 616 (1955).
150. REDDISH, V. C.: Observatory **74**, 68 (1954).
151. OORT, J. H.: Proc. Vatican Conference on Stellar Populations, 1957, p. 415.
152. SPITZER, L.: Astrophysic. J. **127**, 17 (1958).
153. BAADE, W.: Astrophysic. J. **100**, 137 (1944).
154. ARP, H. C., W. A. BAUM and A. R. SANDAGE: Astronomical J. **58**, 4 (1953).
155. SANDAGE, A. R.: Astronomical J. **58**, 61 (1953).
156. POPPER, D. M.: Astrophysic. J. **105**, 204 (1947).
157. BAUM, W. A.: Astronomical J. **57**, 222 (1952).
158. DEUTSCH, A. J.: Principes Fondamentaux de Classification Stellaire, p. 32. Paris 1955.
159. MORGAN, W. W.: Publ. Astr. Soc. Pacific **68**, 509 (1956).

160. Baum, W. A.: Astronomical J. **59**, 422 (1954).
161. Hiltner, W. A., H. L. Johnson and A. R. Sandage: Unpublished.
162. Arp, H. C.: Astronomical J. **60**, 317 (1955).
164. Savedoff, M. P.: Astronomical J. **61**, 254 (1956).
165. Schwarzschild, M.: Harvard College Obs. Circular, No. 437 (1940).
166. Roberts, M. S., and A. R. Sandage: Astronomical J. **60**, 185 (1955).
167. Walker, M. F.: Astronomical J. **60**, 197 (1955).
168. Belserene, E. P.: Astronomical J. **57**, 237 (1952).
169. Salpeter, E. E.: Astrophysic. J. **121**, 161 (1955).
170. Bergh, S. van den: Astrophysic. J. **125**, 445 (1957).
171. Jaschek, C., and M. Jaschek: Publ. Astr. Soc. Pacific **69**, 337 (1957).
172. Bergh, S. van den: Astronomical J. **62**, 100 (1957).
173. Sandage, A. R.: Astronomical J. **59**, 162 (1954).
174. Tayler, R. J.: Astronomical J. **59**, 413 (1954).
175. Sandage, A. R.: Astrophysic. J. **126**, 236 (1957).
176. Oort, J. H.: Groningen Publ. No. 40 (1926).
177. — Bull. Astr. Netherl. **9**, 185 (1941); Astrophysic. J. **116**, 233 (1952).
178. Parenago, P. P.: Astron. J. USSR **27**, 150 (1950).
179. Proc. Vatican Conference on Stellar Populations, 1957, p. 507.
180. Spitzer, L., and M. Schwarzschild: Astrophysic. J. **114**, 385 (1951); **118**, 106 (1953); D. E. Osterbrock: Astrophysic. J. **116**, 164 (1952).
181. Trumpler, R. J., and H. F. Weaver: Statistical Astronomy. University of California Press, Berkeley 1953.
182. Payne-Gaposchkin, C.: Variable Stars and Galactic Structure, Ch. 1. London: Athlone Press 1954.
183. Gondolatsch, F.: Z. Astrophysik **24**, 330 (1947).
184. Mumford, G. S.: Astronomical J. **61**, 224 (1956).
185. Dyer, E. R.: Astronomical J. **61**, 228 (1956).
186. Wehlau, A. W.: Astronomical J. **62**, 169 (1957).
187. Parenago, P. P.: Astron. News Letter, No. 71 and Supplement, 1953.
188. Wijk, U. van: Astronomic. J. **61**, 279 (1956).
189. Roman, N. G.: Astrophysic. J. **112**, 554 (1950); **116**, 122 (1952); Astronomical J. **59**, 307 (1954); Astrophysic. J. Suppl. **2**, 198 (1956).
190. Keenan, P. C., and G. Keller: Astrophysic. J. **117**, 241 (1953).
191. Vyssotsky, A. N.: Publ. Astr. Soc. Pacific **69**, 109 (1957).
192. Delhaye, J.: C. R. Acad. Sci. (Paris) **237**, 294 (1953).
193. Nassau, J. J., and V. M. Blanco: Astrophysic. J. **120**, 464 (1954).
194. Smith, H. J., and E. v. P. Smith: Astronomical J. **61**, 273 (1956).
195. Nassau, J. J.: Proc. Vatican Conference on Stellar Populations, 1957, p. 171.
195a. Vasilevskis, S., and R. A. Rach: Astronomical J. **62**, 175 (1957).
196. Morgan, W. W., S. Sharpless and D. E. Osterbrock: Astronomical J. **57**, 3 (1952).
197. — A. E. Whitford and A. D. Code: Astrophysic. J. **118**, 318 (1953).
198. Weaver, H. F.: Astronomical J. **58**, 177 (1953).
199. Bull. Astr. Netherl. **13**, No. 475 (1957).
200. Baade, W.: Symposium Notes. University of Michigan Observatory 1953.
201. Keenan, P. C.: Astrophysic. J. **120**, 484 (1954).
202. Blanco, V. M., and L. Münch: Bol. Obs. Tonantzintla y Tacubaya **12**, 17 (1955).
203. Irwin, J. B.: Astronomical J. **63**, 46 (1958).
204. Kraft, R. P.: Astrophysic. J. **126**, 225 (1957).
205. Bergh, S. van den: Astrophysic. J. **126**, 323 (1957).

206. MORGAN, W. W., and N. U. MAYALL: Publ. Astr. Soc. Pacific **69**, 291 (1957).
207. HELMBERG, E.: Lund Medd. **II**, No. 128 (1950).
208. STEBBINS, J., and A. E. WHITFORD: Astrophysic. J. **115**, 284 (1952).
209. OORT, J. H.: Astrophysic. J. **91**, 273 (1940).
210. HOLMBERG, E.: Lund Medd. **I**, No. 180 (1952).
211. SCHWARZSCHILD, M.: Astronomical J. **59**, 273 (1954).
212. BAADE, W.: Publ. Univ. Michigan Obs. **10**, 7 (1950).
213. SHAPLEY, H.: Publ. Univ. Michigan Obs. **10**, 79 (1950); The Inner Metagalaxy, Yale University Press, New Haven, 1957, p. 183; Galaxies, p. 216. Philadelphia: The Blakiston Company 1943.
214. ARP, H. C.: Astronomical J. **61**, 15 (1956).
215. HULST, H. C. VAN DE, J. RAIMOND and H. VAN WOERDEN: Bull. Astr. Netherl. **14**, 1 (1957).
216. HEESCHEN, D. S.: Proc. Stockholm Conference on Galactic Structure, 1957.
217. BAUM, W. A., and M. SCHWARZSCHILD: Astronomical J. **60**, 247 (1955).
218. DIETER, N. H.: Publ. Astr. Soc. Pacific **69**, 356 (1957); Astronomical J. **63**, 49 (1958).
219. OORT, J. H. et al.: Unpublished.
220. BUSCOMBE, W., S. C. B. GASCOIGNE, and G. DE VAUCOULEURS: Suppl. Aust. J. Sci. **17**, No. 3 (1954).
221. KERR, F. J., J. V. HINDMAN, and B. J. ROBINSON: Aust. J. Phys. **7**, 297 (1954).
222. HUMASON, M. L., N. U. MAYALL, and A. R. SANDAGE: Astronomical J. **61**, 97 (1956).
223. HEESCHEN, D. S.: Astrophysic. J. **126**, 471 (1957).
224. PAGE, T.: Astrophysic. J. **116**, 63 (1952).
225. STEBBINS, J., and A. E. WHITFORD: Astrophysic. J. **108**, 413 (1948); J. STEBBINS: Monthly Notices roy. Astr. Soc. **110**, 416 (1950).
226. BAADE, W., and H. H. SWOPE: Astronomical J. (in the press).
227. SANDAGE, A. R.: Proc. Vatican Conference on Stellar Populations, 1957, p. 75.
228. ROBERTS, M. S.: Astronomical J. **61**, 195 (1956).
229. HUMASON, M. L.: Astrophysic. J. **83**, 10 (1936).
230. WHITFORD, A. E.: Astronomical J. **58**, 49 (1953); Astrophysic. J. **120**, 599 (1954)
231. Ann. Rep. Washburn Observatory, Astronomical J. **61**, 352 (1956).
232. JEANS, J. H.: Astronomy and Cosmogony, Ch. 13. Cambridge: at the University Press 1929.
233. PAYNE-GAPOSCHKIN, C.: The New Astronomy, Scientific American, p. 107. New York: Simon and Schuster 1955.
234. WEIZSÄCKER, C. F. VON: Z. Astrophysik **22**, 319 (1944).
235. LÜST, R.: Z. Naturforsch. **7a**, 87 (1952); E. TREFFTZ: Z. Naturforsch. **7a**, 99 (1952).
236. HARO, G.: Bol. Obs. Tonantzintla y Tacubaya, No. 14, 16 (1956).
237. MÜNCH, G.: Astronomical J. (in the press).
238. OORT, J. H.: Scientific Amer. **195**, 101 (1956).
239. SPITZER, L., and W. BAADE: Astrophysic. J. **113**, 413 (1951).
240. SANDAGE, A. R.: Hubble Atlas of Galactic Forms, Carnegie Institution of Washington 1960.
241. VAUCOULEURS, G. DE: Handbuch der Physik, Vol. 53, 1958.
242. MORGAN, W. W.: Astrophysic. J. Publ. Astr. Soc. Pacific **70**, 364 (1958).

The Formation of Stars by the Condensation of diffuse Matter

by

F. D. KAHN

With 4 Figures

Contents

I. The observational data 104
  1. Introduction 104
  2. The astronomical time-scale 106
  3. Large scale properties of spiral galaxies 108
  4. Local motion and distribution of stars 117
  5. Local motion and distribution of interstellar matter 120
  6. The surface features of stars 124
  7. Physical properties of interstellar matter 130
  8. Conclusion 135
II. Theories of star formation 136
  1. The origin of stellar associations 136
  2. Contraction and fragmentation problems 146
  3. The motion of expanding associations 154
  4. Conclusion 160
III. Structure and evolution of stars 161
  1. Stationary models 161
  2. Evolution of stars 167
  3. Other differences between old and new stars 171
  4. Conclusion 174
Appendix I: Tests for the reality of associations 174
Appendix II: The synthesis of elements 175
Appendix III: A final summary 176
Acknowledgments 180
Bibliography 181

I. The Observational Data

1. Introduction

Stars are not static systems in a permanent state of equilibrium. This fact was recognized even in the days of KELVIN and of HELMHOLTZ, who were, perhaps, the first scientists to ask themselves what power it is that keeps the Sun (and the stars) shining. In the nineteenth century, nuclear power, and EINSTEIN's relation between mass and energy, were unknown. Gravitational effects provided the most potent source of energy which could be envisaged. The attempt was made to estimate the age of the Sun on the basis of its present luminosity and its present gravitational self-energy. As is well known the resulting time-scale was of the order of 5×10^7 years: even in those days there was geological and paleontological evidence that the age of the Earth itself was far greater than this.

After this century's developments in nuclear physics a more promising approach to the problem became possible. Our present theories of the structure of the stars depend, to a very large extent, on the nature of the process which provides them with their energy. The main reactions possible in stellar interiors are the proton-proton chain

$$H^1 + H^1 = D^2 + \beta^+ + \nu$$
$$D^2 + H^1 = He^3 + \gamma$$
$$He^3 + He^3 = He^4 + 2 H^1,$$

which is probably predominant at low temperatures, and the carbon-cycle [BETHE (1939)]

$$C^{12} + H^1 = N^{13} + \gamma$$
$$N^{13} = C^{13} + \beta^+ + \nu$$
$$C^{13} + H^1 = N^{14} + \gamma$$
$$N^{14} + H^1 = O^{15} + \gamma$$
$$O^{15} = N^{15} + \beta^+ + \nu$$
$$N^{15} + H^1 = C^{12} + He^4 + \gamma,$$

which is more important at relatively higher temperatures. A discussion of these reactions has been given by FOWLER (1953).

The proton-proton chain reaction can occur wherever there are hydrogen nuclei. The BETHE cycle can occur only where one, at least, of the catalysts C^{12}, C^{13}, N^{14} and N^{15} is present. The end result of both chains is the fusion of four hydrogen nuclei into a helium nucleus, together with the emission of two positive electrons, some γ-radiation and two neutrinos. The latter particles escape from the star, and probably never interact with other matter again. The positive electrons interact with negative electrons and add more to the γ-radiation. The radiant energy filters slowly out to cooler parts of the star, changes wavelength as it does so, and is finally emitted from the stellar surface. The conversion of one gram of hydrogen into helium liberates about 6×10^{18} erg of radiant energy.

One very definite statement can now be made about stellar evolution. A star of mass M g and luminosity L erg/s must suffer significant changes in its nature, and particularly in its energy output, within a time shorter than $6 \times 10^{18} M/L$ s $\sim 10^{11}$ years in the case of the Sun; on the other hand a very hot main sequence star may have a mass $20 \odot$ and a luminosity $10^5 L_\odot$. The time limit is then shorter by a factor 5000, and equals 20 million years. If our theory of energy generation is correct — and few modern astrophysical theories are better founded — then the inescapable conclusion is reached that there are stars known to us which must change radically in their nature in a short period even on the geological time-scale. However the maximum life involves no conflict with the known history of the Earth. It clearly makes little difference to terrestrial conditions if a distant O-star flares up, and then goes out again.

These results are fundamental in any discussion of stellar evolution. We are trying to answer the question "What evidence is there that there are still stars in the process of formation even at the present day?" It will be necessary to define "the present day" rather more closely. We therefore must know something of the age of the Universe, of the nature and the distribution of matter in its various galaxies — and particularly in our own — and we require some statistical data of the different varieties of stars.

One unfortunate fact hampers all studies of stellar evolution. A time-scale of 20 million years may be short in comparison with the age of the Galaxy or the Universe, but it is very long compared with the age of an astrophysicist. No detectable evolutionary change has ever been noted for certain in any star (except for the few recorded supernova explosions). The astrophysicist may then be likened to a naturalist presented with a snapshot of some caterpillars, some cocoons and some butterflies, and a lot of unrelated insect life. Like this hypothetical naturalist the astrophysicist can expect to have much difficulty in tracing the correct line of evolution of the species in which he is interested.

2. The astronomical time-scale

A time-scale for the Universe may be determined in two ways. One way is based on the rate of radioactive decay of some elements in the rocks found near the Earth's surface. (A recent review of this method has been given by HOUTERMANS, 1957.) Its particular details need not be discussed here; the results seem reliable, certainly as far as orders of magnitude are concerned. The most recent value found for the age of the Earth is 6×10^9 years; this is, in fact, the lapse of time since the solidification of the surface rocks. On no theory does it seem possible that the Earth is older than the Sun. We therefore know a lower limit to the age of the Sun, or to that of a typical star in our neighbourhood of the Galaxy.

A second, but less compelling, method is based on the observed red-shift in the spectra of extra-galactic nebulae. HUBBLE and HUMASON (1931) were the first to notice that the external galaxies showed the familiar Ca⁺ and Na absorption lines, and that these were shifted systematically to the red — the shift apparently increasing roughly in proportion to the distance between ourselves and the galaxy concerned. The common interpretation nowadays given to this result is that the system of galaxies, in other words the Universe, is in a state of expansion, and that the red shift is due to the Doppler effect. To complete the picture one needs to find the distance to the various galaxies. For the nearer ones such measurements used to be based on the well-known period-luminosity relation of the classical Cepheids: however, it now seems clear that a third variable enters into this relation, and so the measured period of a Cepheid does not define a unique value for the luminosity, but only determines a possible range of values [SANDAGE (1958a)].

In more distant parts of the Universe one must use data based on the brightest visible stars or on the average brightness of galaxies. The velocity-distance relation can now be found only by statistical means and therefore shows rather a large scatter. The latest estimate is that the radial velocity v of a galaxy is related to its distance R from us by $v = HR$, where $H^{-1} = 13 \times 10^9$ years, with a possible error of a factor 2 in H [SANDAGE (1958b)].

The most straightforward explanation of this simple relation is that all galaxies have been moving away from the same small volume of space since a time H^{-1} ago. Those galaxies which set out with the highest speeds have travelled the largest distance. If this is right then conditions in the Universe must have been considerably different at time $t = -H^{-1}$. Some far-reaching theories have been based on this extrapolation from the present data in attempts to discover the effect of a high pressure and density in a possible building-up of the heavier elements from hydrogen. (The various theories dealing with the formation of the elements are discussed in Appendix II.)

Nevertheless there is no immediate justification for the belief that the Universe began its existence in such a small volume at a time $t = -H^{-1}$. Genuine random motions, not due to any observational error, have been found superposed on the systematic expansion, with root-mean-square velocities of the same order as the systematic velocity difference expected between two neighbouring galaxies. Thus the Andromeda Nebula is approaching our Galaxy with about 125 km/s radial velocity[1] and is at present 500 kpc distant [V. D. HULST, RAIMOND and V. WOERDEN (1957)]. On the simple interpretation that there has been no change in galactic velocities the distance between our Galaxy and the Andromeda Nebula must have been about 1630 kpc at time

---

[1] The radial velocity relative to the local standard of rest is 300 km/s, and allowance must be made for galactic rotation.

$t = -H^{-1}$. Let us take this as a lower bound for the "initial" diameter of the Universe.

A numerical investigation may be useful. It is estimated that the Universe contains some 10^{11} galaxies in all, with an average mass of the order of $10^{11}\odot$ each. This leads to an initial mass density of about $5000\odot/\text{pc}^3$. Suppose that all this mass was concentrated into 10^{22} stars, each of solar mass. We shall estimate how many of these stars could possibly have been destroyed by mutual collisions at this early period.

The solar diameter is $D = 1.4 \times 10^{11}$ cm $\approx 5 \times 10^{-8}$ pc; the collision cross-section per star may therefore be estimated to be $\pi D^2 = 7.5 \times 10^{-15}$ pc², and the mean free path for a collision in a medium with 5000 stars per pc³ is 2.5×10^{10} pc, or about 1.6×10^4 times the minimum diameter. This simple model shows that it would have been quite possible for the Universe to have existed in a state of contraction when $t < -H^{-1}$ and for the large majority of stars to have survived this episode unscathed. There is no compelling reason at present for the belief that the age of all stars must be less than H^{-1}*.

If the minimum diameter of the Universe had been smaller by a factor 100 or more, then most stars would not have survived. But it will clearly be difficult on any theory to reconcile such a small initial size with the present random motions of the galaxies.

There are more sophisticated theories [BONDI and GOLD (1948), HOYLE (1948), JORDAN (1948)] in which the continuous creation of matter is postulated to make good the losses due to the general expansion in any particular volume. Model universes can be constructed which are in a quasi-steady state, and no limit can be given to their ages, though different ages can be assigned to the individual galaxies. Other theorists [e. g. FREUNDLICH (1954)] deny that the red-shift is due to the recession of the galaxies, and remove the possibility of determining any age at all.

The very early history of the Universe, if such a concept is permissible, is clearly not yet well understood. But we may be reasonably certain that its nature and its mass density have not changed very significantly in the last 3 or 4×10^9 years. The general properties of the system of galaxies can therefore not have altered very much in a period short compared with this interval, say, in the last 10^8 years. This leads to some rather interesting conclusions.

3. Large scale properties of spiral galaxies

There are two main types of galaxy. Some are elliptical (about 17%), about 80 % have spiral arms, like the Andromeda Nebula, and the

---

* A possible alternative view is that the Galaxy and the Andromeda Nebula belong to a system of negative energy (the Local Group). The consequences are considered by KAHN and WOLTJER (1959).

remainder are said to be irregular. BAADE (1957) points out that the estimated proportion of spiral galaxies may be too large by a factor two owing to observational selection. In this essay we are concerned with spiral galaxies only, for it is commonly supposed that only in them does the formation of stars go on at the present day.

The Andromeda Nebula and our Galaxy have been studied fairly carefully, and seem to be reasonably typical spiral nebulae. We confine ourselves to these two systems.

The study of our Galaxy is seriously hampered by interstellar absorption, particularly so for observations made in the galactic plane. It is therefore often useful to note analogies between its properties and observed features of the Andromeda Nebula, but here a lack of light limits the visual observation of any but the brightest stars. On the other hand a lack of resolving power hinders studies made with the 21 cm line of atomic hydrogen.

The nature of the Galaxy was finally established some thirty years ago through the work of LINDBLAD (1925) and OORT (1928). It had been known previously that the Sun belonged to a stellar system with a central plane of preference, the galactic plane. From a statistical study of the stellar motions in the Sun's neighbourhood it may be found that the Sun is moving at about 20 km/s into the direction $18^h 0^m$, $+30°.0$ relative to the local standard of rest. OORT then noted that, on the average, a star at a distance Δs from the Sun in the direction of galactic longitude l has a radial velocity $\Delta V = A \Delta s \sin 2(l - l_0)$ away from the local standard of rest. In the conventional system of galactic coordinates the best value for l_0 is $327°$. The system also seems to be turning with angular velocity $A \cos 2(l - l_0) + B$ about the local standard of rest. A and B are known as OORT's constants. Some recent values given for them are

$$A = 19.5 \text{ km/s kpc}$$
$$B = -6.9 \text{ km/s kpc}$$

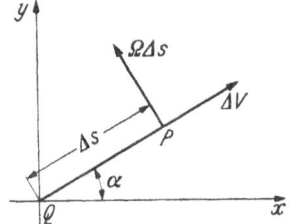

Fig. 1. The co-ordinate system

according to MORGAN and OORT (1951).

To see the significance of OORT's result let us investigate a general flow with velocity components $u = u(x, y)$, $v = v(x, y)$ in the xy-plane. Let Q, P be two points separated by a distance Δs, and let $\sphericalangle QPx = \alpha$. Let Δu and Δv be the difference between the x and y velocity components and Q and at P. (See Fig. 1.)

Then, to the first order,

$$\Delta u = \frac{\partial u}{\partial x} \Delta s \cos \alpha + \frac{\partial u}{\partial y} \Delta s \sin \alpha,$$

and
$$\Delta v = \frac{\partial v}{\partial x}\Delta s \cos\alpha + \frac{\partial v}{\partial y}\Delta s \sin\alpha.$$

The radial velocity of P relative to Q is

$$\Delta V = \Delta u \cos\alpha + \Delta v \sin\alpha$$
$$= \Delta s\left[\frac{\partial u}{\partial x}\cos^2\alpha + \left(\frac{\partial u}{\partial y} + \frac{\partial v}{\partial x}\right)\sin\alpha\cos\alpha + \frac{\partial v}{\partial y}\sin^2\alpha\right]$$
$$= \Delta s\left[\frac{1}{2}\left(\frac{\partial u}{\partial y} + \frac{\partial v}{\partial x}\right)\sin 2\alpha + \frac{1}{2}\frac{\partial u}{\partial x}(\cos 2\alpha + 1) + \frac{1}{2}\frac{\partial v}{\partial y}(1-\cos 2\alpha)\right].$$

Let us choose our coordinate system so that $\alpha = l - l_0$. The x-axis then points to the galactic centre. Comparison of this result with OORT's observation now shows that $\partial u/\partial x = 0$ and $\partial v/\partial y = 0$. The motion of the system of stars, regarded as a fluid, is therefore divergence free.

If Ω is the angular velocity of P about Q, then

$$\Omega\Delta s = \Delta W = -\Delta u \sin\alpha + \Delta v \cos\alpha$$
$$= \Delta s\left[-\frac{\partial u}{\partial x}\sin\alpha\cos\alpha - \frac{\partial u}{\partial y}\sin^2\alpha + \frac{\partial v}{\partial x}\cos^2\alpha + \frac{\partial v}{\partial y}\sin\alpha\cos\alpha\right],$$

and since $\partial u/\partial x = \partial v/\partial y = 0$

$$\Omega = \frac{1}{2}\left(\frac{\partial u}{\partial y} + \frac{\partial v}{\partial x}\right)\cos 2\alpha + \frac{1}{2}\left(\frac{\partial u}{\partial y} - \frac{\partial v}{\partial x}\right).$$

We identify

$$A = \frac{1}{2}\left(\frac{\partial u}{\partial y} + \frac{\partial v}{\partial x}\right)$$
$$B = \frac{1}{2}\left(\frac{\partial v}{\partial x} - \frac{\partial u}{\partial y}\right).$$

Evidently a pair of constants A and B can be chosen for any divergence-free flow. OORT's interpretation of the result goes further and states that the stars in the plane of the Galaxy rotate with an angular velocity $\omega(R)$ about a centre whose distance is R, in the direction $l = l_0$. The form of the results shows that this centre might equally well lie in any of the directions $l_0 + r\pi/2$, $r = 1, 2, 3$; however there are other indications that the Galaxy has a degree of symmetry about $l = l_0$, for instance from the distribution of globular clusters and from the motion of high-velocity stars, which we shall discuss later. It then follows that

$$A = \frac{1}{2}R\frac{d\omega}{dR}, \qquad B = \omega + \frac{1}{2}R\frac{d\omega}{dR}$$

so that

$$\omega = B - A.$$

The observed numerical values of A and B show that ω is negative, so that the local stellar system is moving in the direction of decreasing l.

Since $d\omega/dR$ is positive the absolute value of ω decreases with increasing galacto-centric distance. The inner parts of the Galaxy complete their circles about the centre in shorter periods than the outer parts. The deduced value of $|\omega|$ for the solar neighbourhood is 26.4 km s^{-1} kpc^{-1}; the period for one circuit about the galactic centre is about 2×10^8 years.

The stars which take part in this differential rotation tend to lie near the galactic plane. But other groups of stars, such as the globular clusters and the system of RR Lyrae variables, appear rather to be grouped with more spherical symmetry about the galactic centre. Now spherical symmetry in any system implies the absence of a preferred axis, and therefore the absence of a systematic rotation. By means of radial velocity measurements on globular clusters one can then find the space velocity V of the local standard of rest relative to the galactic centre, and also $R_0 = V/\omega$, the Sun's distance from the galactic centre. A recent value is $R_0 = 8.26$ kpc [V. D. HULST, MULLER and OORT (1954)].

Two useful results follow. Firstly, the period of one galactic rotation is much smaller than the time scale of the Universe. Secondly, since the centripetal acceleration of the stars is due to the gravitational attraction of the Galaxy as a whole, we may estimate its mass roughly from the relation $V^2/R = GM/R^2$, and find $M \sim 10^{11} \odot$. The implicit assumption has been made here that the bulk of the galactic mass is concentrated near the centre. This seems to be the case.

Not all the stars near the galactic plane fit into this simple scheme. Some stars have velocities considerably different from that of the local standard of rest. The maximum speeds observed for these high velocity stars depend on the direction in which they are moving. If their motion is in the direction $l = 57°$, $b = 0$, that is, if they are moving in the direction of the local galactic rotation, then the limiting speed is found to be 60 km/s, approximately. For stars moving in the opposite direction the limiting speed exceeds 300 km/s [OORT (1926)]. This result may be interpreted as follows. No star observed in our neighbourhood is likely to have a speed exceeding the local escape velocity from the Galaxy. The vector sum of the velocity of the local standard of rest and the relative velocity of the star concerned must have a magnitude less than some V_{max}, which depends only on the local gravitational potential. A star moving in the forward direction of the rotation therefore needs the smallest relative velocity in order to escape; a star moving in the backward direction needs the largest relative velocity.

The result confirms that the direction to the galactic centre, $l = 327°$, was correctly chosen by OORT. A statistical study shows that the majority of the nearby high velocity stars have transverse velocities relative to the galactic centre which are smaller than the local circular velocity, and so they move along orbits with diameters much smaller than $2R_0$.

SCHWARZSCHILD (1952) has investigated the distribution of perigalactic distances among the high-velocity stars. With reasonable assumptions concerning the mass distribution in the Galaxy he finds that the majority of them approach the galactic centre to within a distance 0.4—0.6 R_0 at some point in their orbits. The result is not very sensitive to the particular form of variation assumed for the gravitational potential. Even if all the mass of the Galaxy were concentrated at its centre the values of the perigalactic distances would alter by only 15%.

Most high-velocity stars have a direct motion about the galactic centre.

The distribution of matter in the solar neighbourhood is roughly symmetrical with respect to the galactic plane. The thickness of the layer occupied differs for the various constituents. Thus interstellar matter occupies a layer some 220 pc thick [SCHMIDT (1957)], O and B stars a layer of about 500 pc thickness and later type stars fill still thicker layers. We shall consider first some observations of the interstellar matter, for they promise to lead to a great increase in our knowledge of galactic structure.

The best way of studying the interstellar gas is by means of the 21 cm line of hydrogen. With equipment now available it is possible to use this radiation to detect Doppler shifts corresponding to radial velocities of a few km/s, and to resolve angles of some 30 min of arc. 21 cm radiation from distant parts of the Galaxy reaches us without suffering interstellar absorption on the part of the dust grains. Self-absorption in the gas becomes important only when there is little differential motion along the line of sight. This occurs in the directions of the galactic centre and of the anti-centre. WESTERHOUT (1957) has given a comprehensive report of some recent 21 cm studies.

The main difficulty in this work lies in the interpretation of the observations. The recorded results have to be freed first from the blurring effect due to the antenna pattern and then from the smoothing out of the spectral features due to the finite bandwidth of the receiver. Both calculations involve the solution of an integral equation and are liable to serious error if pushed too far. Thus they yield only a set of approximate curves for the intensity of the radiation in terms of the frequency shift at various points in the sky. A few more steps are needed before the position of the hydrogen clouds in the Galaxy can be plotted.

The Dutch astronomers assume that the large-scale motion of the gas is basically a rotation about the galactic centre, the angular velocity being a function of the radial distance only. Random motions must be superposed on any basic rotation, as can be seen from the widths of the profiles observed in the directions of the centre and the anti-centre. The widths are not due to thermal broadening only. They correspond to random velocities rather greater than those in atomic hydrogen gas at 125 deg K, which is its measured (harmonic) mean temperature. The random velocities causing the broadening are due to a supersonic turbulence, with typical speeds of the order of 5 or 10 km/s. It has been considered desirable to remove this turbulent broadening of the spectral lines from the data, and to find the line profiles which would have been seen if the large-scale motion had been present by itself alone. It is not certain that this can be done unambiguously.

But once it has been done a model distribution can be constructed if one assumes that the angular velocity ω about the galactic centre depends only on the radial distance. For parts of the Galaxy further out than the Sun a particular law of variation has to be assumed. For parts closer in than the Sun the law of variation can be determined observationally, provided of course that the assumption is valid. It can then be shown that any given Doppler shift in a particular direction of observation corresponds to one (or two) possible distances from the Sun. The reduced line profiles may now be interpreted to give a map of the hydrogen distribution in the Galaxy.

The survey showed that atomic hydrogen concentrates preferentially into separate arms, and that within these the smoothed-out density is of the order of one atom per cm^3. The density outside the arms is lower by about a factor 3 [see Plate B. WESTERHOUT (1957)].

But some discrepancies are found. Thus the parts of the Galaxy visible from Holland appear to be different from those seen from Australia [KERR, HINDMAN and CARPENTER (1957)]. Next the arms along which the hydrogen lies are rather more circular than the arms of other spiral galaxies. Finally, as WESTERHOUT himself points out, and as can be seen clearly from the line profiles presented by him, there are radial velocities observed in some directions which cannot be fitted at all into a model with rotational symmetry. Any attempt to map the hydrogen distribution by means of the 21 cm line must be based on a simplified model of the Galaxy and is bound to lead to these difficulties, for only half the data can ever be found which are needed to fix the position and velocity of any hydrogen cloud: One can readily measure the galactic latitude and longitude and the radial velocity, but not the parallax nor the two components of the proper motion.

An inspection of WESTERHOUT's Plate B shows the need for great caution; it is easily seen from that picture that the density distribution given there has many features with an apparent, but probably spurious radial symmetry with respect to the Sun. MÜNCH (1957) has drawn attention to this property in some earlier 21 cm plots.

Recent observations by v. WOERDEN, ROUGOOR and OORT (1957) of hydrogen gas near the galactic centre show that the motion there has a pronounced velocity component, of the order of 50 km/s, away from the centre. The intensity of the radiation is rather low, and indicates that there is only a small density of atomic hydrogen.

SCHMIDT (1956) has used the inferred curve $\omega = \omega(R)$ to make a model of the mass distribution in the Galaxy as a whole. He finds a total mass of $7 \times 10^{10} \odot$, and a local mass density 6.3×10^{-24} g/cm^3 near the Sun. Of this he attributes 2.7×10^{-24} g/cm^3 to the nearby stars, 1.4×10^{-24} g/cm^3 to interstellar hydrogen, and leaves the rest unaccounted for.

To gain more insight into the structure of galaxies it seems best now to discuss some observations of the Andromeda Nebula. These can be made in several ways. One can trace the spiral arms from photographs, where they tend to be picked out by the early (O and B) type stars, by H II regions surrounding these stars and by interstellar dust. The general observation has been made, and was first pointed out by BAADE (1944), that in our Galaxy and in many others the O and B stars and the interstellar dust tend to occur rather close to each other. There are two main arms in Andromeda, but BAADE (1956) has been able to trace a rather more detailed structure.

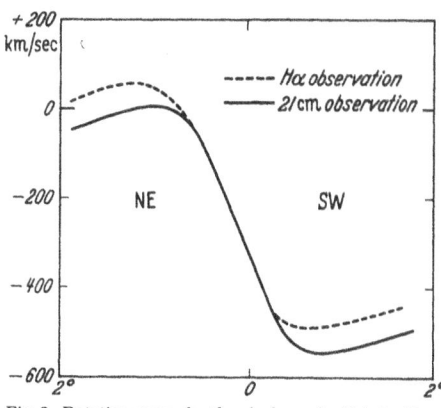

Fig. 2. Rotation curve for the Andromeda Nebula. Data from v. D. HULST, RAIMOND and v. WOERDEN(1957)

The inner parts of Andromeda contain no O and B stars. The brightest stars here are much redder, supergiants of type M.

The outer parts of Andromeda look much the same as the outer part of our Galaxy. The spiral arms lie in a thin disk with a diameter of some 35 kpc. The disk is embedded in a rather larger spherical distribution of stars. BAADE (1957) mentions that the star density falls off with increasing distance from the centre, but members of the distribution can still be detected at an angular distance of about 2°, or about 20 kpc from the centre. The motion of the H II regions in the spiral arms of Andromeda has been investigated by MAYALL (1950). The results are shown in Fig. 2, where they are compared with the recent determination of the motion in the H I regions [v. D. HULST, RAIMOND and v. WOERDEN (1957)]. The latter work was done with the 21 cm line and was rather difficult since the aerial pattern of the radio telescope used was about as wide as the nebula itself. The intensity profiles measured are thus integrated values over considerable areas of the nebula. v. D. HULST and his colleagues found their rotation curve for Andromeda once again with the assumption that the interstellar gas there rotates differentially according to a law $\omega = \omega(R)$. As can be seen from Fig. 2 their curve differs radically from the one by MAYALL. They also found an inconsistency in their interpretation of their own results.

The integrated 21 cm radiation from Andromeda is emitted by a mass of about $2.5 \times 10^9 \odot$ of neutral hydrogen. The rotation curve suggests that the total mass of Andromeda is about $2.5 \times 10^{11} \odot$. The interstellar gas therefore comprises only about 1% of the total mass of that galaxy.

We return to some other determinations of the structure of our Galaxy. The analogy with the Andromeda Nebula shows that early-type stars and interstellar dust tend to occur near each other. Theoretical discussions of the mode of formation of the dust [OORT and V. D. HULST (1946)] make it seem likely that this takes place in interstellar gas clouds. SPITZER (1941) and SAVEDOFF (1955) have shown that the relative motion of the dust and the gas will usually be halted so rapidly that no appreciable separations, larger than 0.5 pc, say, can occur. One expects the gas and the dust to congregate in the same regions of space. In a sense the dust resembles particles of smoke being carried along by a wind.

The reverse is not necessarily true. There may be regions free of dust which nevertheless contain gas. Thus the nucleus of Andromeda shows no trace of dust. However, there certainly is gas in the nucleus of our Galaxy; by analogy we should expect to find it in the nucleus of Andromeda as well.

We therefore assume that dust and gas are present in roughly constant proportions along the galactic spiral arms. It follows that the arms traced out by the early-type stars, and therefore by the dust, should be the same as the arms traced out by the gas. This conclusion can unfortunately not be verified in the case of the Andromeda Nebula owing to the lack of resolving power at 21 cm wavelength. However, MORGAN, WHITFORD and CODE (1953) have found the general trend of the distribution of O and B stars within 1400 pc of the Sun. These stars occupy a roughly elliptical region, whose major axis points in the direction $l = 20°$. This contrasts with the direction $l = 45°$ determined for the local gaseous arm by means of the 21 cm line. The discrepancy may be real or it may be due to the insufficiency of the model used for interpreting the hydrogen line measurements.

The following then is a tentative picture of our Galaxy and of Andromeda. There is a bright nucleus consisting of a multitude of stars, none very brilliant or very blue, which gradually fades out, after about 3 kpc, into a spherical halo with roughly the same character. Within this halo is embedded a disk-like distribution of stars, including some of very early spectral type, of gas and of dust. The inner radius of the disk is about 3 kpc, its outer radius about 16 kpc; its thickness is different for its various components, but may be as large as one kpc. The gas, the dust and the youngest stars in the disk seem to trace out two spiral arms, with many subsidiary features. The stars in the disk revolve about the galactic centre, the rotation being differential, with the largest angular velocities at inner points and the smallest ones outside.

The last result leads to a well-known difficulty. The angular velocities in the inner and the outer parts of the disk may differ by a ratio as large as 3 : 1. The typical rotation period for a galaxy is 10^8 years. If the spiral arms follow the general law of rotation then they will lose their characteristic shape in about 10^8 years, a period small compared with the typical life-time of a galaxy. But spiral galaxies are very common objects, and we cannot assume that they are peculiar to the present epoch of the Universe. Therefore, either a spiral arm must be an excitation moving through the interstellar matter, or the motion of the gas in the arms has been wrongly determined. This point may be examined further.

We have mentioned that the atomic hydrogen density within a spiral arm is larger by about a factor three than outside an arm. One might, nevertheless, imagine that the gas has a roughly uniform density in the galactic disk, but that it is largely molecular outside the arms, and so escapes detection by the 21 cm line. If the gas takes part in the general differential rotation while the excitation moves with uniform angular velocity then we must find regions in the Galaxy where it has a speed of the order of 100 km/s relative to the gas. Measured along an arc of a circle of 8 kpc radius, concentric with the Galaxy a spiral arm is some 10 kpc wide. The excitation therefore must take about 10^8 years to sweep past a given mass of gas. If such a picture is to be taken seriously then it would need to be shown that dust particles can grow and that atomic hydrogen gas can become largely molecular in such a period.

We do not suggest that this simple model explains why spiral arms exist, but mention it only to show one conceivable way in which an excitation mechanism might explain the observations. Even so it is clear that the spiral shape cannot persist without a large-scale organising field of force, e. g. a galactic magnetic field.

One might take the alternative view that the gas in a spiral arm actually moves with the arm, and therefore with a roughly uniform angular velocity. This is approximately true in the case of the arm which runs through the solar neighbourhood. The motion of the gas must then differ from that of the nearby stars, and so an extra set of forces must act on the gas. Once again a magnetic field seems to be the only possibility. A very crude estimate may be made to find an order of magnitude for its intensity. The magnetic energy density in a field of strength H is $H^2/8\pi$; the kinetic energy density of the gas moving at a speed V relative to the local standard of rest is $\varrho V^2/2$. We may expect that

$$\frac{H^2}{8\pi} \sim \frac{1}{2}\varrho V^2.$$

With

$$\varrho \sim 2\times 10^{-24}\,\text{g/cm}^3, \quad V \sim 100\,\text{km/s}$$

we find

$$H \sim 5\times 10^{-5}\,\Gamma.$$

There exists other indirect evidence for the presence of magnetic fields. Observations of the intensity and the isotropy of cosmic rays suggest that the Galaxy has a storage factor of at least 10^3 for these high energy particles. This implies a presence of a magnetic field of the order of $10^{-5}\,\Gamma$. Further, observations of radio noise from the galactic halo suggest that this is due, at least in part, to the radiation emitted by fast electrons moving through a magnetic field. Finally, the interstellar polarization seen on stars in the galactic plane again suggests the presence of a magnetic field of about $10^{-5}\,\Gamma$ field strength, provided the effect is explained by the theory of DAVIS and GREENSTEIN (1951). The variations in the degree of polarization for stars in different galactic longitudes have lately been plotted by HILTNER (1956). Using the DAVIS-GREENSTEIN theory he finds that the mean direction of the field is roughly in the direction $l = 10°$. For comparison the direction of the local spiral arm was found to be $l = 20°$ according to the method of locating the O and B stars, and $l = 45°$ according to 21 cm line measurements.

It is generally agreed that the formation of new stars takes place exclusively in the gaseous spiral arms of galaxies. Our discussion shows that there still is much uncertainty concerning the fields of force and the mechanical conditions which prevail there.

4. Local motion and distribution of stars

The presence of stellar associations is a most important feature in the local structure of the Galaxy. Following AMBARTSUMIAN (1954) one may define these as follows: An association is a relatively compact group of stars of a rare type, endowed with a positive amount of mechanical energy. Occasionally groups of rare stars are found having a negative total energy, but these cannot readily disperse into space. The existence of an association is, however, remarkable, for, as AMBARTSUMIAN points out, it is bound to dissolve owing to the random velocities of its members and owing to the tidal effect due to the gravitational field of the Galaxy as a whole.

A number of associations of O and B stars have now been studied and the proper motions of their members noted. [See, for example,

BLAAUW (1952), BLAAUW and MORGAN (1953), KOPYLOW (1954).] In most cases associations are found to be expanding, though it is never very clear just where one is to place the centre of expansion in any particular case. Whatever centre is picked there is usually a considerable dispersion in the ratio of radial distance to outward velocity for the various stars. It is hard to say whether the spread is real or is mainly due to errors of measurement.

BLAAUW (1952), and others, have estimated the ages of some associations by a comparison of the presently observed mean speeds of expansion with the present linear dimensions. The ages usually found are of the order of a few million years. Typical member stars tend to have radial velocities of the order of 10 km/s.

Many associations cover a fairly large area of the sky. The Lacerta aggregate, for example, spreads over approximately $15° \times 15°$ [BLAAUW and MORGAN (1953)]. The process of isolating them is therefore necessarily rather difficult. In addition the genuine members may be intermingled with back- or foreground stars which have to be discarded. MEURERS (1957) has expressed doubts about the reality of associations. He has made experiments by placing random dots on pieces of paper and has convinced himself that the human eye has a natural tendency to arrange these dots into groups. To judge by the diagrams given in his paper he found that there is a relatively good chance of picking out groups with about eight members each and separated by distances of the same order as the typical group diameter. Some comments on this result will be found in Appendix I.

But most of the stellar associations studied so far seem to have twenty or more members each; they should therefore be real groups. In particular the earliest type stars are rarely found outside an association.

Some associations contain interstellar gas and dust as well. MENON (1958) has studied the interstellar hydrogen surrounding the Orion nebula and has found that it is expanding radially at about 10 km/s. On the other hand the association ζ Persei contains dust, but RAIMOND (1957) failed to find any neutral hydrogen gas there, nor was any hydrogen detected near the association I Lacertae. MÜNCH (1957) observed large irregularities in the Doppler shifts of the interstellar Ca^+ and Na lines in the neighbourhood of some associations.

Finally BLAAUW and MORGAN (1954a, b) and BLAAUW (1956) have drawn attention to the remarkable trio of early type stars AE Auriga, μ Columbae and 53 Arietis, which all have exceptionally high space velocities relative to the local standard of rest. The estimated values of their speeds are 128 km/s, 126 km/s and 73 km/s, respectively.

AMBARTSUMIAN (1954) has also pointed to the existence of T-associations, so called because they consist of large groups of irregularly variable stars, mainly of T-Tauri type. Nothing is known about their expansion or their possible lifetime. They, too, are believed to be gravitationally unstable; AMBARTSUMIAN holds that the reason why their member stars are irregular variables is that they are newly formed bodies, still in an unstable state. HERBIG (1957) has stated that T-associations always occur in the neighbourhood of bright and/or dark interstellar nebulosities, and some early type stars. It has therefore been suggested that the passage of interstellar clouds of non-uniform opacity in front of these stars may be the cause of their irregular variability. BÖHM (1956) has shown that this hypothesis is hardly tenable.

In a typical T-association the density of stars is some five to fifteen times higher than is the space density of stars in the same luminosity range near the Sun.

It has been claimed [HERBIG (1957)] that two Herbig-Haro objects in a T-association in Orion had become visible for the first time in 1955 and had then been photographed in a position where no star had been seen about eight years before. It is hard to say whether these were really newly-formed stars, particularly since such objects are known to be irregular variables. In a group of even 1000 stars, evolving on a time-scale of millions of years, one can expect a completely new star to become visible perhaps once every few thousand years. In the case of these two objects the mean luminosity must have increased at a rate faster than three magnitudes in eight years. None of the T-Tauri like objects of this particular association are much brighter than the two "newly-formed" stars. We should be forced to conclude that such stars can brighten at more than one-third of a magnitude per annum and can then suddenly attain a steady state.

FESSENKOV and ROSHKOVSKI (1952) have stated that in regions where interstellar matter is abundant one may often find chains of stars, roughly similar in colour and luminosity, linked by filaments of glowing gas. STRUVE (1954), quoting some observations by MAYALL, has claimed that the chains are mere optical illusions, and has not been able to trace them himself on any photograph. But RING (1955) has photographed some of the regions observed by FESSENKOV through an interference filter, and has confirmed the existence of the chains. The gaseous matter linking neigbouring stars was found to emit the $H\alpha$ line.

Groups of stars with negative mechanical energy are also found in our part of the Galaxy. They are called galactic clusters, and are often rather loose structures of irregular shape. Among the better-known clusters are the Pleiades, the Hyades, h and χ Persei, etc. A cluster can be recognized by the proximity of its members one to another. The study of the luminosity-colour diagrams of different clusters is extremely relevant in any attempt to construct a theory of stellar evolution, for it is commonly held that all the member stars of any one cluster must have been formed at about the same epoch in the past.

Stars with a common origin can also be picked out by their present peculiar velocities. EGGEN (1957) has described such a group in the solar neighbourhood. Its luminosity-colour diagram is an important datum for us.

In the outer parts of the Galaxy there are the globular clusters. These are more nearly spherical in shape and are much richer in stars than the galactic clusters. Their motion and general nature indicate that they are older than most of the galactic clusters. The luminosity-colour diagrams are known for several of them.

Finally an interesting kinematic effect has been noted for stars in the solar vicinity which are not involved in any particular cluster or association. It is found [see, for instance, VYSSOTSKY (1957)] that there are systematic differences between the motions of stars of different spectral type. If the local standard of rest is defined in terms of stars of any one spectral type, then the various standards, corresponding to different colours, may differ by as much as 8 km/s. The dispersion of velocities among stars of a given colour also depends markedly on the colour chosen, and increases systematically as the spectral type gets later. For B stars it is only 8 km/s; for dM stars it may be as high as 28 km/s. Evidently this observation explains why the stars of earliest type define the thinnest galactic disk: they just do not have enough energy to rise very far against the gravitational attraction of the galactic plane.

5. Local motion and distribution of interstellar matter

The local density and velocity distributions of the interstellar matter have considerable influence on its physical properties, and are largely dominated by the radiation field of the nearby stars. In the present problem they are of fundamental importance.

Near O and B stars the interstellar gas is almost fully ionized and forms the so-called HII regions. The size of these regions depends on the luminosity of the exciting star (or stars); typical linear dimensions are measured in parsecs. Quite often an HII region appears to be nearly spherically symmetrical, as in the case of the Monoceros Nebula [MINKOWSKI (1949)], with only a small proportion of its characteristic $H\alpha$ radiation coming from the central regions and the bulk from the outer parts. The $H\alpha$ radiation is emitted during the recombination of some protons and electrons, and the intensity emitted per unit volume varies as the square of the particle density and only inversely as the square root of the temperature. There is good reason to believe that the temperature throughout an HII region is practically uniform [cf.

SPITZER (1954)]: the distribution of Hα intensity therefore reflects to some extent the density distribution in the region. The observed concentration of matter to the outer parts of an HII region is very reminiscent of the mass distribution in an expanding bubble of gas following a violent explosion [TAYLOR (1946)].

The light of the HII regions is produced directly by the recombination of an ionized gas. But there are other clouds, called reflection nebulae, which appear bright against the background of the sky, but which only shine by the reflected light of nearby stars. The statistics of the association of these nebulae with stars has been used to show that clouds capable of scattering starlight occupy about 10% of the interstellar space inside a spiral arm [STRÖMGREN (1948)]. Reflection nebulae have spectra similar to those of the stars exciting them, and are easily distinguished from HII regions.

HII regions also radiate in the continuum. This was at one time taken to be evidence that they contain solid particles capable of scattering starlight. But it is possible that two-quantum transitions from the 2s to the 1s state of the hydrogen atom take place often enough to explain the continuum emission [SPITZER and GREENSTEIN (1951), KIPPER (1952)] and it is not known whether any solid particles are present there [SEATON (1955)].

An HII region is generally in a state of highly non-uniform internal motion, shown up by the relative Doppler shifts of radiation from its different parts. COURTÈS (1955) and MÜNCH (1958) have studied this turbulence. Velocity differences of up to 25 km/s may occur within a second of arc, corresponding to a separation of about 0.0025 pc in the Orion Nebula. The statistical rate of increase of velocity differences with distance of separation does not follow the same law as in incompressible turbulence, where $\langle |\Delta v| \rangle \propto l^{1/3}$. This is not surprising: incompressible turbulence is said to occur when the velocity differences are much below the local speed of sound. In fully ionized hydrogen gas the speed of sound is $0.17 \sqrt{T}$ km/s at temperature T deg K. Only if $T \gg 20\,000$ deg K could the turbulence be incompressible. MÜNCH attributes the velocity differences to shock-waves in the gas brought about by an interaction between the ionized matter and non-uniformities in the cool, non-ionized gas surrounding it.

The outer parts of HII region frequently run up against masses of material which look dark in front of the bright background of the glowing gas. Often the dark matter fills an elongated shape — nicknamed an elephant's trunk — pointing towards a nearby bright star. Occasionally the matter fills small convex pockets, called globules.

POTTASCH (1956, 1958) has systematically analysed the shapes of a number of trunks and globules, and has found that in almost all cases there exists a very pronounced bright rim on the side near the closest bright star. [Bright rims were first noticed by DUNCAN (1920).] The rims are relatively thin, perhaps 0.1—0.01 pc across, in comparison with their associated trunks or globules. Their spectrum is the same as that of the background H II region, but one can infer that the rate of emission of light per unit volume is much larger in the rims. The gas density in a rim is some fifteen times higher than in the H II region as a whole.

POTTASCH has found a systematic variation in the general shape of a typical elephant's trunk depending on its position in the H II region as a whole. Structures with gentle curvature tend to lie on the outside, far from the exciting star. Trunks with sharper curvature lie nearer the centre, the globules closer still. This is thought to show up a sequence of evolution in the nature of the trunks as the background H II region grows. The innermost trunks have had most time to develop, the outermost ones have only just begun to interact with the ionized gas and the ionizing radiation. The time-scale of the evolution is not known. H II regions can be observed by radio telescopes in continuum radiation, usually in emission and sometimes in absorption. In the latter case a temperature estimate may be made. Thus MILLS, LITTLE and SHERIDAN (1956) found a temperature of about 6500 deg K for NGC 6357.

About 10% of the interstellar matter is contained in H II regions. Most of the matter exists in the so-called H I regions, where the hydrogen gas is neutral. STRÖMGREN (1939) first pointed out that a very sharp separation is to be expected between H I and H II regions owing to the large absorption cross-section of the H atom for radiation in the Lyman continuum and owing to the very low density of matter in interstellar space. Intermediate regions probably occupy a negligibly small volume.

H I regions can be seen mainly by means of the 21 cm line, by the interstellar absorption lines of Ca$^+$ and Na, and by the absorption and scattering effects due to interstellar dust grains. The sodium and calcium lines show mean Doppler shifts corresponding to radial velocities about half as large as those of the stars against which they are seen. The curves-of-growth and the doublet ratios for both sets of lines are best interpreted by means of a model in which the matter is concentrated into separate clouds. It is estimated that some eight to twelve clouds intersect a line of sight one kiloparsec in length [BLAAUW (1952)]. The clouds have random velocities; the probability of finding a cloud with a line-of-sight velocity v appears to vary as $\exp(-|v|/\eta)$, where $\eta \sim 5$km/s.

The neutral hydrogen gas is observed by means of the 21 cm line. Data obtained in this way have a higher signal:noise ratio than those obtained from the calcium or the sodium lines. However, existing radio telescopes have large aerial patterns, about 30 min of arc in angular dimensions in the case of the best instrument now available (at Dwingeloo in Holland). It has therefore not been possible to isolate 21 cm radiation from individual clouds in emission, and as a rule one observes the combined spectra of several of them at the same time. But the results can be reconciled with the cloud picture, and with the values of the random velocities deduced from Ca^+ and Na line studies. The harmonic mean temperature in the gas is found to be at least 125 deg K [WESTERHOUT (1957)]. Locally, in regions where there is a high density of interstellar dust and, presumably, also of gas, the temperature may be as low as 60 deg K [DAVIES (1956)].

It has not yet been confirmed observationally that the local distribution of the interstellar absorbing material is always similar to that of the gas. In particular, observations of a region in Taurus showed little correlation between 21 cm line intensities and the obscuration of background stars in the same part of the sky [V. D. HULST, MULLER and OORT (1954)]. In other parts of the Galaxy some correlation may be observed [BOK (1955)].

Interstellar reddening is due to a selective scattering of starlight which depends on colour, blue light being more strongly scattered than red. It is well correlated with interstellar absorption. Interstellar polarization is due to a selective scattering which depends on the direction of the electric vector of the incident light. It is fairly well correlated with interstellar extinction in some parts of the sky but not in others [HILTNER (1956)].

The material which causes the general interstellar extinction is presumably also responsible for the appearance of elephants' trunks and globules in front of H II regions, but this assumption has not been checked by observation. Neither the trunks nor the globules are large enough for detection by means of the 21 cm line, and they are too opaque to observe by the Ca^+ or Na absorption lines. Their great opacity also makes it difficult to find out any details about the nature of the solid grains causing the extinction of light there.

The available data indicate the approximate size of an average interstellar cloud in the following way. Suppose that the interstellar material is concentrated into spherical clouds of typical radius a, there being N clouds per unit volume. The material occupies one-tenth of the space available to it, so that

$$\frac{4\pi}{3} N a^3 = \frac{1}{10}, \tag{1}$$

and ten clouds are, on the average, seen along a line of sight one kiloparsec in length, so that

$$\pi N a^2 l = 10, \tag{2}$$

where $l = 1$ kpc.

On dividing (1) by (2) we deduce that

$$a \sim 0.0075\, l = 7.5 \text{ pc}$$

is a typical cloud radius, while

$$N \sim 60000 \text{ clouds/kpc}^3.$$

Neither value should be taken too seriously nor should the existence of separate clouds be believed in too literally. One should instead picture an interstellar medium with rather violent fluctuations in density, so that the mean square value $\langle n^2 \rangle$ of the particle density is some ten times larger than the square $\langle n \rangle^2$ of the mean density of particles. A maximum in the material density is, in general, surrounded by a region some five or ten parsecs in linear dimensions, in which the density exceeds the interstellar average. Within such a region one expects only a small spread in gas velocities. About one-tenth of the length of any line of sight would be expected to lie in regions where the density is above the average.

There are also larger agglomerations of interstellar matter. The Orion Nebula is an example, with an estimated mass of about $10^5 \odot$ [SAVEDOFF (1956)].

6. The surface features of stars

The superficial properties of different stars are best exhibited on a Hertzsprung-Russell (HR) diagram. This gives a plot of the absolute magnitudes of different stars, measured at a particular effective wavelength ($\lambda \approx 5290$ Å), against their spectral types. The latter are defined by the presence of some features in the spectrum and the absence of others: most of the differences between the spectra of different stars can readily be explained in terms of temperature differences. Stellar surface temperatures can also be gauged by observations of the colour index, which is effectively the difference in the apparent magnitudes measured at different wavelengths ($\lambda\lambda$ 5290, 4250 Å), but this is to some extent affected by interstellar reddening. If the stars radiated like black bodies the interpretation of the colour index in terms of a temperature would be immediate. In stellar atmospheres the variation of opacity with wavelength is unfortunately such that different effective photospheres exist for different colours. One therefore finds such phenomena as the Balmer discontinuity, where the intensity of the radiation suffers a downward jump with decreasing wavelength. The longer wavelength radiation comes from a deeper and, therefore, hotter layer than does the radiation of shorter wavelength, which undergoes additional absorption in the Balmer continuum. The Balmer jump is observed only in the hotter stars, but similar effects, due to other absorbing elements, seem to occur at all the common stellar temperatures.

To place a nearby star on an HR diagram one measures its apparent magnitude and converts this to an absolute magnitude from a knowledge of the star's parallax. Such a star is usually bright enough for its spectrum to be observed in some detail, and it can readily be classified according to spectral type. The colour index can also be measured: from a series of measurements on different stars the correlation of the colour index with the spectral type can then be determined.

Parallax measurements cannot be made on the more distant stars; these can only be plotted on the HR diagram according to their apparent magnitudes. To obtain a meaningful result one must ensure that the differences between the apparent magnitudes of the stars entered equal the differences in absolute magnitude. In other words it is best to study only the HR diagrams of compact stellar groups so that all the stars are at approximately the same distance from us and have suffered the same amount of interstellar extinction. Measurements at several colours [JOHNSON and MORGAN (1953)] provide a check on the importance of the extinction effect in particular cases. It is easier to measure the colour index for a distant star than it is to find its spectral type; as we have mentioned, the correlation between spectral type and colour index may be found from nearby stars. A large number of HR diagrams have now been made for a variety of different star groups. The stars of any one group usually lie within a magnitude or so of a curve in the luminosity-colour index plane (or LC plane); with the conventional orientation for such diagrams the bottom half of the picture, occupied by less luminous stars, usually looks similar for all groups, while there are wide divergences among the top halves of different diagrams. The common lower envelope of all the LC curves belonging to the disk population is called the main-sequence. The conversion from apparent to absolute magnitude is made by first removing the colour excess, due to interstellar reddening, and then choosing the absolute magnitude scale so that the lower sections of the different HR diagrams coincide. It seems generally agreed that this is a valid procedure for the younger star groups in the galactic disk (henceforth called population I); it is probable that the lower ends of the sequences traced by the spherical population of the Galaxy (population II) should be 0.5 to 2.0 magnitudes below those of population I sequences [ARP and JOHNSON (1955)]. There are also some LC curves for clusters of population I whose upper parts follow the main sequence all the way but whose lower halves break away from it at some point, and then lie one or two magnitudes above it. WALKER (1956, 1957) has given LC curves for two such clusters.

Some different LC curves are sketched in Fig. 3. Data are included from different stellar groups of population I, from M 67 and from some nearby high-velocity stars, to represent the intermediate population, and from the globular cluster M3 to represent population II. The various results come from papers by the authors listed on the caption of the diagram.

HR diagrams are extremely useful: all astrophysicists agree that a comparison of the LC curves traced out by stellar groups of different ages will one day lead to a correct theory of the evolution of stars. It is

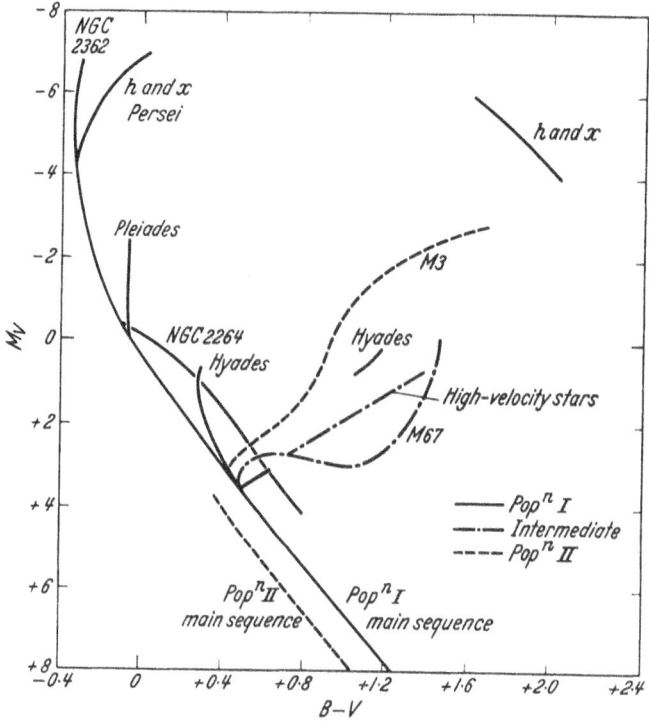

Fig. 3. Some typical LC curves. Data taken from SANDAGE (1958a), EGGEN (1958), REIZ (unpublished) and WALKER (1957) (read h and χ Persei instead of h and x)

therefore particularly interesting to study the diagrams provided by groups of stars of roughly uniform age such as are found in different associations, in galactic and in globular clusters.

The age of a group of stars is usually estimated by finding where the upper end of its LC curve turns away from the main sequence. The theory of the method is sketched out in part III of this essay. The determination is possible only if the initial chemical composition of the group is known. REIZ has found that the exact age corresponding to the turn-off point is higher for star groups deficient in the heavier elements (STRÖMGREN, private communication).

A large variety of LC curves exists for population I groups. Qualitatively there is much less difference between the curves for M 67 and the high-velocity stars, which are intermediate between populations I und II, and those for globular clusters (population II). JOHNSON and SANDAGE (1955), however, believe that there are significant differences between the upper ends of the curves for the globular clusters M 3 and M 92, on one hand, and for M 67 on the other.

The HR diagram displays stars classified by luminosity and colour. For stars earlier than type F 5 another classification has been used by CHALONGE and Mlle. DIVAN [CHALONGE (1955)]. Here the stars are arranged by means of three parameters λ_1, D and φ_b. λ_1 is the effective wavelength of the Balmer jump in the stellar spectrum; its displacement from the series limit determines the electron density in the effective photosphere of the star. D is the height of the Balmer jump and may be used to find the number of H atoms above the photosphere which are in the second quantum state. For a star of known composition these two data lead to some possible values of the stellar radius and mass in the following way.

The photospheric electron density n_e can be calculated from a knowledge of λ_1; the quantity D is a function of the temperature T in the photosphere and of the number of H atoms in a column of unit cross-sectional area above it.

Let N be the number of H atoms in the second quantum state per unit volume of photosphere, and let h be the local scale height. Let n be the number of H atoms per unit volume. If the atmosphere consists predominantly of hydrogen $N = N(n, T)$ and

$$D = D(N h) = D[h N(n, T)]. \tag{1}$$

This is the first equation. The second equation

$$\lambda_1 = \lambda_1(n_e) \tag{2}$$

determines the photospheric electron density. The local scale height is given by

$$h = \frac{kT}{\mu g} \tag{3}$$

where g is the stellar surface gravity; μ is the mean molecular weight and is determined by the stellar composition, and by the ionization equilibrium, so that

$$\mu = \mu(C; n, T). \tag{4}$$

In (4) C stands for composition. The absorption coefficient $\varkappa$ per unit length is also a function of the composition and ionization equilibrium and so

$$\varkappa = \dot\varkappa(C; n, T). \tag{5}$$

Finally the photosphere occurs at optical depth unity, approximately, so that

$$\varkappa h \sim 1 \qquad (6)$$

and the equation of ionization equilibrium gives that

$$n_e = n_e\,(C\,;\,n,\,T)\,. \qquad (7)$$

These seven equations determine the unknown quantities n, n_e, T, μ, $\varkappa$, h and g, provided that λ_1, D and the stellar composition are known. Usually a multivalued solution is found, but with a knowledge of φ_b the right set of numbers can be chosen.

To find the appropriate values for the mass, radius and luminosity of the star one has to use the empirical result that, for a stellar group with the same initial composition, the position of a star on the Chalonge-Divan diagram together with a knowledge of T corresponds to one possible position on the HR diagram. The value of L can therefore be found, and the mass and radius follow because

$$L = \pi R^2 f(T) \quad \text{and} \quad g = GM/R^2$$

where $f(T)$ is a known function of T.

If all the stars in the group have the same chemical composition then the values of λ_1, φ_b and D lie on a single surface. This provides a check on the assumptions made in the above calculation. CHALONGE (1958) has reported that the particular surface on which the values lie changes continuously when one varies the stellar population from which the sample is drawn. High-velocity stars and stars of population II are systematically too blue to fit on the population I surface. One interpretation is that the relative excess of blue light is due to a scarcity of metallic elements, with a consequent reduction in continuum absorption at shorter wavelengths [cf. JOHNSON and SANDAGE (1955)].

The weakness in the suggested method for finding M and R or, for that matter, n and T is that the atmospheric composition must be known beforehand. But the composition can only be found by an interpretation of the stellar spectrum, and this requires a knowledge of n and T. The vicious circle can be broken in a number of ways, as explained by UNSÖLD (1957); the best solution is probably to construct model atmospheres with various possible assumed compositions and to see which fits the data best. UNSÖLD (1955) suggests that the error in such a computation may be about $\Delta \log_{10} \sigma = \pm\, 0.3$, where σ is proportional to the abundance of the particular element considered.

It is clear that the physical characteristics of a star are only found at the end of a lengthy and delicate process. In every case one has to assume that the stellar atmosphere is in hydrostatic equilibrium. More complex effects, such as convection and granulation, undoubtedly occur in most stars, but it has not yet been possible to include them in any calculation.

The results of such a mass determination are thus rather indefinite. On the other hand a direct method for finding the stellar mass exists only for binary stars. A complete determination can be made for visual binaries, which are rather rare and are all close to the Sun, or for eclipsing binaries. In the case of the latter one can infer not only the stellar masses but also the stellar radii and moments of inertia. Yet such stars are never quite typical. They are necessarily very close to one another and may strongly disturb each other's equilibrium configurations. In many cases (KOPAL's semidetached systems) these gravitational effects are so strong that a point of zero gravity exists on the surface of the secondary where matter can escape freely [KOPAL (1957)].

All the same two results derived from measurements on eclipsing binaries have been used frequently. They apply to stars on the main sequence and state that

$$L \propto M^{3.9}, \quad R \propto M^{0.75};$$

these are the well-known mass-luminosity and mass-radius relations.

The Chalonge-Divan and the Hertzsprung-Russell diagrams show up some differences between stars of the various populations. These can probably be explained in terms of the abundances of the heavier elements. There are some other observational results which can also be accounted for by the same hypothesis. One can for instance explain in this way why the absorption lines of heavier elements are generally weaker in F and G high-velocity stars than in the corresponding population I stars [SCHWARZSCHILD and SCHWARZSCHILD (1950), ROMAN (1950), AX (1953)]. One can also see why the CH bands are strengthened relative to the CN bands in high-velocity G and K giants. According to SCHWARZSCHILD, SPITZER and WILDT (1951) this is due simply to the differences in the chemical and ionization equilibria which result from the different atmospheric composition. It is suggested that the metal abundances in high-velocity stars are one-third and the carbon, nitrogen and oxygen abundances one half of those in typical population I stars. DE JAGER and NEVEN (1957) find even larger composition differences. But one must not conclude that these results show more than a general trend in the typical chemical composition of stars as one moves from population I towards population II. There is no evidence that all population I stars have exactly the same abundances for the various elements, and that all high-velocity stars have another set of abundances. In fact there is a considerable dispersion in the degree of line weakening observed in different high velocity stars [KEENAN and KELLER (1953)].

Not much information is available about the spectra of genuine population II stars. An excess of ultraviolet radiation is observed among the stars in some globular clusters (NGC 4174, M 3, M 92) but not in all (e. g. the excess is not large in M 13) [JOHNSON and SANDAGE (1955)]. This may again be due to a deficiency of metallic elements. Still further notable differences, possibly due to the same cause are:

(1) In a period-luminosity diagram the classical Cepheids (population I) fill a strip which differs from that filled by the cluster type variables (population II).

(2) Novae and planetary nebulae belong to the spherical component of the Galaxy.

(3) The more violent (class I) supernovae seem to be members of population II. The Crab Nebula (M 1) is an example of a remnant of one of these. The composition of the outermost regions of M 1 appears to be the same as that of a population I star, however, with possibly a slight hydrogen deficiency [WOLTJER (1958)]. The bulk of the material is in the inner parts of the nebula and emits no spectral lines.

(4) Finally there is one difference which most probably has no connexion with the stellar composition. Only stars of population I ever show high angular velocities [SLETTEBAK (1955)]. Even in population I no star later than type F rotates appreciably fast.

7. Physical properties of interstellar matter

For lack of better data it is generally assumed that interstellar matter has the same composition as do the atmospheres of population I stars. Observed interstellar abundance ratios of hydrogen to calcium, sodium and oxygen [SEATON (1951)] indicate that the assumption is good. Most elements are, however, too sparsely distributed in space or do not possess resonance lines in a suitable part of the spectrum. They have therefore escaped detection up till now. Observations from an Earth satellite of interstellar resonance lines in the ultra-violet will become feasible soon and should considerably increase our knowledge.

The problem is further complicated because the heavier atoms in interstellar space may, to some extent, be locked up within the dust grains. The best one can do at present is to adopt the model composition given in Table 1. Only the more important elements are listed, together with their relative abundances, atomic weights and ionization potentials. The data are quoted from UNSÖLD (1955).

The great relative abundance of hydrogen leads to the division of space into H I and H II regions and this profoundly affects all mechanical and physical properties of the gas.

In an H II region almost all the matter is ionized. In an H I region the abundance of electrons is effectively equal to that of C, Mg, Si and Fe,

Table 1. *Abundances of elements*

Z	Element	First ionization potential: χ_i	Relative abundance of atoms	
			by number: ε_i	by mass: μ_i
1	H	13.54 Volt	25,100	25,100
2	He	24.46	4,570	18,200
6	C	11.22	6.31	75.9
7	N	14.48	13.5	190
8	O	13.55	24.5	389
10	Ne	21.47	28.8	575
12	Mg	7.61	1.55	38.0
14	Si	8.12	1.41	39.8
26	Fe	7.83	2.34	132
			$\Sigma \varepsilon_i = 29{,}750$	$\Sigma \mu_i = 44{,}760$

about 5×10^{-4} of the particle density according to Table 1. This follows because these four elements are the commonest substances present which can be ionized, since there is no radiation in an HI region with a wavelength shorter than $\lambda = 912$ Å, the Lyman limit. There are two major consequences.

The gas in HI and HII regions has so low a density that it can gain energy from the radiation field only when an atom is ionized. The ionizing photon in general has more energy than is required to detach an electron; the remainder goes to heat the gas. But in equilibrium the ionization rate can only balance the recombination rate. At a given gas density the charged particle density in an HII region is some 2000 times higher than in an HI region; it follows that the recombination rate, and therefore the rate of photo-electric heating, is some 4×10^6 times larger in the former than in the latter.

Interstellar gas loses heat in a number of ways. In the first place some thermal energy is re-radiated when ions and electrons recombine. If this were the only means by which the gas could cool, and if absorption of heat on ionization were the only way in which the gas could be warmed, then eventually an equilibrium temperature would be reached comparable with that of the star exciting the gas. But there are other mechanisms of re-radiation. In an HII region the most important one is radiation by O^+ ions, excited by electron impact; in HI regions there can be radiation from low lying levels of the Si^+ and C^+ ions, also excited by electron impact, radiation by H_2 molecules excited by atomic impacts, and thermal radiation from solid grains [SPITZER (1954), SEATON (1955b), KAHN (1955a, b)].

It is calculated that the temperature in the HII regions generally settles down somewhere between 8000 and 10000 deg. K; in the HI regions it would settle down to about 50 deg. K at a gas density of 10 atoms/cm³, if heating by photo-ionization were the only effective process.

This is not the case because the more intensely heated HII regions acquire a greater gas pressure and expand. Nearby HI clouds are therefore pushed away, and thus there is set up the irregular motion with strong density fluctuations which is observed in HI regions [OORT (1955), SCHLÜTER (1955)].

The energy of this motion decays at a rate of the order of u^3/l, where u is the typical cloud velocity and l the mean free path for collision between clouds. The decay leads to a rate of heating in HI clouds larger by two or three orders of magnitude than that due to photo-ionization [KAHN (1955a)]. It would indeed have been surprising if some heat had not leaked out in this way from the HII regions, for these absorb between them about 4×10^5 times more heat owing to the photo-electric effect than do the HI regions.

A particular HI cloud will be involved in a collision about once in 10^7 years and will then be heated to about 4000 deg. K. In the interval between successive collisions the gas cools off; the estimated rate of loss of heat by the radiation of H_2 molecules and by Si^+ and C^+ ions is probably large enough to explain the observed harmonic mean temperature of 125 deg. K in HI regions [SEATON (1955b), KAHN (1955b)]. There is additional cooling by the dust grains, which are believed to have a very low surface temperature, of the order of 20 deg. K [SPITZER (1954)]. In one collision with a grain an atom will lose most of its thermal energy. But the mean free path of an atom for a collision with a grain is about the same as the mean free path for the extinction of a light quantum by a grain, and equals about 1 kpc for an average line of sight in interstellar space, or 100 pc in a cloud with a density of 10 atoms/cm³. The root-mean-square speed of a hydrogen atom is 1.7 km/s at 125 deg. K, and so a given atom will strike a grain perhaps once in 60 million years, under typical conditions. During this time the average gas cloud will have been heated and cooled several times. Cooling by dust particles is therefore significant only in rather quiescent conditions and at relatively high densities.

The high concentration of ions and electrons in HII regions and their low concentration in HI regions lead to a further difference. If there is a magnetic field present almost all the particles in an HII region are virtually bound to the lines of force. In an HI region only the ions and electrons are thus bound; the neutral atoms can only be affected indirectly by the magnetic field after collisions with charged particles. In certain conditions the neutral particles may be able to slip past the charged particles and move more or less independently of any magnetic field. Such a case has been studied by MESTEL and SPITZER (1956), whose work on the condensation of an interstellar cloud in a magnetic field will be considered later.

At the third symposium on cosmical aerodynamics (held in June 1957, reported in Rev. Mod. Phys. No 3, 1958) there took place a lengthy discussion on the possible reduction, in the presence of magnetic fields, of the rate of loss of kinetic energy in cloud collisions. It was generally agreed that such a reduction can be at most by a factor two or three. No thorough quantitative work appears to have been done on this problem.

Inferences about the nature of the interstellar dust grains have been made from the following data. The grains need a rather high albedo to explain the existence and brightness of reflection nebulae, and it has been concluded that they are composed of dielectric rather than metallic material [v. D. HULST (1949)]. The probable composition of a grain can be deduced from the assumed abundances of the elements in interstellar space. The most common substances, H and He, cannot condense into solid matter at the low densities and relatively high temperatures found in space. It is therefore likely that the typical grain consists largely of H_2O, CH_4 and NH_3 in solid form, with some metallic impurities.

Our table of abundances shows that only a small mass is available for condensation into dust, perhaps one per cent of that of all interstellar matter. There must be enough solid matter to explain the average extinction of 1 mag/kpc in the galactic plane. If the typical grain has linear dimensions a, if there are N grains per unit volume made of material with a specific gravity ϱ, and if Q is the average extinction efficiency, the mass of interstellar dust per unit volume of space is $(4\pi/3) N a^3 \varrho$, and we have

$$\frac{4\pi}{3} N a^3 \varrho \leq 2.4 \times 10^{-26} \text{ g/cm}^3, \tag{1}$$

while the optical cross-section per unit length of a line of sight is $N\pi a^2 Q$ and so

$$N\pi a^2 Q l \sim 1, \text{ when } l = 1 \text{ kpc.} \tag{2}$$

From (1) and (2) it follows that

$$\frac{a}{l} \leq \frac{3}{4} \times 2.4 \times 10^{-26} Q$$

or

$$a \leq 5.4 \times 10^{-5} Q \text{ cm} = 5400 \, Q \text{ Å}.$$

Q is of order unity when the grain radius a is about the same as the wavelength of the light being extinguished [v. D. HULST (1949)], and is very much smaller when a/λ is small. The radius of a grain is therefore of the order of 5000 Å; this agrees with estimates of grain sizes made from observations of interstellar reddening. The matter in space seems to be barely sufficient to explain interstellar extinction effects. If the grains were much larger than they seem to be then they could not together present a big enough cross-sectional area; if they were much smaller, the extinction efficiency Q would be severely reduced.

The occurrence of interstellar polarization shows that some, at least, of the dust grains must be considerably anisotropic. It may be that the effect is due to particles which are of ordinary composition (H_2O, CH_4, NH_3) but are considerably elongated in shape [v. D. HULST (1955)]. But it has also been suggested that the presence of small flakes of graphite can be responsible [CAYREL and SCHATZMAN (1955)]. There is general agreement that a magnetic field of about 10^{-5} Γ is required to explain the alignment of the polarizing particles.

The surfaces of grains probably act as catalysts in the formation of interstellar molecules, such as CH, CN and H_2. It is thought that any atom which strikes a grain is adsorbed onto the surface and, in general, evaporates later. The grains are expected to be covered in a film of hydrogen. If, say, an oxygen atom strikes this film it has a high probability of combining with the hydrogen present to form a water molecule, and to freeze onto the body of the grain. This model leads to an estimate of the rate of growth of a grain.

The impact of each oxygen atom (atomic weight 16) adds one water molecule (molecular weight 18) to the mass M of the grain. If the grain is assumed to be spherical

$$\frac{dM}{dt} \approx 4\pi a^2 \cdot \frac{18}{16} \cdot \varrho_0 \overline{C}_0,$$

where a is the grain radius, ϱ_0 the mass of oxygen per unit volume in the gas cloud and $\overline{C}_0$ the mean thermal speed of the oxygen atoms. If ϱ_G is the specific gravity of the grain material

$$4\pi a^2 \varrho_G \frac{da}{dt} \approx 4\pi a^2 \cdot \frac{18}{16} \cdot \varrho_0 \overline{C}_0$$

or

$$\frac{da}{dt} \approx \frac{9}{8} \frac{\varrho_0}{\varrho_G} \overline{C}_0.$$

Typical values are

$$\varrho_0 = 2.4 \times 10^{-25} \text{ g/cm}^3 \text{ in a cloud}$$

$$\varrho_G = 1 \text{ g/cm}^3 \text{ for ice}$$

$$\overline{C}_0 = 4 \times 10^4 \text{ cm/s}$$

and so

$$\frac{da}{dt} = 10^{-20} \text{ cm/s}.$$

To reach a radius $a = 5 \times 10^{-5}$ cm takes about 5×10^{15} s, or 170 million years.

Presumably the growth of grains stops when the material is exhausted. On the other hand it is likely that grains are destroyed when they are immersed in H II regions.

It is not known why there are just the right number of centres of growth for grains to form with just the required extinction properties. If more growth centres had been available a larger number of grains of smaller size would have formed, and quite different extinction effects would have been caused.

8. Conclusion

It may be useful to summarise the data collected thus far.

There is good evidence that the Earth is about 5 or 6×10^9 years old. Neither the Sun nor the Galaxy can be younger than this. On the other hand the distribution of red-shifts among the extragalactic nebulae shows that the Universe passed through a stage of relatively high density about 13×10^9 years ago (the estimate being liable to revision). Possibly no form of matter existed before that time, but this is not certain.

Most of the visible matter in the Universe is concentrated into spiral nebulae. The physical and dynamical properties of stars in these galaxies are correlated. The hottest and brightest stars seem always to be found in the rotating and disk-like parts and close to concentrations of interstellar gas and dust; they are typical members of BAADE's population I. The spherical population (BAADE's population II) of a spiral galaxy contains no dust, probably only a little gas and no very bright stars. Certain easily recognizable types of star or star-groups are associated with one or other of the populations: thus classical Cepheids and galactic clusters are found in population I, globular clusters, planetary nebulae and RR Lyrae variables in population II. Possibly many of the differences between populations arise because elements heavier than helium are less abundant in population II than in population I. There appears to be an intermediate population, the high-velocity stars, which takes part to some extent in galactic rotation and whose members have physical properties resembling those generally ascribed to population II.

Member stars of a given cluster are usually assumed to be equally old. Comparison of Hertzsprung-Russell diagrams for different clusters picked from population I shows up considerable differences between luminosity-colour curves for the brighter stars; with a few exceptions the dimmer stars appear to lie on much the same curve in all clusters. LC curves for high-velocity stars and for globular clusters resemble, to some extent, one extreme variety of the LC curves found for population I.

In the galactic disk there exist gravitationally unstable associations of early type stars (OB associations) and of irregular variables (T associations). Both types tend to dissolve within 10^6 or 10^7 years. Associations probably consist of newly formed stars. The LC curve for a typical OB association resembles the other extreme in the family of such curves for population I.

Masses and radii of stars are known indirectly from details of their spectra; alternatively they may be measured in eclipsing binaries, but such stars are rather a-typical. Exact values are therefore unknown, but even if the estimated stellar masses were wrong by a factor five, and this is most unlikely, it would still follow from the observations that there are early-type stars with luminosities so high that their energy sources cannot support them in their present state for even 10^8 years. This is a very generous upper limit; the actual life-time is believed to be much shorter. Such bright stars are usually found in unstable associations and are thought to be relatively newly formed. The argument concerning the life-time assumes that stars gain their energy largely from the conversion of hydrogen into helium, with a yield of about 6×10^{18} erg/g.

The association of early-type stars with interstellar matter suggests that there is a genetical connexion between them. The total mass of interstellar matter in our Galaxy is about $10^9 \odot$, only about one per cent of the total galactic mass, but comparable with the mass of the disk population. Interstellar gas seems to take part in the general rotation of the galactic disk: its composition is thought to resemble that of a population I star. The gas is divided in the ratio of about 1 : 10 between H II regions, where it is highly ionized and hot, with $T \approx 10^4$ deg. K, and H I regions, where it is only lightly ionized and cool, with $T \approx 125$ deg K. Both H I and H II regions are often found to be in a state of supersonic turbulence.

These are the basic data for theories of the formation and evolution of stars.

Part II: Theories of star formation
1. The origin of stellar associations

AMBARTSUMIAN (1955) has proposed the simplest theory to explain the formation of new stellar associations. In his view they originate from dense, "pre-stellar" matter, whose nature is not further specified but which is endowed with enough energy to form an expanding group of stars. The theory does not explain why stellar associations, and in particular O and B stars, occur only in the spiral parts of most galaxies, nor why they are usually close to concentrations of interstellar dust and gas. It is difficult to discuss the hypothesis since no way has been found of verifying it, either by observation or by theory.

There are two other principal theories concerning the formation of early type stars; we deal first with the theory of accretion [HOYLE and

LYTTLETON (1939)], according to which interstellar matter can condense onto an existing smaller star to form a large star. It predicts that a star of mass M moving with a speed V through a stationary diffuse medium of density ϱ will sweep up all the matter within a distance of about $r = GM/V^2$. The rate of increase of its mass is therefore

$$\frac{dM}{dt} \approx \pi \varrho V \frac{G^2 M^2}{V^4}. \tag{1}$$

To explain the formation of an association one must assume that its various members have all passed through the same diffuse cloud and have accreted a sufficient mass of material to change them into early type stars. Let us work out some consequences.

If ΔM is the mass accreted onto a typical star whose final mass is M_f, then $\Delta M/M_f$ must be of order unity, or else there would be no great change in the stellar luminosity. Let V_f be the final velocity of the star, and let t be the time it takes to pass through the cloud from which it accretes matter, then

$$\Delta M \approx \pi \varrho V_f t \frac{G^2 M^2}{V_f^4}$$

$$\approx \pi \varrho L \frac{G^2 M^2}{V_f^4}, \tag{2}$$

when the cloud has linear dimensions L. It follows that

$$\frac{\Delta M}{M_f} \approx \pi \varrho L \frac{G^2 M_f}{V_f^4} \sim 1. \tag{3}$$

An association usually contains some dozens of stars and so the accretion radius of each of them must be much smaller than L, otherwise all the matter would condense onto one star. Hence

$$\frac{G M_f}{V_f^2} \ll L. \tag{4}$$

But unless the cloud is supported by its own gas pressure against its gravitational self-attraction, it will collapse within a time of the order $(G \varrho)^{-1/2}$. If the collapse takes place in less time than it takes for the accreting star to move through the cloud, then clearly accretion is only of secondary importance. If accretion is to be dominant it is necessary that

$$\frac{1}{\sqrt{G \varrho}} \gg \frac{L}{V_f},$$

or that

$$L \ll \frac{V_f}{\sqrt{G \varrho}}. \tag{5}$$

Insertion of this inequality into (2) leads to the condition

$$\frac{\pi \varrho \, G^2 M_f}{\sqrt{G \varrho V^3}} \gg 1$$

or

$$\frac{G M_f}{V_f^2} \gg \frac{1}{\pi} \frac{V_f}{\sqrt{G \varrho}}. \tag{6}$$

Comparison with (5) shows that therefore

$$\frac{G M_f}{V_f^2} \gg L, \tag{7}$$

which contradicts condition (4).

The members of a group of stars cannot therefore increase their masses by large amounts in any accretion process from the same cloud, except possibly if such a cloud is stabilised against its self-attraction by gas pressure. To consider that case let a be the speed of sound in the gas; if stability occurs, then

$$G \varrho L^2 \sim a^2. \tag{8}$$

When a is small compared with V_f, the stellar velocity, then this condition leads to another contradiction. For it follows from (3) and (4) that

$$\frac{\pi G \varrho L^2}{V_f^2} \gg 1, \tag{9}$$

or, by means of (8), that

$$a^2 \gg V_f^2. \tag{10}$$

If a and V are comparable the accretion radius is given, approximately, by $GM/(a^2 + V_f^2)$ [BONDI (1952)]. Once again this length must be smaller than L. Following through the same working one finds that the condition

$$\frac{\pi G \varrho L^2}{V_f^2 + a^2} \gg 1 \tag{11}$$

replaces (9), and leads to the self-contradictory inequality

$$a^2 \gg V_f^2 + a^2. \tag{12}$$

It seems that the accretion theory, whatever its value may be in other astronomical applications, cannot possibly explain the formation of O and B associations.

The remaining possibility is that a group of new stars can be formed after the collapse of a large cloud of interstellar matter under its own self-attraction. These are the principal problems to be considered:

(i) Is the average interstellar cloud likely to be unstable under its gravitational self-attraction?

(ii) What mass, dimensions and temperature may be expected in an unstable cloud?

(iii) How does external gas or radiation pressure affect the instability?
(iv) What is the effect of the presence of a magnetic field?
(v) How does the further collapse take place?
(vi) How long does it take?
(vii) Why are stars formed with the masses which commonly occur?
(viii) Why do associations expand? Can the usual speeds of expansion be explained?

The general state of motion and the space distribution of interstellar matter have been inferred from the data discussed in some previous sections. There are many uncertainties and ambiguities involved, and all conclusions are therefore rather vague. For instance one cannot say for sure whether the average rate of interstellar extinction is one or two magnitudes per kpc, whether the harmonic mean temperature of the gas is 125 or 150 deg K, whether the random motion of the interstellar clouds has a mean line-of-sight velocity component of 4 or 6 km/s, and so on. All arguments which can usefully be applied must be based on orders of magnitude rather than on exact values. Individual numbers which may be quoted should not be taken seriously.

Let us consider first the gravitational stability of an average interstellar H I cloud, of radius $R = 7.5$ pc and with a mean density of $n = 10$ atoms/cm³. According to the usual rather rough criterion of the virial theorem the cloud will expand or contract according as

$$2T + V \gtrless 0, \tag{13}$$

where T is the sum of the kinetic and thermal energies[1] of the cloud and V is its gravitational self-energy. In a quiescent cloud at temperature Θ

$$\left.\begin{aligned} T &= \frac{3k}{2\mu}\Theta M \\ \text{and } V &= -\frac{3}{5}\frac{GM^2}{R}, \end{aligned}\right\} \tag{14}$$

where M is the mass, supposed to be uniformly distributed within a sphere of radius R, and μ is the mean molecular weight. It follows from (13) and (14) that expansion or contraction takes place according as

$$\frac{k\Theta}{\mu} \gtrless \frac{1}{5}\frac{GM}{R}, \tag{15}$$

or since

$$M = \frac{4\pi}{3} n \mu R^3,$$

according as

$$k\Theta \gtrless \frac{4\pi}{15} Gn\mu^2 R^2. \tag{16}$$

---

[1] For a monoatomic gas. Contributions from rotational and excitational degrees of freedom are not included in T.

With $\mu = 2.4 \times 10^{-24}$ g, corresponding to a mean molecular weight 1.5, the criterion becomes

$$k \, \Theta \gtrsim 10^{-15} \text{ erg}$$

or

$$\Theta \gtrsim 7 \text{ deg K}. \tag{17}$$

No part of an interstellar cloud ever reaches a temperature as low as 7 deg K; in general the average cloud is therefore in a state of expansion. This conclusion could be wrong only if the average cloud were strongly condensed towards its centre, so that our assumption of a uniform density n would be inappropriate and the absolute value of V would be much larger. But the statistical interpretation of interstellar absorption lines is always based on the assumption that the average cloud has a roughly uniform density distribution. If the typical cloud were highly condensed, the statistics would lead to a mean cloud radius much smaller than R. It seems that in fact gravitational collapse can take place only in HI clouds whose density or whose mass is much larger than the average, or in smaller masses subject to a high pressure from the medium surrounding them. In the former case the collapse might eventually lead to the formation of an association; the latter case might, perhaps, be a suitable description of a globule or an elephant's trunk immersed in an HII region.

Some work on the collapse of HI clouds has been done by EBERT (1955), by MCCREA (1957) and by ROUSKOL (1955). EBERT and MCCREA follow the evolution of a cloud through a succession of stationary states, ROUSKOL studies a dynamical model of a collapsing cloud with spherical symmetry. In all cases more complex dynamical effects, such as turbulence or rotation, are left out of account, nor is the possible influence of magnetic fields considered. EBERT points out that for a large range of densities a gas cloud will tend to a more or less constant temperature, and this seems correct as long as heating due to viscous effects can be neglected, that is as long as the motion goes on slowly enough. He therefore studies a model sequence of isothermal gas spheres in equilibrium under external pressure. MCCREA remarks that the gas pressure may have been applied suddenly to a gas cloud not in equilibrium, for instance in the case of a globule, and works out some of the numerical differences which occur. He remarks that a slight improvement is needed in the work by EBERT, who uses JEANS's criterion for the instability of a mass of gas. In fact one can give a relation between the mass M of a gas sphere, its radius R and fixed temperature Θ_0, and the external pressure Π. It is found that a given isothermal mass of gas can support at most an external pressure Π_0 which is attained when the radius R takes the value R_0. For other radii the external pressure in an equilibrium state must be less than Π_0. Therefore as the external pressure is raised gradually the gas sphere will contract until it reaches a radius R_0. A collapse must follow if the pressure increases still further, for a suitable configuration is now impossible. McCrea reasons that all clouds are brought to the verge of instability in this way; there seems to be no way in which a cloud can reach the condition where it satisfies Jeans's criterion for instability without its previously having satisfied McCrea's condition.

Let us find an expression for the pressure Π at the boundary of a spherical mass M of gas in gravitational equilibrium at a temperature Θ_0. The condition of hydrostatic equilibrium leads to

$$\frac{k\Theta_0}{\mu}\frac{d\varrho}{dr} \equiv \frac{dp}{dr} = \varrho F = -\frac{GM(r)\varrho}{r^2}, \tag{18}$$

where $M(r)$ is the mass enclosed within a radius r, while

$$\frac{dM}{dr} = 4\pi r^2 \varrho. \tag{19}$$

Equations (18) and (19) lead to

$$\frac{d^2u}{dr^2} + \frac{2}{r}\frac{du}{dr} + \frac{4\pi G \varrho_0}{a^2} e^u = 0, \tag{20}$$

which is Emden's equation for an isothermal cloud. In (20) $\varrho \equiv \varrho_0 e^u$, and $a^2 \equiv k\Theta_0/\mu$. With $x = (r/a)(4\pi G \varrho_0)^{1/2}$, (20) may be written

$$\frac{d^2u}{dx^2} + \frac{2}{x}\frac{du}{dx} + e^u = 0. \tag{21}$$

The mass enclosed to (dimensionless) radius x is, by (18),

$$M(x) = -\frac{k\Theta_0}{G\mu}\frac{r^2}{\varrho}\frac{d\varrho}{dr} = -\frac{a^3 x^2}{G^{3/2}(4\pi \varrho_0)^{1/2}}\frac{du}{dx}; \tag{22}$$

the pressure at radius x is

$$p(x) = \frac{k\Theta_0}{\mu}\varrho = \varrho_0 a^2 e^u. \tag{23}$$

The relation of p with M is expressed by (22) and (23) in parametric form; from Emden's tables [see Eddington (1926)] the function $u(x)$ is known.

Instability sets in when the further contraction of a gas sphere of given mass M_0 requires no further increase in the boundary pressure. Assume that in a small contraction the central density changes from ϱ_0 to $\varrho_0 + \delta\varrho_0$ and the dimensionless radius from x to $x + \delta x$. Taking logarithmic derivatives in (22) and noting that $\delta M_0 = 0$ we find

$$\frac{\delta\varrho_0}{2\varrho_0} - \frac{u''\delta x}{u'} - \frac{2\delta x}{x} = 0, \tag{24}$$

and with Π as the boundary pressure we find from (23) that

$$\frac{\delta \Pi}{\Pi} = \frac{\delta \varrho_0}{\varrho_0} + u'\,\delta x. \tag{25}$$

When the critical condition is reached, $\delta \Pi = 0$, and so

$$\frac{u'}{2} + \left(\frac{u''}{u'} + \frac{2}{x}\right) = 0,$$

or since

$$u'' + \frac{2u'}{x} = -e^u$$

it follows that

$$e^u = \frac{1}{2} u'^2. \tag{26}$$

From EMDEN's tables it can be seen that this condition is satisfied when

$$u = -2.65, \quad x = 6.5 \quad \text{and} \quad e^u = 0.070.$$

The corresponding value of the radius R for a sphere of given mass at a given temperature can now be found from the relations (21) and (23), and from the definition of x in terms of r.

Both EBERT and MCCREA point out that this mode of collapse cannot occur in two- or in one-dimensional configurations. Two-dimensional configurations are either stable to symmetrical disturbances for all pressures, or else they are unstable for all pressures. One-dimensional configurations always become stable at a high enough compression.

Let us now see, very roughly, how a large interstellar cloud may possibly form and how it can be brought to a state of collapse by a mechanism similar to MCCREA's and EBERT's.

Following OORT (1955) we adopt the following picture. Every now and again an interstellar cloud rather more massive or more dense than the average will form somewhere in the interstellar medium. In general two clouds of about the same size involved in a collision tend to break up [KAHN (1955a)]. However an unusually dense cloud will tend to gain mass whenever it collides with another cloud of smaller density. We therefore consider a mass $M(t)$ of gas with radius $R(t)$ in gravitational equilibrium under an effective outside pressure $\frac{1}{2}\bar{\varrho}\,\overline{V^2}$, where $\bar{\varrho}$ is the mean density of interstellar gas and $\overline{V^2}$ the mean square random velocity component of interstellar clouds. The pressure is exerted on the large mass by the collisions of other clouds with its outer parts. The factor $\frac{1}{2}$ occurs because the collisions are inelastic. The sphere gains mass at a rate

$$\frac{dM}{dt} = 4\pi R^2 \bar{\varrho}\,\overline{|V|}. \tag{27}$$

According to the inferred distribution of interstellar cloud velocities [BLAAUW (1952)] the probability that a cloud have a radial velocity component in the range $(V, V+ dV)$ is

$$p(V)\, dV = \frac{1}{2\eta} e^{-|V|/\eta}\, dV,$$

with $\eta \sim 5$ km/s. It follows that

$$\overline{|V|} = \eta \quad \text{and} \quad \overline{V^2} = 2\eta^2.$$

If the growing mass of gas is in equilibrium under its own gravitation and the outside pressure, then, as before [equation (23)]

$$\Pi = \varrho_0 a^2 e^u = \bar\varrho\, \eta^2 \tag{28}$$

at the boundary of the sphere, so that

$$\varrho_0 = \frac{\bar\varrho\, \eta^2}{a^2 e^u}. \tag{29}$$

If the boundary of the sphere occurs where $x = x_0$ it now follows from (22) that

$$M(x_0) = -\frac{a^3 x_0^2}{(4\pi G^3 \varrho_0)^{1/2}} \left(\frac{du}{dx}\right)_{x=x_0} = -\frac{a^4 x_0^2 e^{u/2}}{\eta\, (4\pi G \bar\varrho)^{1/2}} \left(\frac{du}{dx}\right)_{x=x_0}. \tag{30}$$

With $x_0 \equiv (R/a)(4\pi G \varrho_0)^{1/2}$ we deduce from (27) that

$$\frac{dM}{dt} = \frac{4\pi a^2 x_0^2}{4\pi G \varrho_0} \bar\varrho\, \overline{|V|} = \frac{a^4 x_0^2}{G\eta} e^u, \tag{31}$$

and so

$$\frac{1}{M}\frac{dM}{dt} = -(4\pi G \bar\varrho)^{1/2} \left(\frac{e^{u/2}}{du/dx}\right)_{x=x_0}. \tag{32}$$

In these equations x_0 is to be regarded as a function of t. As the cloud grows x_0 increases, and when the critical value $x_0 = 6.5$ is reached a gravitational collapse sets in. Equation (32) leads to an estimate of the time required for the formation of an unstable cloud. The table given by EDDINGTON (1926) p. 90 shows that $e^{u/2}(du/dx)^{-1}$ is an increasing function of x. (EDDINGTON uses z to denote his independent variable.) The percentage rate of increase of mass is therefore least in the last stages of the growth: the time scale of the development of an instability is thus approximately equal to

$$\left[\frac{M}{dM/dt}\right]_{x_0=6.5} = -\frac{(du/dx)\, e^{-u/2}}{(4\pi G \bar\varrho)^{1/2}} = -\frac{1}{(2\pi G \bar\varrho)^{1/2}}, \tag{33}$$

and depends only on $\bar\varrho$, the mean density of interstellar matter. Adopting $\bar\varrho = 2.4 \times 10^{-24}$ g/cm³ one finds

$$\left[\frac{M}{(dM/dt)}\right]_{x=6.5} \sim 10^{15}\text{ s} \sim 3.3 \times 10^7 \text{ years}, \tag{34}$$

a relatively short time for the formation. Equation (30) leads to an estimate of the mass of the unstable cloud. With an assumed mean molecular mass 2.4×10^{-24} g, as before,

$$(M)_{x \approx 6.5} = 3.3 \times 10^{32} \, \Theta_0^2 \text{ g} \\ \sim 400 \; \odot, \text{ if } \Theta_0 = 50 \text{ deg K}, \\ \sim 1600 \; \odot, \text{ if } \Theta_0 = 100 \text{ deg K}. \quad (35)$$

When instability sets in the cloud radius is

$$R = \frac{6.5 \, a}{(4 \pi G \varrho_0)^{1/2}} = \frac{6.5 \, a^2 \, e^{u/2}}{\eta \, (4 \pi G \bar{\varrho})^{1/2}} = 1.4 \times 10^{17} \, \Theta_0 \text{ cm} \\ \sim 2.5 \text{ pc, if } \Theta_0 = 50 \text{ deg K} \\ \sim 5 \text{ pc, if } \Theta_0 = 100 \text{ deg K}. \quad (36)$$

The unstable sphere of gas is comparable in size with an average interstellar cloud if $\Theta_0 = 100$ deg K, and a little smaller if $\Theta_0 = 50$ deg K. Its mass is rather above average at 100 deg K. and about the same as the average mass at 50 deg K. In either case the distribution of material in the sphere shows a strong concentration towards the centre.

This description is in no way complete. Many effects have not been considered. In particular the turbulent pressure is exerted only intermittently by the interstellar gas: there will, in fact, be occasional collisions between the growing gas sphere and other clouds, whose actual density exceeds ϱ by a factor ten, or so. The collisions set up shockwaves, which we have ignored; in between collisions the growing mass is subject to less than average pressure and is free to expand. Finally spherical symmetry is unlikely to occur, and the cloud may be rotating. But it is at least satisfactory that the suggested process leads to the formation of gravitationally unstable clouds, with masses of the right order.

About 15 years ago, when interstellar data were even more incomplete than now, SPITZER (1941) and WHIPPLE (1946) described a rather beautiful mechanism. They suggested that differences in the radiation intensity coming from various directions might produce forces on dust grains in clouds which would ultimately lead to a collapse. SPITZER and WHIPPLE supposed that the background radiation field of the Galaxy was spherically symmetrical, and that an asymmetry was introduced locally because the grains cast shadows. The spherical symmetry is not essential; the mathematics can be formulated thus:

Let $I(\mathbf{r}, \mathbf{a}) \, d\Omega$ be the intensity of radiation at position $\mathbf{r}$ coming from a solid angle $d\Omega$ around the unit vector $-\mathbf{a}$. A flux vector $\mathbf{F}(\mathbf{r})$ may be defined by

$$\mathbf{F}(\mathbf{r}) = \int I(\mathbf{r}, \mathbf{a}) \, \mathbf{a} \, d\Omega \quad (37)$$

integrated over the direction sphere. Let an interstellar grain have cross-sectional area S, scattering efficiency Q and albedo γ, and let the forward directivity g of the scattered radiation be defined by

$$g = \int f(\mathbf{b}, \mathbf{a})(\mathbf{b}\cdot\mathbf{a})\, d\omega,$$

where $f(\mathbf{b}, \mathbf{a})\, d\omega$ is the fraction of the scattered light entering a solid angle $d\omega$ around the unit vector $\mathbf{b}$ when the incident light has come from the direction $-\mathbf{a}$. A dust grain exposed to the radiation flux $\mathbf{F}$ experiences a force

$$\frac{1}{c}[(1-\gamma g)\, S\, Q]\, \mathbf{F}. \qquad (39)$$

The force on a group of n particles in a unit volume is

$$(n/c)\,(1-\gamma g)S\, Q\, \mathbf{F}.$$

Alternatively let the mass density of the interstellar matter be ϱ, and let $\varkappa\varrho$ be the extinction coefficient per unit length. Then $n\, S\, Q \equiv \varkappa\varrho$, and the force per unit volume becomes

$$\frac{\varkappa\varrho}{c}(1-\gamma g)\mathbf{F} \equiv \varrho\, \mathbf{f}_r,$$

say.

A fraction $(1-\gamma)$ of the incident stellar radiation is absorbed by a grain and later re-emitted as thermal radiation, presumably without any preferential direction, and at a much greater wavelength. It will therefore produce no recoil on the grain where it originates, nor will it be scattered again elsewhere, since the dust particles have a low optical cross-section at large wavelengths.

Now consider any volume V bounded by a surface S. Let E be the energy density due to stellar radiation; the amount of energy absorbed in the volume element dV is $(1-\gamma)\varkappa\varrho c E\, dV$. If the volume V contains no source of radiation — i. e. no star — the sum of the net flux of radiation out of the volume and of the radiant energy absorbed within it must be zero. Hence

$$(1-\gamma)\varkappa c\int \varrho E\, dV = -\int \mathbf{F}\cdot\mathbf{n}\, dS, \qquad (40)$$

and

$$\int \mathbf{f}_r\cdot\mathbf{n}\, dS = \frac{\varkappa}{c}(1-\gamma g)\int \mathbf{F}\cdot\mathbf{n}\, dS = -\varkappa^2(1-\gamma)(1-\gamma g)\int \varrho E\, dV. \qquad (41)$$

There is an evident similarity between formula (41) for the acceleration $\mathbf{f}_r$ due to radiation pressure, and that for $\mathbf{f}_g$ due to gravitational attraction; thus Gauss's theorem states that

$$\int \mathbf{f}_g\cdot\mathbf{n}\, dS = -4\pi G\int \varrho\, dV. \qquad (42)$$

The vector $\mathbf{f}_g$ is known to be irrotational. The vector $\mathbf{f}_r$ in general has a rotational part $\mathbf{f}_{r,r}$ and an irrotational part $\mathbf{f}_{i,r}$. Only the latter helps in any possible condensation of material.

Now the density of radiant energy in the interior of the cloud will never exceed E_b, its value on the boundary. A comparison of (41) and (42) then shows that

$$\frac{|f_{i,r}|}{|f_g|} \leq \varkappa^2 \frac{(1-\gamma)(1-\gamma g)}{4\pi G} E_b . \qquad (43)$$

Some possible values of γ and g, determined by observations in the Cygnus region, have been quoted by v. D. Hulst (1955). From his diagram (Fig. 3 in his paper) it is seen that, although γ and g cannot be found independently, the largest value of $(1-\gamma)(1-\gamma g)$ compatible with observations is about 0.7. For grains of impure ice its value is about 0.1.

Let us make a numerical estimate. The average interstellar mass density is $\bar{\varrho} = 2.4 \times 10^{-24}$ g/cm^3; optical depth unity is reached when the line of sight is one kpc long so that $\varkappa \bar{\varrho} l = 1$, when $l = 1$ kpc $\sim 3 \times 10^{21}$ cm. This gives $\varkappa \sim 140$ cm^2/g. The density of radiant energy E_b due to starlight is about 6×10^{-13} erg/cm^3 and so

$$\frac{|f_{i,r}|}{|f_g|} \leq \frac{1}{10}(1-\gamma)(1-\gamma g) . \qquad (45)$$

At best the irrotational component of the acceleration due to radiation pressure is, perhaps, seven per cent of the gravitational acceleration, but probably it is as little as one per cent. If there happens to be a star within the dust cloud, equation (40) no longer holds, and the attraction due to radiation pressure is even less, while the gravitational attraction is increased.

Finally, since the dust particles cannot move far without being stopped by the friction of the interstellar atoms [Savedoff (1955)] it is most unlikely that any preferential separation of solid matter can be brought about by the selective radiation pressure.

If $\varkappa$ were larger the relative importance of radiation pressure would be greater. It is easy to see that $\varkappa$ will be overestimated if $\bar{\varrho}$ is underestimated. Fifteen years ago the value of $\bar{\varrho}$ was not at all well known and presumably this is why it was then thought that this process would work. It is a great pity that it does not.

2. Contraction and fragmentation problems

Rouskol (1955) has described the mode of contraction of a sphere of gas and dust under its own gravitation. He considers a cloud with a mass $M = 10 \odot$, containing about one per cent, by mass, of dust, and he assumes that any excess thermal energy generated in the contraction will be transferred to the solid grains and radiated by them. Rouskol treats the case in which only radial motions occur, and he does so by solving the equations of continuity and of energy balance, and Poisson's equation. These are integrated numerically.

The cloud just satisfies JEANS's criterion for instability at the beginning of the motion. ROUSKOL finds that the outer parts of the cloud, containing about 40% of the mass, will expand, that the inner parts contract and that the model departs more and more from the equilibrium state. Thus it is quite inappropriate to seek a solution of the problem in terms of a sequence of homologous equilibrium states. The collapse goes on for about 10^5 years; the cloud then ceases to be transparent to its own thermal radiation and so its internal pressure builds up and impedes the further collapse. It is estimated that another 10^7 or 10^8 years are needed for the formation of a star, but no further calculation is given.

ROUSKOL's model is clearly quite different from MCCREA's, for his gas sphere is not subject to any pressure from outside. It is therefore not surprising that its outer parts expand on being released, but, since the conditions postulated by ROUSKOL are not likely ever to be exactly fulfilled, this effect is not of great importance. It would, however, be extremely interesting to study the further evolution of a gas cloud brought to the verge of instability after MCCREA's fashion. Some relevant preliminary details of such a calculation have been discussed.

In the first place MESTEL and SPITZER (1956) have worked out the possible influence of the presence of a magnetic field. They consider a spherical mass on the verge of instability according to JEANS's criterion, which they have suitably modified to include magnetic effects and which now states that a collapse will not or will occur according as

$$2T + \mathfrak{M} + V \gtreqless 0 \tag{1}$$

in our notation, where $\mathfrak{M}$ is the magnetic field energy. If the gas is a good conductor of electricity so that the lines of force are bound to the charged particles, then the magnetic energy $\mathfrak{M}$ must increase in any symmetrical contraction. This requires that the charged ions are numerous enough to prevent the neutral atoms and molecules from slipping past them. When both conditions hold, the conservation of magnetic flux requires that

$$H R^2 = H_0 R_0^2, \tag{2}$$

where H and R are, respectively, the magnetic field strength and the cloud radius, and the suffix zero denotes their initial values. It follows that

$$\mathfrak{M} = \frac{H^2}{8\pi} \cdot \frac{4\pi}{3} R^3 = \frac{H_0^2 R_0^4}{6R} = \frac{\mathfrak{M}_0 R_0}{R}. \tag{3}$$

If initially $\mathfrak{M}_0 + V_0 > 0$ then $\mathfrak{M} + V > 0$ always, with the present assumptions. On the other hand a cloud may contract without hindrance by magnetic forces if $\mathfrak{M}_0 \ll |V_0|$. In a uniform cloud this leads to

$$\frac{1}{6} H_0^2 R_0^3 \ll \frac{3}{5} \frac{GM^2}{R_0}. \tag{4}$$

With $M = 10^3 \odot$ and $H_0 = 10^{-6}\,\Gamma$ this means

$$R_0 \ll 10 \text{ pc};$$

with $H_0 = 10^{-5}\,\Gamma$,

$$R_0 \ll 3 \text{ pc}.$$

The typical unstable clouds which we considered earlier are borderline cases in which magnetic effects may be important.

Alternatively one can argue as follows. When a cloud is on the verge of collapse the absolute value of its gravitational self-energy is of the same order as its kinetic energy content. If the magnetic field energy content is also of the same order, then clearly the kinetic energy density and the magnetic energy density are of the same order. In other words, if there is any tendency towards equipartition of kinetic and magnetic energies then magnetic forces are bound to be significant in the subsequent collapse.

It should be noted that $\mathfrak{M}$ and V remain proportional to one another only during a symmetrical collapse. If the gas cloud contracts along the lines of force $\mathfrak{M}$ decreases while $|V|$ increases.

MESTEL and SPITZER later consider to what extend the plasma and the neutral gas must move together. This problem occurs also in a study of H I regions in general, as we mentioned earlier. Only collisions of ions and neutral atoms are likely to influence the gas motion. An electron is too light to transfer an appreciable amount of momentum. One might call the plasma and the neutral gas well coupled if the Mach number of their relative motion is forced to remain small. Consider the forces acting on the neutral atoms in a small volume. With the usual notation the gravitational acceleration acting on them will be GM/R^2; the retardation due to friction against ions is about $v\,a/l$, when the relative velocity is v, when l is the mean free path for a collision of an atom with an ion and when a is the speed of sound. It follows that the relative Mach number v/a can be small only if

$$1 \gg \frac{v}{a} \sim \frac{GMl}{R^2 a^2}.$$

But $a^2 \sim GM/R$ when the collapse occurs. To satisfy the condition of good coupling requires therefore that l/R shall be small, or that the mean free path for a collision with an ion shall be small in comparison with the linear dimensions of the cloud. A numerical illustration may be useful.

There should be about 5×10^{-4} ions to each neutral atom in a typical H I cloud. If n is the density of atoms

$$5 \times 10^{-4}\, n\, l\, \sigma \sim 1 \tag{5}$$

where $\sigma\ (\sim 10^{-16}\ \text{cm}^2)$ is the collision cross-section. With $n\,\mu = \varrho$, and $\mu = 2.4 \times 10^{-24}$ g, (5) leads to

$$\varrho\, l \sim 4.8 \times 10^{-5}\ \text{g/cm}^2. \tag{6}$$

The opacity of this depth of cloud due to interstellar extinction is about

$$\tau = \varkappa\, \varrho\, l \sim 7 \times 10^{-3} \tag{7}$$

if we adopt the value $\varkappa = 140\ \text{cm}^2/\text{g}$ from an earlier estimate. In other words the plasma and the neutral gas are quite well coupled in a cloud which is only one per cent opaque. The average interstellar cloud is, perhaps, 10% opaque; the typical cloud on the verge of collapse is denser and should be rather well coupled.

We should be faced with a serious problem if we were now to conclude that the collapsing cloud drags the magnetic field along all the time until a star has been formed. If this were so then newly formed stars would have much stronger magnetic fields than they do possess. MESTEL and SPITZER therefore point out that in very opaque clouds ions will readily attach themselves to dust grains, and will be removed from the plasma. This attachment is usually prevented because the grains acquire a positive charge by the photo-electric emission of electrons. It is hard to say how opaque such a cloud needs to be; MESTEL and SPITZER consider a cloud with an optical depth $\tau = 40$ in their calculation.

When ion attachment occurs freely it necessarily has a time-scale rather shorter than the time of free fall. Let v_m be the thermal speed of a typical neutral atom, v_i the speed of an ion. Let μ_m, μ_i be, respectively, the mean masses of the neutral atoms and the ions in the cloud. Let R be the linear dimensions of the cloud. Then we have that the time of free fall is given by

$$t_{free} \sim (4\pi\, G\, \bar{\varrho})^{-1/2} = \left(\frac{R^3}{3\, G\, M}\right)^{1/2} \tag{8}$$

and, by the virial theorem,

$$\overline{v_m^2} \sim \frac{G M}{R}. \tag{9}$$

Hence

$$t_{free} \sim \frac{R}{(3\overline{v_m^2})^{1/2}} = \left(\frac{8}{9\pi}\right)^{1/2} \frac{R}{\bar{v}_m} \sim 0.5\, \frac{R}{\bar{v}_m}, \tag{10}$$

if the velocity distribution is Maxwellian. In that case we also have that

$$\mu_i^{1/2}\, \bar{v}_i = \mu_m^{1/2}\, \bar{v}_m,$$

and so
$$t_{free} = 0.5 \left(\frac{\mu_m}{\mu_i}\right)^{1/2} \frac{R}{\bar{v}_i}. \tag{11}$$

The cross-sectional area of the dust grains is such that the mean free path for a photon is of the order $1/\varkappa\,\bar{\varrho}$. If the optical extinction efficiency of the grains is Q then the mean free path for an ion before collision with a grain is $Q/\varkappa\,\bar{\varrho}$. The typical time required for an ion to attach itself to a grain is then

$$t_{att} = \frac{Q}{\varkappa\,\bar{\varrho}\,\bar{v}_i},$$

and using (11) we now find that

$$t_{att} = 1.8 \left(\frac{\mu_i}{\mu_m}\right)^{1/2} \frac{Q}{\varkappa\,\bar{\varrho}\,R}\, t_{free}.$$

Since $\varkappa\,\bar{\varrho}\,R = \tau$ is the optical depth of the cloud

$$t_{att} = 1.8 \left(\frac{\mu_i}{\mu_m}\right)^{1/2} \frac{Q}{\tau}\, t_{free}. \tag{12}$$

With $(\mu_i/\mu_m)^{1/2} = 3$ and $Q \sim 2$ we find that

$$t_{att} = \frac{11}{\tau}\, t_{free}. \tag{13}$$

Thus when ion attachment does occur it will do so rather suddenly. After the removal of the ions the neutral gas can move independently of the magnetic field, but only so long as the kinetic energies of the gas atoms remain low enough that ionization by collision may be neglected. (I owe this remark to Mr. RAY WEYMANN.) Let us briefly consider what this means.

When the cloud radius is R the typical kinetic energy of a gas atom is

$$\varepsilon = \frac{GM\mu}{R}. \tag{14}$$

The optical depth then is

$$\tau = \varkappa\,\bar{\varrho}\,R = \frac{3\varkappa M}{4\pi R^2} \tag{15}$$

and so

$$\varepsilon = G\mu \left(\frac{4\pi\tau}{3\varkappa} M\right)^{1/2}. \tag{16}$$

Using the usual values for $\varkappa$ and μ and taking $\tau = 40$ we find

$$\varepsilon = 7.2 \times 10^{-15} \left(\frac{M}{\odot}\right)^{1/2} \text{erg} = 4.5 \times 10^{-3} \left(\frac{M}{\odot}\right)^{1/2} \text{eV}. \tag{17}$$

If $M = 10^3 \odot$, the mean particle energy ε is about 0.14 eV, much below the ionization potential (7.83 eV) of iron, the element most likely to cause trouble. In this case it seems that the cloud will be able to contract appreciably before collisional ionization sets in. The value of ε will be larger in a more massive cloud, but one would have to make a rather careful calculation if one wanted to see how massive a cloud must be for collisional ionization to be important when $\tau = 40$.

At some time the contracting mass must break into smaller units, some of which ultimately become stars. We have seen that typical unstable clouds have masses of the order $10^3 \odot$ and the average early type star has a mass of about $10 \odot$. The study of globules and of elephants' trunks gives independent evidence of the occurence of such fragmentation. Thus POTTASCH (1958) has been able to show, from data concerning their evolution, that these structures do not form because of a Taylor instability in the system; it can also be shown that they are not due to an instability of an ionization front separating a growing H II region from the neutral hydrogen gas surrounding it [KAHN (1958)]. It seems more likely that they appear when an ionization front, produced by radiation from a newly-born star, sweeps across compact regions of higher density in the surrounding non-ionized cloud, and this indicates that fragmentation must have occured previously.

HOYLE (1955) first considered the possibility of fragmentation. MESTEL and SPITZER (1956) have also studied it, using a model cloud which passes through a sequence of quasi-equilibrium states, and therefore ignoring any possible dynamical effects. In their particular mechanism the collapse of a cloud goes on until the density becomes so high that a mass of half the cloud dimensions is unstable according to JEANS's criterion. The further collapse is then assumed to take place around several centres, not just around one. The smallest sub-unit which can be formed in this way has a mass $\odot/2$.

There are major difficulties here. The fragmentation is assumed to occur after the beginning of the first collapse. MCCREA's work shows that there are no suitable equilibrium states immediately after the collapse begins, and it is doubtful whether one can justify the neglect of the dynamical effects in the ensuing mass motion.

SCHATZMAN (1954a) has studied the effect of turbulence in the collapsing cloud and has found a possible distribution of angular momenta for the resulting fragments. He proposes that the new stars form from rotating eddies in the condensing gas. When the turbulence is homogeneous and isotropic it is found that

$$p(\omega)\, d\omega = C\, \omega^2 \exp[-3\, \omega^2/2\, \overline{\omega^2}(l)]\, d\omega$$

is the probality that an eddy of linear dimensions l shall have an angular velocity between ω and $\omega + d\omega$ where $\overline{\omega^2}(l)$ is the mean square angular velocity for eddies of that size, and C is a constant. It is assumed that each eddy conserves its angular momentum as it condenses further.

SCHATZMAN also seeks to find the frequency function $f(M)$ for the initial masses M of newly formed stars. To do so he investigates the rotational stability of these bodies, and he notes that a star cannot form if its angular momentum is such that the centrifugal force at the stellar surface exceeds gravity. He adopts FESSENKOV's theory of stellar evolution [FESSENKOV (1954)] to find a frequency function for the masses of stars which have evolved to some extent and which have emitted a certain amount of corpuscular radiation. This leads him to a distribution of stellar rotational velocities u which agrees quite well with the values of $\overline{u^2}/(\overline{u})^2$ determined observationally by BÖHM (1952) and SLETTEBAK (1949, 1954). SCHATZMAN also finds that stars of a type later than F0 will have negligible speeds of rotation.

In studying the behaviour of a turbulent condensing cloud SCHATZMAN has probably come closer to the real physical problem than anyone else has done. On the other hand his conclusions, and the comparison of the predicted with the observed values of u, stand or fall by FESSENKOV's theory, which we shall discuss later. There is no doubt that the observed distribution of rotational velocities among early type stars gives us a clue to the state of motion in the condensing pre-stellar medium. But any interpretation must use a correct theory of stellar evolution.

SCHATZMAN (1954b) has also discussed how a diffuse cloud can condense by its self-gravitation and then form an association in which the stars, regarded as mass-points, have a positive total energy. It should be noted that there is a considerable reservoir of negative energy within the stars when these are regarded as self-gravitating gas spheres; the total energy of the association is the sum of the energies of the stars, regarded as mass-points, and of their internal energies. The sum is clearly negative even in expanding associations. No one seems as yet to have found out whether there can be a transfer of the internal energy of one condensing proto-star to the kinetic energy of motion of another. One might imagine the following process. As a proto-star shrinks the conservation of angular momentum demands that its rotational energy should increase. Perhaps a suitably oriented collision of two rapidly rotating proto-stars can increase the translational velocities of their centres of mass.

SCHATZMAN does not investigate this possibility. Instead he discusses, by an order-of-magnitude argument, whether there can be any fragmentation and gravitational condensation in a turbulent cloud with positive total energy. The turbulence is assumed to be homogeneous, incompressible and isotropic, and it is assumed that the further condensation takes place in the smallest eddies, which are free of internal movement. The size of these eddies can be calculated from details of the motion. The suggested process is found to be impossible.

SCHATZMAN also points out that no fragmentation can occur in a uniformly expanding cloud, having rediscovered the well-known result that in such a case there can either be a condensation of all possible sizes — and then the system has negative energy — or else no condensation can occur.

In all these studies it has been assumed that the fragmentation occurs before any early type star has been formed and begun to radiate. OORT (1954b) supposes, on the contrary, that no fragmentation occurs until a central star has started to shine and to ionize part, at least, of the surrounding neutral gas. We shall discuss his description of the process later.

SHAJN and HASE (1952) have studied a mode of collapse with axial symmetry in an attempt to explain the formation of the occasionally observed chains of early type stars. They envisage a contraction of a cylinder of interstellar gas to a density high enough for a collapse into stars to take place. (They mention a value 10^{-19} g/cm^3.) They suggest that hydromagnetic effects may be important. It seems that there are three possible configurations.

(i) A magnetic field with lines of force along the cylinder will tend to hinder any contraction. If there is sufficient ionized matter present to couple the motion of the gas to that of the field then the conservation of flux requires the field strength H to remain proportional to the gas density ϱ. The transverse magnetic pressure varies as H^2, while in an isothermal contraction the gas pressure only varies directly as ϱ, and therefore as H. Now both EBERT (1955) and MCCREA (1957) have shown that the force per unit volume due to gravitational self-attraction in a condensing isothermal filament always remains proportional to the pressure gradient, and have concluded that if a given filament is to be unstable at any radius it must be unstable at all radii. But since the magnetic pressure increases much more rapidly than the gas pressure in a contraction, it must ultimately become large enough to stabilise the filament against its self-gravitation. The collapse can occur only if the neutral gas is not well coupled to the field, but then the hydromagnetic effects are not important, by definition.

(ii) If the magnetic field is due to a current discharge along the filament a pinch effect makes the column contract, but the system becomes unstable to lateral deformations [KURCHATOV (1957)]. Such a column discharge has a strong tendency to lengthen itself and this is, in part, the reason for the instability. The magnetic field will in this case certainly tend to resist any possible condensation into stars.

(iii) Calculations seem to show that no stable configurations exist in which a filament contains a longitudinal magnetic field and a current discharge, unless the system is enclosed in a rigid conducting cylinder of relatively small radius [TAYLER (1957)]. There can be no such cylinder in interstellar space, but possibly the gravitational self-attraction may help to prevent the instability. This last mechanism seems not to have been studied yet.

We conclude that magnetic fields are probably not important in the formation of star chains.

Finally, McVITTIE (1956) has followed through the contraction of a model gas cloud into a star. In its initial state the gas is at rest and is distributed within a sphere; to describe the ensuing motion McVITTIE uses the equation of motion

$$\frac{\partial q}{\partial t} + q \frac{\partial q}{\partial r} = -\frac{1}{\varrho}\frac{\partial p}{\partial r} + 4\pi G \frac{\partial \psi}{\partial r},$$

the equation of continuity

$$\frac{\partial \varrho}{\partial t} + \frac{1}{r^2}\frac{\partial}{\partial r}(r^2 \varrho q) = 0,$$

and Poisson's equation

$$\frac{1}{r^2}\frac{\partial}{\partial r}\left(r^2 \frac{\partial \psi}{\partial r}\right) = -\varrho.$$

Only radial motions are considered. There are three equations for the four unknown functions q, p, ϱ and ψ, representing the radial velocity, pressure, density and gravitational potential, respectively, and there is no unique solution. McVITTIE now simplifies his equations by the use of a similarity variable ζ, and derives three equations in terms of ζ and the quantity τ, which represents a reduced time variable. He seeks the simplest polynomial form of a function $R(\tau)$ which describes the time variation of the radius of the model, and which satisfies some plausible boundary conditions. $R(\tau)$ is found to be a fifth-order polynomial in τ. The contraction time is of the order $(4\pi G \varrho)^{-1/2}$, as was to be expected.

This treatment ignores the equation of energy balance. The internal thermal energy distribution will strongly influence the mode of contraction of the gas sphere, and is itself determined by the rate at which heat can flow from the warm interior towards the outside and by the rate at which nuclear reactions at the centre can liberate energy [SALPETER (1954)]. It is very hard to give a full description of this process and yet it is an essential feature of the contraction. If it is left out of a calculation one can hardly be sure of finding the right answer at the end.

3. The motion of expanding associations

It is commonly held that associations of newly formed stars are in a state of expansion, but PETRIE (1958) and WOOLLEY and EGGEN (1958) have concluded that the observed motions of some associations look random rather than systematic. In particular they suggest that one cannot give a reliable date for the birth of an association by tracing backwards the present velocities of its members.

Now whatever the answer to this particular problem may be it does seem well established that an association usually contains a collection of unusual stars which could not have come together purely by chance. Further these stars usually move fast enough in some direction to give the association as a whole a positive total energy. Yet in order to be able to condense at all in the first place, the cloud of diffuse matter from which the association was formed must have had a negative total energy. It was suggested in the previous section that possibly some of the gravitational energy which is lost in the condensation of individual stars might be transformed into kinetic energy of motion of other stars in the system. SCHATZMAN (1954b) has also considered whether it is possible that an association can consist of stars which have left a compact cluster, with negative total energy, after acquiring a speed exceeding the escape velocity.

It is SCHATZMAN's basic assumption that a Maxwellian velocity distribution is set up among the stars in the cluster within a period called the relaxation time. If a star with an exceptionally high speed has escaped, it will be replaced by another within a time t_r. A numerical estimate shows some of the difficulties involved in the process.

Let there be N stars in the cluster, each having a mass M_*. Let R be such that the typical gravitational potential at a star within the cluster is GM_*N/R; the mean square speed $\overline{v^2}$ of the cluster stars has the same order of magnitude. Significant energy exchanges occur when stars approach so close that r, their distance of least separation, satisfies

$$\overline{v^2} < \frac{GM_*}{r} ; \qquad (1)$$

the collision cross-section for such encounters is

$$\pi r_0^2 \sim \pi \frac{G^2 M_*^2}{(\overline{v^2})^2}, \qquad (2)$$

where

$$\overline{v^2} = \frac{GM_*}{r_0}.$$

If there are n stars per unit volume on the average, the mean free path for these collisions becomes

$$L = (n \pi r_0^2)^{-1} = \frac{(\overline{v^2})^2}{n \pi G^2 M_*^2}. \qquad (3)$$

If the cluster is in a secularly steady state then, by the virial theorem,

$$\overline{v^2} = \frac{GM_* N}{R}, \qquad (4)$$

and

$$L = \frac{N^2}{n \pi R^2}. \qquad (5)$$

The time which elapses before some star in the cluster is involved is about

$$\frac{L}{N(\overline{v^2})^{1/2}} = \frac{N}{\pi n R^2 (\overline{v^2})^{1/2}}. \tag{6}$$

This is a minimum value for t_r. We may say that

$$t_r > \frac{N}{\pi n}(GM_* NR)^{-1/2}$$

$$= \frac{1}{\pi \varrho}\left[\frac{M_* N}{GR^3}\right]^{1/2} = O\left[(G\varrho)^{-1/2}\right],$$

where ϱ is the mean mass density. If, say, twenty stars are to escape within 10^6 years we require

$$t_r < 5 \times 10^4 \text{ years} \quad \text{or} \quad O\left[(G\varrho)^{1/2}\right] > 10^{-12}$$

or $O(\varrho) > 1.6 \times 10^{-17}$ g/cm$^3 \sim 2 \times 10^5$ ⊙/pc^3.

The clusters from which the stars escape must have a very high space density of matter if the process is to go on sufficiently fast. Such clusters do not seem to exist near associations, but SCHATZMAN (1956) has suggested that possibly the systems with negative energy should be identified with close multiple stars. After the escape of one or more stars from a multiple system there may be a pair of close binaries left behind. Stellar associations do, in fact, contain a large number of binary star systems. Whether a multiple stellar system can really eject members in this way is one of the unsolved problems of stellar dynamics.

Proponents of other theories try to explain the formation of expanding associations by suggesting ways in which the nuclear energy, stored in the stars, may be turned into the kinetic energy of mass motion. The following are the successive stages in the theory of OORT and of SPITZER [OORT (1954b), OORT and SPITZER (1955)], which we have mentioned before. First a massive cloud of gas and dust collapses at its centre and forms a new O star there. This star produces an HII region around itself, and then interesting gas dynamical effects may occur at the boundary of the growing HII region and the surrounding HI region. The gas is heated on passing through the ionization front, which divides the two regions; if the ionized gas streams away towards the exciting star a type of recoil action may raise the gas pressure in the HI region and then send a shock-wave ahead. As a result the gas in the HI region is compressed and pushed away.

The compressed gas can be cooled rapidly by radiation from H_2 molecules and from various ions, and may thus become very dense, with up to 10^6 atoms/cm^3. It is supposed that, when this has happened, the gas becomes gravitationally unstable, and breaks up into the stars of an expanding association. OORT and SPITZER realize that such a process cannot explain how the exceptionally fast-moving stars, such as AE Aurigae, have acquired their present speeds. But their proposal is subject to some further objections.

A mass of gas, i. e. the compressed HI cloud, cannot be made gravitationally unstable by compression in one dimension only. In such a process the gas density ϱ varies as R_0/R, where R is the present thickness and R_0 the initial thickness of the layer; the smallest unstable sphere has a radius r given by

$$G \varrho r^2 = \text{const.} \, k \, \Theta_0 \,, \tag{7}$$

and so

$$r \propto \left(\frac{k \Theta_0}{G \varrho}\right)^{1/2} \propto \left(\frac{R}{R_0}\right)^{1/2}. \tag{8}$$

If the layer of gas is to be unstable then $r < R$, but inspection of (8) shows that r decreases less rapidly then R during a compression, and in fact the HI layer is made more stable thereby. The same result is found for a shell of gas which is compressed in a radial direction. A condensation into stars can take place only if the HI cloud which is being overtaken contains considerable fluctuations in density; these might lead to the formation of the elephants' trunks mentioned before. When the advancing HII region overtakes a convex region in the HI cloud with exceptionally high density we might expect the formation of a curved ionization front. The rocket effect will then act at right angles to this front at all points and will further compress the existing region of high density. A three-dimensional compression like this may well make the matter in this denser region unstable against its own self-gravitation.

Rather stringent conditions must be satisfied if the rocket effect is to occur at all [KAHN (1954)]. Let there be J photons/cm²/s incident on the ionization front, with an average energy $\frac{1}{2} \mu_H Q^2$ in excess of the Lyman limit, and let ϱ_0 be the density of the undisturbed neutral gas, assumed to consist almost entirely of atomic hydrogen. μ_H stands for the mass of the H atom. A discussion of the stability conditions then shows that a shock-front will form if, and only if,

$$\frac{3 a^2}{4 Q} < \frac{\mu_H J}{\varrho} < \frac{4 Q}{3}, \tag{9}$$

where a, the sonic speed in the undisturbed gas, is assumed to be much smaller than Q. This means that if the flux of radiation is excessive then the gas is ionized so fast that no pressure wave can precede the I-front. If the flux of radiation is too low, the rocket effect is too feeble, and the non-ionized gas tends to expand towards the source of radiation. The inequalities of relation (9) set an upper limit to the speed which can be given to the neutral, compressed gas. The highest possible value, for an

O 5 star, is about 40 km/s; average values, however, tend to lie between 5 km/s and 10 km/s.

The speeds are rather low in comparison with those observed in associations. Possibly a given H I cloud can be accelerated several times while an association is being formed. Thus the first star to light up would produce its own H II region and would set up a rocket effect at the boundaries of several other incipient condensations. Some of these then collapse, become early-type stars, and emit more ionizing radiation, which then produces a second rocket-effect on the boundaries of still further condensations, and so on. A particular mass of gas may thus be accelerated some ten or twenty times. The various accelerations do not all point into the same direction. This may explain why the expansion of an association is never very clear-cut.

A mass of gas which has been given ten or twenty separate impulses may reach a final speed above 5 km/s or 10 km/s, but no detailed calculations have been made yet to illustrate the sequence of events.

It is interesting to note here that POTTASCH (1958) estimates that the gas densities inside the more highly evolved elephants' trunks are large enough to make them gravitationally unstable.

GOLDSWORTHY (1958) has studied the formation and expansion of symmetrical H II regions in much more detail and has taken into account the thermal balance and the effect of recombinations in the ionized gas. He confirms that the surrounding neutral gas can be accelerated when the intensity of the ionizing radiation lies in a certain range. He has not yet considered what effect is produced by inhomogeneities in the neutral gas nor how new stars can condense.

SAVEDOFF (1956) has used a similar model but applied a less rigorous method of calculation to it. He has given reasons for ignoring any possible rocket effect in his case. The energy derived from the radiation of his central O star mainly goes to heat and to re-ionize the gas in the surrounding H II region; the high pressure thus set up leads to the pushing away of the H I envelope, in the manner originally proposed by BIERMANN and SCHLÜTER [SCHLÜTER (1955)]. A comparison is made of the energy developed by the Orion Nebula in this way with the amount transferred to the neutral gas in Barnard's ring, and it is estimated that the latter has a mass of some $10^5 \odot$, moving at 14 km/s. SAVEDOFF suggests that a collapse into O and B stars can take place in the compressed H I envelope; in this respect his views resemble those of OORT and SPITZER, and the same comments may be made about them.

SAVEDOFF has also discussed a proposal by ÖPIK (1953) that a supernova explosion within the Orion Nebula may be the cause of the present expansion in the surrounding H I region. Supernovae eject matter with a speed of the order of 1000 km/s; ÖPIK supposes that this fast-moving matter collides with the surrounding neutral hydrogen and pushes it outwards. He tacitly assumes that in this process momentum is conserved in any given solid angle centred on the supernova. The mass originally ejected is relatively small, and so the final speed of the neutral matter is much lower than 1000 km/s.

SAVEDOFF objects, pointing out that violent supernovae belong to population II, while Orion type formations belong to population I. If momentum were conserved in every solid angle a supernova with a mass exceeding 500 ☉ would be needed to explain the expansion in the neighbourhood of the Orion Nebula. Supernovae are believed to occur when an exhausted star explodes, yet no stars are known with such large masses. Finally it would be rather a coincidence for a supernova to occur inside an exceptionally large gas cloud.

The conservation of momentum in every solid angle would lead to a great inefficiency in the conversion of the kinetic energy of the supernova remnants into the kinetic energy of motion of the neutral gas. If a small mass m with a high speed V collides inelastically with a large mass M, the combined mass moves on with a speed $mV/(m+M)$, and therefore has a kinetic energy $\frac{1}{2} m^2 V^2/(m+M)$, much smaller than the initial kinetic energy $\frac{1}{2} m V^2$.

It is by no means necessary for momentum to be conserved in this way, since the system has spherical symmetry. In such a collision most of the kinetic energy of the ejected remnants would rapidly become thermal energy, and a region filled with very hot gas would form around the site of the explosion. The subsequent mechanical effects would be rather like those which occur when an HII region is formed near an O star immersed in an HI region; in fact they would closely resemble the situation described by SAVEDOFF himself.

A comparison of the relevant energies is now somewhat less unfavourable to ÖPIK's theory. The expanding shell of gas enclosing the Orion Nebula has about 10^7 ☉ km²/s² kinetic energy; this is the same as the energy of a mass of 20 ☉ expanding at 1000 km/s. The mass of the supernova is still rather large.

Probability arguments do not favour ÖPIK's hypothesis. It has been estimated that the rate of occurrence of supernovae is about three per galaxy per 1000 years. The stars of population II in our Galaxy occupy a sphere with a radius of about 10 kpc, or a volume of about 4×10^{12} pc³. The frequency of population II supernovae is then about 7.5×10^{-13} outbursts/pc³/1000 y. In general one would expect to have to wait some 1.3×10^{12} years before the explosion of a supernova in a given volume of 1000 pc³. Possibly ÖPIK's process leads to the formation of a very few associations; clearly it cannot be responsible for the majority of them.

MCCREA (1955) has described still another variant of OORT and

SPITZER's process. He also believes that new stars are formed during the collapse of a gravitationally unstable cloud of gas and dust, but he assumes that the larger part of the diffuse matter remains uncondensed. When the new stars begin to radiate they will ionize and heat the remaining diffuse gas, and will give it enough thermal energy to escape from the gravitational field of the system. The gravitational force on each newly formed star is then much reduced. If the original unstable cloud was in a state of turbulent motion, the typical speed of the newly formed stars must be comparable with the typical speed of the eddies from which they condensed. If this is fast enough the newly formed association expands after the expulsion of the diffuse matter.

To discuss this suggestion in detail one needs to know much more about the mechanics of compressible turbulence in a gaseous medium where gravitational effects are not negligible. At present the properties of such a gas cannot be described even approximately. Only one definite conclusion can be drawn.

The typical eddy speeds cannot be smaller than the typical speeds of stars in associations. The mean square turbulent speed $\overline{v^2}$ will approximately equal GM/R, where M is the mass and R a typical linear dimension of the turbulent cloud. Since stars in associations have speeds of the order of 10 km/s, it follows that $M/R \gtrsim 2.5 \times 10^4 \odot/\text{pc}$. If the condensing cloud has a mass of $10^3 \odot$, its radius in the final stages would have to be less than 0.04 pc to satisfy this condition. A cloud with a mass of $10^5 \odot$, however, could have a radius as large as 4 pc. This mechanism, like all the others which have been proposed, also leaves unexplained the exceptionally high speeds of AE Aurigae, μ Columbae and 53 Arietis.

4. Conclusion

Attempts to explain the formation of associations have been faced with the two major difficulties that only vague data are known about the physical state and properties of the interstellar matter and that the dynamics of all the processes involved is highly complex. One does not know how rapidly a given mass of gas will radiate away its thermal energy; this would be a fundamental datum in any theory of gravitational condensation. The calculations by McVITTIE and by GOLDSWORTHY show how complex a satisfactory mechanical theory will be, particularly when one remembers that these two authors dealt only with specially selected models possessing spherical or cylindrical symmetry.

All present theories therefore make many appeals to physical intuition and none can give very exact answers. Some definite statements can however be made.

The existence of associations cannot be explained by accretion theory nor by any theory involving a supernova explosion. The formation of stars in chains, though probably a real effect, seems to have little to do with hydromagnetic forces.

A longitudinal magnetic field can provide the axial symmetry needed, but it resists compression perpendicular to its length. A circumferential field promotes compression, but tends to break up the column.

In general large clouds can possibly grow, and can be held together by the turbulent pressure of the interstellar medium until a gravitational collapse sets in. There are indications that the collapse leads to a fragmentation into smaller masses; it is almost certainly out of place to use a quasi-equilibrium theory to describe this process. The presence of magnetic fields may hinder the collapse, but only if the degree of ionization is high enough to couple the neutral gas to the lines of force, and if the field energy density is large enough. The coupling is probably very weak in highly obscured regions.

A collapse can occur only in a cloud with negative energy, and the system of stars thus formed should become a cluster with negative energy. It is almost certainly impossible that the occasional escape of fast moving stars from the cluster can then lead to the formation of an expanding association.

According to other theories the motion in an association results indirectly from the heating of the gas in the collapsed (or collapsing) cloud after the ionization of the hydrogen there by one or more newly formed stars. A rocket-like effect may occur, in which non-ionized gas is pushed away from the centre of the expansion. Possibly this accelerates the eddies which are about to condense into stars; possibly the expulsion of a large mass of gas reduces the gravitational potential and allows the system of new-born stars to expand. But the existence of a few stars with very high speeds is still quite unexplained.

Several key problems need to be solved before the processes of star formation can be better understood.

Three of them are:

(i) Do the observations really confirm that inhomogeneities exist in the neutral gas before ionization? And if so

(ii) How does fragmentation occur in a turbulent medium collapsing under its own gravitation?

(iii) How long does it take for eddies of various sizes to turn into stars?

III. Structure and evolution of stars
1. Stationary models

Theories of stellar structure ought to be based on very accurately known physical data. Only the outer layers of a star, containing perhaps one part in 10^{12} by mass are known to us. This means that one has to solve the equations for the structure of the star with a knowledge of the outer boundary conditions only, and without the possibility of a check to see whether the right equations are, in fact, being solved. The equations themselves are fortunately not very complex in nature: for a non-rotating star they can all be expressed in terms of one independent variable. Non-linear dynamical effects, such as supersonic flow, do not commonly occur inside stars. Only small velocities are usually encountered, for instance in convective regions. The treatment required for a study of stellar interiors is clearly quite different from that needed for an investigation of the properties of interstellar matter.

We shall first consider the structure of a non-rotating star of uniform composition. This is, of course, an idealized model: if a star has shone for some time some of the hydrogen in its interior will have turned into helium, and will have made its composition non-uniform to a certain extent. We shall see later how this influences the stellar structure.

The following equations must be satisfied: Relation between the density ϱ and the mass M included in a sphere of radius r

$$\frac{dM}{dr} = 4\pi r^2 \varrho ; \tag{1}$$

relation between the gravitational acceleration F, and M and r

$$F = -\frac{GM}{r^2} ; \tag{2}$$

equation of hydrostatic equilibrium relating the pressure p to ϱ and F

$$\frac{dp}{dr} = \varrho F ; \tag{3}$$

an expression for the pressure in terms of the kinetic pressure and the radiation pressure

$$p = nkT + \tfrac{1}{3} a T^4, \tag{4}$$

where n is the particle density, k is BOLTZMANN's constant, a is STEFAN's constant and T the temperature;
and finally expressions for the particle density and the mass density in terms of the mass μ_i and particle densities n_i of the individual species present

$$n = \Sigma n_i ; \quad \varrho = \Sigma n_i \mu_i . \tag{5}$$

The particle density for the ith species depends on the ionization equilibrium, and can be expressed in terms of the abundances of the elements, the mass per unit volume and the temperature. This quantity can be tabulated if the stellar composition is known.

There are also equations for the rate of transfer and for the rate of generation of energy. At the densities and temperatures which are expected in stellar interiors both diffusion and thermal conductivity are usually very slow because the particles have only small mean free paths. An exception occurs when the electron gas becomes degenerate, and then conductivity is important. In other cases heat is transported by radiation and, given suitable conditions, by convection. The opacity of the stellar material must therefore be known, and two processes are found to be important here.

At the temperatures to be expected (of the order of 10^7 deg K) the outer electron shells of atoms will have been largely stripped off. The mean thermal energy of particles at 10^7 deg K is of the order of 10^3 eV; this is approximately the ionization potential for the last electron of an element with atomic number $Z = 9$. Deep inside a star the continuous absorption coefficient is therefore important only for the heavier elements. KELLER and MEYEROTT (1955) have described some calculations, which they feel can be trusted to within 10%, if the abundances of the heavier elements are accurately known. This is, of course, not the case. We have seen that abundances of elements other than hydrogen and helium cannot, at present, be determined to better than a factor two; in addition the composition of a star probably depends on the epoch when and the place where it was formed. Different values of the opacity therefore apply to different stars.

Scattering in spectral lines, as opposed to continuum absorption, can generally be neglected. Free-free transitions by electrons in ionic fields are usually unimportant as well, but THOMSON scattering by electrons may be significant at high temperatures, when most of the atoms present have been completely ionized, or in stars with a low abundance of the heavier elements. The contribution of this effect to the opacity is readily estimated. It is almost certain that at least 99% by mass of stellar material consists of hydrogen and helium, both fully ionized. If a fraction X by mass is hydrogen and a fraction Y is helium then there are $N(X + \tfrac{1}{2} Y)$ free electrons present per gram of matter, where $N (= 6 \times 10^{23})$ is Avogadro's number. Since $X + Y \approx 1$, $N(X + \tfrac{1}{2} Y) \approx \tfrac{1}{2} N (1 + X)$. The electron scattering cross-section is $8 \pi \varepsilon^4 / 3\, m^2 c^4$, where ε and m are, respectively, the electronic charge and mass [UNSÖLD (1955), p. 178].

When the opacity $\varkappa$ is known the equation for the rate of transfer of energy is given by

$$\frac{1}{r^2} \frac{d}{dr} \left[\frac{r^2 c}{3 \varkappa \varrho} \frac{d}{dr} (a T^4) \right] = U, \qquad (6)$$

where U is the rate of generation of energy per unit volume, and the other symbols have the same meaning as before.

It may turn out, when a stellar model has been constructed, that the temperature gradient in some parts is larger than the adiabatic gradient.

Where this occurs convective instability sets in [LEDOUX (1954)]. A very small excess above the adiabatic gradient can maintain the convection. Therefore the stellar model must be reconstructed so that dT/dr satisfies equation (6) when $|dT/dr| < |dT/dr|_{ad}$ or else $|dT/dr| = |dT/dr|_{ad}$; the subscript "ad" here stands for "adiabatic". Convection is found to occur in the central regions of the hotter population I stars; the Sun is near the line dividing the stars which have convective cores from the stars which do not. Stars deficient in the heavier elements are less likely to have convective cores because in them the opacity is lower and the energy generation less strongly concentrated towards the centre. Several processes contribute to the rate of generation of energy. If the star contracts the decrease in its volume leads to an increase in thermal energy. In a static model this effect is necessarily negligible, but it may be important in a study of the early evolution of a star. Energy can also be released by reactions involving some light elements such as lithium [SALPETER (1954)], though these too probably only occur early on.

Most of the energy must be derived from the reactions which involve the proton, that is from the proton-proton chain or from the carbon-nitrogen cycle. Reaction rates can be calculated if the relevant nuclear cross-sections and the abundances of the elements are known. Cross-sections are usually found by extrapolation from laboratory data or by calculations based on reasonably good theory [FOWLER (1954)]. Probable errors seem not to be quoted anywhere in the literature.

At higher temperatures, of the order of 10^8 deg K, there may be a three-body reaction $3\,He^4 \rightarrow C^{12}$; only if hydrogen is completely absent can stellar material get as hot as this, for otherwise the proton chain or the carbon cycle would produce energy at a prohibitively large rate. For this reason the helium reaction probably only becomes important in the central parts of more highly evolved stars.

In general one can therefore specify the rate of energy output U in terms of the local density and temperature, provided the composition of the stellar material is given. We may write

$$U = U(\varrho, T).\tag{7}$$

There remains the problem whether any of the physical processes discussed are seriously influenced by the high densities found inside stars. A few examples will help to show how likely this is.

(i) Energy generation: The reactions take place between nuclei separated by distances of the order of nuclear dimensions, say 10^{-12} cm. Average distances between neighbouring particles in a gas at a density N are of the order $N^{-1/3}$; only if the density is as high as 10^{36} particles per cm^3 will three-body collisions be at all frequent.

(ii) Photo-electric absorption: The typical ionization potential is of the order of 10^3 eV, or about 3 e.s.u. The electrostatic potential at any point due to the nearest electron is of the order $N^{1/3}\varepsilon$, where ε is the electronic charge. The photo-electric effect will be seriously disturbed by stray electric fields only if

$$N^{1/3}\varepsilon \gtrsim \frac{kT}{\varepsilon} \sim 3 \qquad \text{(for } T = 10^7 \text{ deg K)},$$

or if the density exceeds 5×10^{29} particles/cm^3. This is higher than the density usually found in stellar interiors.

(iii) Electron degeneracy: The equation of state for the stellar material as well as the heat conduction equation are changed significantly by the exclusion principle if the density is too high. Degeneracy becomes important when

$$N \gtrsim (2\pi m kT)^{3/2}/h^3.$$

With $T = 10^7$ deg K the limit is approximately 10^{26} electrons/cm^3 [cf. BIERMANN (1954)]. Partial or total degeneracy occurs near the centres of more highly evolved stars [SCHWARZSCHILD (1958)].

(iv) Refractive index: The combined action of the electrons changes the refractive index of the medium, and this effect may be more significant than THOMSON scattering. The refractive index n at a frequency ν is given by

$$n^2 = 1 - N\varepsilon^2/\pi m \nu^2.$$

At a temperature T, $\nu \sim kT/h$ and

$$\frac{N\varepsilon^2}{\pi m \nu^2} \sim \frac{N\varepsilon^2 h^2}{\pi m (kT)^2} \sim 1.6 \times 10^{-27} N,$$

with $T = 10^7$ deg K. The refractive index will be seriously changed at the frequencies which carry most of the energy only if N exceeds about 10^{27} particles/cm^3.

(v) Cooperative effects in electron scattering: Long range interactions between charged particles introduce a partial ordering of the electron positions in a plasma. As a result the THOMSON scattering is reduced for radiation whose wavelength exceeds the DEBYE length λ_D [KAHN (1959)]. The effect is therefore important when

$$\lambda_D \sim \left(\frac{kT}{N\varepsilon^2}\right)^{1/2} \lesssim \lambda = \frac{c}{\nu} \sim \frac{hc}{kT},$$

or when

$$N \gtrsim \frac{k^3 T^3}{\varepsilon^2 h^2 c^2} = 3 \times 10^{23} \text{ cm}^{-3},$$

again with $T = 10^7$ deg K. This condition is satisfied near the centres of most stars. However the rate of energy transfer will be seriously altered only if the KRAMERS opacity is negligible in comparison with that due to electron scattering. This may be so near the centres of hot main sequence stars of population I, or near the centres of stars deficient in the heavier elements.

The equations for a static stellar model are now complete. The set is together of the fourth order and four boundary conditions must therefore be specified. The equations can be solved in terms of the temperature $T(r)$, and of $M(r)$ and $L(r)$, which respectively denote the mass contained and the energy generated within a sphere of radius r. If a star has radius R_*, mass M_* and luminosity L_* the boundary conditions are

$$M(R_*) = M_*, \quad (dM/dr)_{r=R_*} = 0, \quad T(R_*) = 0 \quad \text{and} \quad L(R_*) = L_*,$$

and the equations can now be integrated inwards from the outer boundary.

For a given stellar composition the solution will in general describe a star with a singularity at its centre: as a rule it is found that the model so constructed has $M(0) \neq 0$ and $L(0) \neq 0$. But the composition of a star can be expressed in terms of two parameters X and Y, the fractional abundances of hydrogen and helium, respectively; the remaining fraction $1 - X - Y$ is then assumed to be made up of a standard mixture of the other elements. By making a suitable choice of X and Y one can now ensure that $M(0) = 0$ and $L(0) = 0$ for any given set of values M_*, L_* and R_*. A knowledge of the mass, radius and luminosity of a star can therefore be used to derive some information about its composition.

Alternatively one can construct models of stars with a non-homogeneous composition, which is given as a function of M. The same argument again shows that the four boundary conditions are needed for a solution: to find a model without a central singularity one can this time adjust the values of L_* and R_* until $L(0) = 0$ and $M(0) = 0$. In other words, if the composition of a star is given as a function of M and if the total mass M_* is known, then the values of L_* and R_* are determined by the equations of stellar structure. SCHWARZSCHILD (1958) has recently given a very comprehensive description of the best way to construct stellar models.

It is usually found that the brighter stars have cores which are in convective equilibrium and which include from 15 to 35%, by mass, of the star [TAYLER (1954)]. But the size of the convective core depends sensitively on the value of the opacity $\varkappa$ because convection occurs only when the transfer of energy by radiation is too slow. At the same time the outward appearance of a star also depends sensitively on $\varkappa$. If the outer layers of a star were suddenly made more opaque, the flow of radiation would be dammed up and the gas pressure would increase. The whole star would then swell and its surface temperature would fall.

2. Evolution of stars

A static model of a star serves for a short period of time. Since nuclear reactions are taking place there must be changes in composition. If M_e is the mass contained in the region where hydrogen is effectively being turned into helium then the important time-scale for evolution is $6 \times 10^{18} M_e/L$ s; L denotes the stellar luminosity, in erg/s, and M_e is expressed in grams. The transmutation of hydrogen into helium goes on most rapidly near the stellar centre, and an inhomogeneous star tends to form whose inner parts are richer in helium than the outer parts are.

Two possibilities must now be considered:

(i) The composition differences so produced may be smoothed out relatively rapidly, in a time much smaller than $6 \times 10^{18} M_e/L$ s, by convection currents throughout the star.

(ii) The composition remains uniform in the convective core (if this exists), but changes there with time. The material in the radiative envelope is forced to stay where it is and its composition gradually becomes non-uniform.

The convection in the core of a star never seems to affect more than about one-third of the stellar material by mass; this is shown quite well by the trend in the Tables 28.1 to 28.8, given by SCHWARZSCHILD (1958). There may also be convection due to a thermal instability in the outer layers of a star [UNSÖLD (1931)], but it seems that this cannot reach down to the regions where the nuclear reactions take place, except possibly in the very coolest stars.

Only one other means has been discussed for transporting material through a star, and that is convection induced by rotation. No rotating star whose composition is homogeneous near any equipotential surface can transfer energy entirely by radiation and remain in equilibrium. This follows from a theorem due to MILNE (1923) and v. ZEIPEL (1924).

The proof is simple. Let φ be the potential of the combined gravitational and centrifugal fields of force, then

$$\operatorname{grad} p = \varrho \operatorname{grad} \varphi,$$

p and ϱ being pressure and density, respectively. Surfaces of constant p therefore coincide with surfaces of constant φ, and so $p = p(\varphi)$; it follows that

$$\operatorname{grad} p = \frac{dp}{d\varphi} \operatorname{grad} \varphi,$$

and that

$$\varrho = \frac{dp}{d\varphi},$$

and is a function of φ only. In a region where the stellar composition is uniform the mean weight μ of the particles is also uniform, and there the temperature T is a function of p and ϱ only, and therefore of φ. The temperature must therefore be the same at all points on an equipotential surface, and if radiation pressure may be neglected

and
$$p = k \varrho T/\mu$$

$$T/\mu = F(\varphi), \qquad (1)$$

say.

The equation of heat conduction has the form

$$\mathbf{H} = -f(\varrho, \mu, T) \operatorname{grad} T$$

where $\mathbf{H}$ is the energy flux. If μ has the same value at all points on each equipotential surface in a given region then

$$f(\varrho, \mu, T) \operatorname{grad} T \equiv f(\varrho, T) \frac{dT}{d\varphi} \operatorname{grad} \varphi = g(\varphi) \operatorname{grad} \varphi, \qquad (2)$$

say, and so

$$\operatorname{div} \mathbf{H} = -g'(\varphi) (\operatorname{grad} \varphi)^2 - g(\varphi) \nabla^2 \varphi. \qquad (3)$$

Now

$$\nabla^2 \varphi = -4\pi G \varrho + 2\Omega^2$$

in a rotating star, and is a function of φ only. But on any equipotential surface $(\operatorname{grad} \varphi)^2$ is large near the axis of rotation, where the surfaces are close together, and small near the equatorial plane. Unless $g'(\varphi) \equiv 0$ it follows that div $\mathbf{H}$ cannot vanish, so that energy tends to accumulate in some parts (near the axis, as a rule) and a defect occurs elsewhere. Convection currents then begin to flow to redistribute the energy [SWEET (1950), ÖPIK (1951), MESTEL (1953)]. It seems that the currents are too slow to mix the stellar material well except in stars earlier than type F [see Table 21.2 quoted by SCHWARZSCHILD (1958)].

The essential problem now is this. Does the convection current produced change conditions in the star so as to bring it to a new equilibrium state, or does it take the star further away from equilibrium? MESTEL, in his discussion, points out that the current rising along the rotational axis of the star will bring up material richer in helium than the average; the mean molecular weight at points near the axis is therefore increased relative to that near the equator, and this promotes stability. On the other hand the temperature also rises near the axis, since T/μ must have the same value at all points on any equipotential surface. The increase in μ further implies a decrease in the electron density, if ϱ remains constant; the changes in electron density and temperature lead to a drop in the opacity. It now depends on the structure of the star concerned whether, as a result of all these effects, the value of div **H** actually gets closer to zero. If it does then the circulation is choked back; if it does not then it is amplified. MESTEL finds that the circulation tends to stop, though he does not include the change of electron density in his calculation.

Thus the whole problem is essentially altered. The effect of the circulation current is simply to provide an initial disturbance. The stability of a star against convection does not depend on the particular value of the angular velocity, and, by going to the limit $\Omega \to 0$, one can convince oneself that a non-rotating star could be unstable in just the same way. The only exception would occur if the star were in neutral equilibrium with respect to the μ-currents. In that case the circulation would go on as described by SWEET.

In order to follow the evolution of a star it is essential to know how rapidly the material within it can be transported from place to place. The procedure is as follows. A star is supposed to form from interstellar material with an essentially homogeneous composition. The initial radius R_* and luminosity L_* can therefore be determined for a star whose mass is M_*. The rate of transmutation of hydrogen can then be calculated, and if the rate of transport of material is known then the composition is known for all shells of the star at time δt later. A new model, with a radius $R_* + \delta R_*$ and luminosity $L_* + \delta L_*$, can then be constructed, and so on.

The following is a rough physical explanation of the development of a star. If a star can remain well-mixed throughout, the gradual increase in the mean molecular weight will cause the pressure to decrease everywhere, at a given temperature. The star therefore shrinks slowly and heats up slightly, its rate of energy generation increases and it becomes more luminous and bluer. In other words it moves up the main sequence.

But if only an inner core of the star remains well-mixed, or if there is no mixing at all, then the mean molecular weight increases only in the region near the centre. Now the slow contraction and subsequent heating only take place near the centre; the outer portions remain essentially unchanged in composition, and they are made to expand by the greater flux of energy. The star is found to increase in brightness, but its surface temperature falls. In the Hertzsprung-Russell diagram it moves upwards and to the right, away from the main sequence.

The mode of evolution has been calculated by SANDAGE and SCHWARZSCHILD (1952), by TAYLER (1954) and by KUSHWAHA (1957). There are differences between the sets of results found, but there is qualitative agreement.

A paper by HÄRM and SCHWARZSCHILD (1955) contains a warning about all such calculations. Here it is shown that a great difference, amounting to as much as one order of magnitude in radius, can be made by incorrect assumptions about the abruptness of changes in the composition.

However the calculations based on models with limited mixing fit in well enough with the different observed LC diagrams of various clusters. Evolution away from the main sequence is fastest for the brightest and most massive stars, and these tend to leave first. The sequence of diagrams observed for different groups illustrates how the LC curve changes with time (see Fig. 4). The further the upper branch of its LC curve is removed from the main sequence, the greater is the age of a particular group. In principle the position of the "knee" in its LC curve can be used to find the age of a given set of stars [cf. SANDAGE (1958a)].

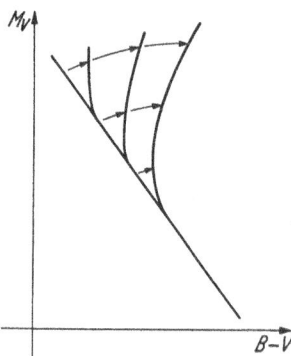

Fig. 4. To illustrate how the LC curve of a group of stars changes with time

It has been found by REIZ (unpublished work) that stars deficient in the heavier elements do not evolve in quite the same way. Initially they lie on an LC curve rather below the main sequence and even a star which is not well mixed will tend to evolve along this sequence for a considerable time before turning off. This means that star groups belonging to different populations cannot be dated by means of the same HR diagram. In fact it even raises the question whether the age of M 67 was found correctly by JOHNSON and SANDAGE (1955). According to their calculation this cluster is 5×10^9 years old, but very probably the composition of stars formed as long as 5×10^9 years ago differs from more recent compositions.

Finally, according to the theory due to FESSENKOV (1954) the stars evolve down the LC curves of different clusters. Here it is postulated that all stars satisfy the mass-luminosity relation $L \propto M^{3.9}$ throughout their career, and it is assumed that they remain well-mixed at all times. To satisfy the equations of stellar structure a star must continually emit corpuscular radiation in such a way that its rate of loss of mass is given by

$$\frac{dM}{dt} = -0.07 \left(\frac{L}{L\odot}\right) \odot, \tag{4}$$

where t is in units of 10^9 years. The existence of a variety of different LC curves for the different clusters is not held to be an age effect and can only be explained if each cluster has a different composition [RUBIN and MASSEVICH (1957)]. This might be hard to reconcile with the observation that stellar associations, which are known to be young, all have very similar LC curves. No mechanism is suggested which might produce the corpuscular emission; the rate of loss predicted for the Sun by equation (4) is equivalent to about 10^{-7} g/cm^2/s, and is intolerably high.

3. Other differences between old and new stars

Three main differences have been noted between old and new stars.

(i) Stars of later spectral type seem to have higher random space velocities.

(ii) The stars of population I are richer in the heavier elements.

(iii) The typical variables belonging to population II (the cluster variables) have a range of period-luminosity relations which differs from the range of such relations for the population I variables (the classical Cepheids).

Let us consider some possible interpretations.

(i) The average age of the late type stars exceeds that of the early type stars; this follows because the early type stars evolve more rapidly and acquire late type spectra as they do so.

SPITZER and SCHWARZSCHILD (1951, 1953) have suggested that the older stars have gained their larger random speeds by passing through irregularities in gravitational fields, caused by the presence of other stars or of massive complexes of interstellar matter.

Now VYSSOTSKY (1957) gives velocity dispersions as large as 30 km/s for dM stars. Let the dispersion be due to successive encounters of the stars with other bodies of large mass M, having a space concentration N. If V is the speed of a star relative to a large mass, the cross-section for a collision which appreciably changes the velocity is of the order of $\pi G^2 M^2/V^4$; in a time T the expected number of collisions is $\pi N G^2 M^2 T/V^3$. If most stars are to be deflected appreciably this quantity must be of order unity; with $T \sim 6 \times 10^9$ years and $V \sim 30$ km/s this implies that $NM^2 \sim 10^{16}$ g^2/cm^3. The stars in the galactic disk provide a space density $NM \approx 0.08$ ☉/pc$^3 \approx 6 \times 10^{-24}$ g/cm^3, The typical mass of a deflecting star must therefore be at least 1.6×10^{39} g $= 8 \times 10^5$ ☉. Such massive stars do not exist. Little is gained by taking account of distant collisions between the star and the various gravitating centres. This only changes the numerical values by a small logarithmic factor.

One must argue a little differently if the dominant irregularities in the gravitational field are due to fluctuations in the density of interstellar matter. Let R be the typical linear dimension of a complex of interstellar matter and let L be the distance between complexes. Let v_r^2 be the typical mean square random speed of the material in the complex;

this term includes both the turbulent and the thermal speeds. If a complex, of mass M, is not to collapse under its self-gravitation then

$$v_r^2 \gtrsim GM/R. \tag{1}$$

A star passing with a speed V spends a time of the order of L/V in a gravitational field with an acceleration of the order of GM/L^2, and so experiences a change of speed of the order of

$$\Delta V \sim \frac{GM}{L^2} \cdot \frac{L}{V} = \frac{GM}{LV}. \tag{2}$$

The mean square change in speed due to a passage through or past N such complexes is of the order of

$$N (\Delta V)^2 \sim \frac{NG^2M^2}{L^2V^2}. \tag{3}$$

If there have been N passages in the time T during which the star has existed then

$$N \leq \frac{VT}{L} \tag{4}$$

and

$$V^2 \sim N (\Delta V)^2. \tag{5}$$

From (3), (4) and (5) it follows that

$$V^3 \sim \frac{G^2 M^2 T}{L^3} = \left(\frac{GM}{R}\right)^{3/2} \left(\frac{GM}{L^3}\right)^{1/2} \left(\frac{R}{L}\right)^{3/2} T,$$

or, using (1),

$$\left(\frac{V}{v_r}\right)^3 \lesssim (G \bar\varrho T^2)^{1/2} \left(\frac{R}{L}\right)^{3/2}, \tag{6}$$

since the mean density M/L^3 per complex is less than $\bar\varrho$, the mean density of interstellar material. With $\bar\varrho \sim 2.4 \times 10^{-24}$ g/cm^3, $T \sim 6 \times 10^9$ years and $R/L = \varepsilon$, say, it now follows that

$$\frac{V}{v_r} \lesssim 4 \sqrt[]{\varepsilon}. \tag{7}$$

If v_r is the typical thermal speed in an HI cloud, i. e. about 1 km/s, then clearly V cannot be larger than 4 km/s, and the suggested mechanism fails. On the other hand if one identifies the complexes with the spiral arms of the Galaxy and v_r with the speed of a typical interstellar cloud, then one might have $v_r \sim 10$ km/s and $\varepsilon \equiv R/L = 1/3$, say, so that

$$V \lesssim 23 \text{ km/s}. \tag{8}$$

This is rather more promising, but it will clearly be difficult to account for speeds which are much larger. One would also need to show that the system of spiral arms produces a sufficiently irregular gravitational field.

According to an entirely different hypothesis the high speeds of the older stars reflect the more turbulent conditions of the interstellar material at the time of their formation [OORT (1954a)]. The greater thickness of the disk occupied by the older stars can be explained in the same way.

(ii) Two reasons have been suggested why the composition of population I stars differs from that of older stars. Either the heavier elements condensed more readily into the members of population I, or the relative abundance of heavier elements in interstellar space has increased since the early days when most population I stars were formed.

A mechanism of the first type has been discussed by SPITZER (1951). Here the interstellar dust grains are concentrated preferentially into newly formed proto-stars by the action of radiation pressure. The grains are thought to be relatively rich in the heavier elements; unfortunately however the condensation by radiation pressure is not likely to occur.

In an alternative theory [SCHWARZSCHILD and SPITZER (1953), HOYLE (1954)] the heavier elements are synthesized in aging stars, mainly of population II, and are ejected into space by supernova explosions. Theories of element formation are briefly discussed in appendix II; the main problem here is whether there can have been enough supernova explosions in the past, and whether a good enough mixing process exists to carry the heavier elements to the outer parts of the Galaxy.

It has been estimated that supernovae occur about once every 300 years per galaxy. It is not known how much of the substance of a supernova is turned into heavier elements; WOLTJER (1958) finds them to be present with quite normal abundances in the outer filaments of the Crab Nebula. But no definite conclusions can be reached because the mass and composition of the inner, amorphous parts are not known.

The population I stars and interstellar matter in the Galaxy together have a mass of $5 \times 10^9 \odot$. Perhaps one or two per cent of this mass is in the form of elements heavier than helium, and 5×10^7 or $10^8 \odot$ of them must have been produced by the $3 \times 10^{-3} T$ supernovae of the past T years. If the average supernova has produced one solar mass, then T equals 1.7 or 3.4×10^{10} years. Of course supernovae may have been rather more frequent in the past. A check on this possibility can be made from a study of their rates of occurrence in different types of galaxies.

Most supernova outbursts would be expected to occur near the galactic centre where population II stars are most plentiful, and the heavy elements formed there would have to diffuse through the galactic disk. With a typical length l (= 100 pc) and a typical speed U (= 10 km/s) the random motions of interstellar gas clouds will cause the matter to diffuse through a distance of the order of $(l\, U\, T)^{1/2} = (10^{-3}\, T)^{1/2}$ pc in time T. It takes some 6×10^{10} years to spread the heavy elements throughout a region 8 kpc in diameter; if diffusion takes place only along a spiral arm an even longer time will pass before the solar neighbourhood is reached.

The main difficulty seems to be that there may not have been time enough for the process to have worked well. But we have seen that the velocity distribution of older stars indicates that interstellar turbulence may have been more violent in the past, and this affects our estimates.

In any case it would be interesting to see whether there is any systematic variation of the abundance of the heavier elements with position in the Galaxy.

(iii) The fundamental reason for the existence of variable stars is not yet understood. But it is believed that at least one other parameter enters into the period-luminosity relation: possibly this third variable is the age of the star, possibly it is the abundance of heavier elements in it. SANDAGE (1958a) has given a brief discussion of the problem.

4. Conclusion

Modern theories seem to describe quite adequately the general structure of the stars and the means by which they generate their energy. The way in which a star evolves depends very much on the degree to which its composition remains homogeneous as the hydrogen changes into helium. Probably non-uniformities do develop in time.

In principle one can follow the evolution of a star step by step; however small errors made early in the calculation can magnify as time goes on and it seems that the necessary physical data and initial conditions are not yet known quite accurately enough. This should not affect the qualitative conclusions which have been drawn, but makes it extremely hazardous to estimate the age of a group of stars from the present appearance of its LC curve. The present abundance of the heavier elements can be understood in terms of the evolution of a past generation of stars; the newly-formed substances were probably mixed with the remaining diffuse matter in the Galaxy by the turbulence of the interstellar gas. The present-day velocity distribution of the stars can probably also be explained by the past mechanical behaviour of the interstellar medium.

Appendix I
Tests for the reality of associations

Early type stars seem to occur in groups; the different groups are separated by distances comparable with their mean diameters. But MEURERS (1957) found experimentally that the human eye tends to arrange randomly spaced points into groups, with about eight members on the average.

MEURERS was probably right to study this problem by means of an experiment. The idea of a group is too loosely defined to be suitable for a mathematical treatment, and in any case the identification of groups involves a great deal of personal choice by the observer. However mathematical treatments can be given to decide whether a given distribution of points is random or whether it has a tendency to cluster [cf. NEYMAN, SCOTT and SHANE (1954)].

A simple method might be as follows. Let a predetermined grid, with meshes of area S, be superimposed on the field being studied. The average number $\bar{n}$ of stars per grid area S can then be found. The quantities n follow a Poisson distribution; the probability that there be n stars within a mesh, in the absence of clustering is

$$p(n) = \frac{\bar{n}^n}{n!} e^{-\bar{n}}.$$

It can then be shown that

$$\overline{n^2} = n + (\bar{n})^2,$$

and the mean square deviation in n is

$$\overline{\Delta n^2} = \overline{(n-\bar{n})^2}$$
$$= \overline{n^2} - 2(\bar{n})^2 + (\bar{n})^2 = \bar{n}.$$

The expected sum of the square deviations for the whole grid is $\Sigma \overline{\Delta n^2} = \Sigma \bar{n} = N$, the total number of stars. If the stars have a tendency to cluster there will be a large number of meshes with too many stars in them, and a corresponding number with too few. The counted value of Σn^2 will be too large, and so will $\Sigma (\Delta n)^2$. The standard error in any determination of $\Sigma (\Delta n)^2$ is $\sqrt{N}$; if $\Sigma (\Delta n)^2$ exceeds $N + 2\sqrt{N}$, say, then one may conclude that the stars do show a significant tendency to cluster.

Appendix II

The synthesis of elements

Several theories attempt to explain the observed abundances and distribution of the elements. In most cases the basic assumption is that the primeval matter of the Universe consisted of hydrogen nuclei and electrons (or possibly just of neutrons); however some theories, such as that of MAYER and TELLER (1949), postulate a primary substance which is essentially a very dense nuclear fluid. This rapidly splits into smaller drops which subdivide until nuclei of the more familiar kind are formed.

But the great observed abundance of hydrogen makes it seem more likely that the heavier elements have been, in fact, formed from lighter ones, rather than the reverse. In one group of theories describing such a building-up process it is assumed that the synthesis of the elements took place at the earliest epoch, when the Universe was in a highly condensed state [ALPHER, BETHE and GAMOW (1948)], or in super-dense stars which then existed [KLEIN, BESKOW and TREFFENBERG (1946)]. A primary difficulty is that no elements heavier than atomic weight 4 can be formed in this way, except in regions where hydrogen is absent and the density is large. But clearly there was hydrogen in the Universe in those days.

According to an alternative theory the formation of elements heavier than helium takes place in the cores of stars where all the hydrogen has been burnt. A comprehensive review of this process has been given by BURBIDGE, BURBIDGE, FOWLER and HOYLE (1957). Some elements heavier than helium, such as C^{12}, O^{16}, Ne^{20} up to Fe^{56} can now be formed by the successive addition of α-particles, at temperatures of the order of 10^8—10^9 deg K. Nevertheless hydrogen nuclei must be kept nearby, in the outer layers of such stars, so that those nuclei may be formed whose atomic weights are not multiples of four. This formation occurs in a final catastrophe when the core of the star becomes so hot that reverse reactions of the type $Fe^{56} \rightarrow He^4$ can occur; because these are highly endothermic the star collapses, the outer, hydrogen-rich layers are heated and the whole mass disintegrates in a supernova.

Elements heavier than iron are supposed to be formed during the last few seconds of a star's existence. It is evident that the elements lighter than iron are formed in an equilibrium process; the heavier elements form when a star is far from equilibrium. A division like this is suggested by the statistics of the frequency with which different nuclei are found and should occur in all theories.

Much ingenuity has gone into all this work. Of course a great deal of it depends on values of nuclear reaction cross-sections which are still uncertain. Again some of the mechanics and astrophysics has not been thoroughly studied. For instance, how much hydrogen in a star can be kept sufficiently cool for it not to turn into helium, when the temperature at the stellar centre has reached 10^9 deg K? Is it necessary for the continued existence of the hydrogen that the star have an extended envelope of relatively low density? If so, can the final collapse and the supernova explosion be fast enough?

Appendix III

A final summary

The four charts which follow give a brief survey of the evidence on which theories of stellar evolution are based.

Galaxy

	Accepted data	Evidence from Observation	Evidence from Theory	Implications
Age	$\sim 10^{10}$ y		From age of Earth, expansion of Universe	Brightest stars cannot be as old
Mass	$\sim 10^{11}$ ⊙		Inferred from rotation data	
Size	Diameter between 25 and 35 kpc	Analogy with M 31	Distance to galactic centre from kinematics	
Stars	Majority (pop. II) in spherical, brightest (pop. I) in flat distribution	Distribution in sky. Analogy with M 31		Population I and, in particular, the brightest stars have a genetic connexion with interstellar gas and dust
Gas	Mostly in flat distribution. Some near galactic centre	21 cm line work		
Dust	In flat distribution	Interstellar absorption. Zone of avoidance of galaxies		
Abundance of Population I	(i) Disk pop. 5×10^9 ⊙ (ii) Gas $\sim 10^9$ ⊙ (iii) Dust $\sim 10^7$ ⊙	(ii) 21 cm work	(i) Dynamics based on 21 cm work (ii) direct 21 cm observation (iii) Abundances of elements	Affects possible interchange between diffuse and stellar matter
Magnetic Field	$\sim 10^{-5}$ Γ in spiral arms		Interstellar polarization, cosmic rays, elongated dark clouds	Affects motion of gas
Kinematics and Structure	Disk in differential rotation. Contains spiral arms	Analogy with M 31	Local stellar velocity distribution, 21 cm work, positions of associations	Can spiral arms retain shape indefinitely?

Die Entstehung von Sternen

178   Appendix III: A final summary

Stars

	Accepted data	Evidence from Observation	Evidence from Theory	Implications
Surface Temperatures	For visible stars between 3000 and 70000° K	From atmospheric spectra and from colour		
Absolute Magnitude	From —5 to +20	From apparent magnitudes and parallaxes	Some distances inferred from proper motions	Luminosity L depends sensitively on mass M. In main sequence $L \propto M^4$
Mass	From 0.1 to 10 ⊙, approximately	From motion of binary stars		
Atmospheric Composition	Mainly H, He. Other elements less than 2% by mass	From spectra. By interpretation of curves of growth with model atmospheres		Older stars tend to be poorer in the heavier elements
Energy Generation	Mainly by $4H \to He$ (6×10^{18} erg/g), possibly also $3He \to C$, gravitation		$4H \to He$ is best and most suitable reaction	Restricts lifetime of bright stars
Uniformity of Composition	Stellar composition initially uniform stops being so as $4H \to He$		Calculation suggests no mixing outside central core	Affects structure
Mode of evolution	(i) Star brightens and reddens, or (ii) evolves down main sequence	From luminosity colour curves	(i) Evolution with little mixing. (ii) Belief in relation $L \propto M^4$	
Lifetime	10^6 years for brightest stars. Up to 10^{12} years for dimmer ones		Calculations of structure; available energy sources	Different phases in evolution should be evident
Origin	From interstellar matter	Occurrence together of young stars and gas	Only means apparently possible	Affects balance between diffuse and stellar matter

Appendix III: A final summary

Interstellar matter

	Accepted data	Evidence from Observation	Evidence from Theory	Implications
Local Density	10^{-24} g/cm³ on average	From 21 cm work	Upper limit is set by gravitational effect	Cloud structure implies a non-equilibrium state. Typical cloud is not gravitationally bound (cf. temperature). Turbulence decays in collisions
Local Distribution	Concentrated into clouds	Reflection nebulae; patchiness of extinction	From curves of growth of interstellar Na and Ca⁺ lines; from their doublet ratios	
Local Motion	Supersonic turbulence superposed on galactic rotation	From 21 cm work		
State of ionization	H ionized near O B stars; H neutral but C, Fe ionized elsewhere	H II regions (where H⁺ exists) seen by Hα line, radio continuum	From properties of radiation field	H II regions strongly heated by $H^+ + e \leftrightarrow H + h\nu$. Much less photoelectric heating in H I regions. Hence pressure difference, interstellar turbulence
Temperature	8000°K in H II regions; 125°K elsewhere (in H I regions)	H II temperature from radio continuum, H I from 21 cm line	H II temperature from thermal balance	
Composition	Like atmospheres of population I stars	Rather weak, only few elements observed		Abundance of C, N, O, Si, Fe and so on determines abundance of dust.
Proportion of Dust	Between 0.3 and 3% by mass		From possible grain sizes and required extinction effects	Dust has small mechanical influence, possibly catalyses formation of molecules
Interaction with Magnetic Field	Direct in H II regions, via C⁺, Fe⁺ ions in H I regions	Filamentary structures	High conductivity implies strong interaction	May cause spiral structure and stabilize large masses of gas

Stellar associations

Effect	Explanation	Reason for belief in explanation	Consequence
Existence of OB and of T associations	Associations are newly formed from interstellar matter	Coincidence of so many O B or T Tauri stars is improbable. Such stars must be young. Accretion by old stars is unimportant	An association is probably formed after collapse of a large cloud of gas
Formation and collapse of a large mass of gas	Radiation of energy keeps cloud more or less isothermal. External pressure cannot be supported after certain stage		The rate of radiation determines the rate of collapse
Expansion of associations	Newly formed O B stars heat surrounding gas. General expansion follows	Observed outward motion near Orion Nebula, for instance. Observed consequences, e.g. trunks and globules	Only moderate speeds (~ 10 km/s) can be given to stars. No explanation yet for fast-moving stars like AE Aur
Luminosity-colour curves for associations are close to main sequence	Associations contain young stars only	Newly formed stars are believed to be on the main sequence	Possibly sequence of stellar evolution can be studied by observation of partially dissolved associations

Acknowledgments

This essay was written in Manchester. The final revision was made during a stay at the Institute of Advanced Study, Princeton, N. J.

The author has learnt much about stellar evolution and related problems during discussions with Drs. M. SCHWARZSCHILD, L. SPITZER and B. STRÖMGREN. The author thanks Dr. ROBERT OPPENHEIMER and the Institute for Advanced Study for their hospitality and for financial support. He also thanks the Fulbright Commission for a travel grant.

Bibliography

In this bibliography many references are made to papers read at one or other of the recent astronomical colloquia. The following abbreviations are used:

Gas Dynamics	= I. A. U. Symposium No. 2, Cosmical Gas Dynamics (ed. J. M. BURGERS and H. C. V. D. HULST) North-Holland, 1955.
Les molécules	= Les molécules dans les astres, Liège, 1957.
Non-stable Stars	= I. A. U. Symposium No. 3, on Non-stable Stars (ed. G. H. HERBIG) Cambridge U. P., 1957.
Particules solides	= Les particules solides dans les astres, Liège, 1954.
Principes fondamentaux	= Principes fondamentaux de classification stellaire, Paris, 1955.
Procéssus nucléaires	= Les procéssus nucléaires dans les astres, Liège, 1955.
Stellar Populations	= Ricerche Astronomica Vaticana, vol. 5, 1958.
3rd Symposium	= Report of 3rd Symposium on Cosmical Aerodynamics, Rev. Mod. Phys 30, No. 3, 1958.

ALPHER, R. A., H. A. BETHE and G. GAMOW: Physic. Rev. 73, 803 (1948). — AMBARTSUMIAN, W. A., Trans. I. A. U. 8, 665 (1954); Observatory 75, 72 (1955). — ARP, H. C., and H. L. JOHNSON: Astrophysic. J. 122, 171 (1955). — AX, J.: Z. Astrophysik 32, 257 (1953).

BAADE, W. A.: Astrophysic. J. 100, 137 (1944); Mitt. Astr. Ges. 1955, 51; Observatory 77, 165 (1957). — BETHE, H. A.: Physic. Rev. 55, 434 (1939). — BIERMANN, L.: Procéssus nucléaires, p. 130, 1954. — BLAAUW, A.: Bull. Astr. Netherl. 11, 405 (1952a); Bull Astr. Netherl. 11, 459 (1952b); Astrophysic. J. 123, 408 (1956). — BLAAUW, A., and W. W. MORGAN: Astrophysic. J. 117, 256 (1953); Bull. Astr. Netherl. 12, 76 (1954a); Astrophysic. J. 119, 625 (1954b). — BÖHM, K.-H.: Z. Astrophysik 30, 117 (1952); Astrophysic. J. 123, 408 (1956). — BOK, B. J.: Particules solides, p. 480, 1955. — BONDI, H.: Monthly Notices 112, 205 (1952). — BONDI, H., and T. GOLD: Monthly Notices 108, 252 (1948). — BURBIDGE, E. MARGARET, G. R. BURBIDGE, W. A. FOWLER and F. HOYLE: Rev. mod. Physics 29, 547 (1957).

CAYREL, R., and E. SCHATZMAN: Particules solides, p. 601, 1955. — CHALONGE, D.: Principes fondamentaux, p. 55, 1955; Stellar Populations, p. 345, 1958. — COURTÈS, G.: Particules solides, p. 453, 1955.

DAVIES, R. D.: Monthly Notices 116, 443 (1956). — DAVIS, L., and J. L. GREENSTEIN: Astrophysic. J. 114, 206 (1951). — DUNCAN, J.: Astrophysic. J. 51, 4 (1920).

EBERT, R.: Z. Astrophysik 37, 217 (1955). — EDDINGTON, A. S.: Internal Constitution of the Stars. Cambridge U. P. 1926. — EGGEN, O. J.: Observatory 77, 229 (1957); Monthly Notices 118, 154 (1958).

FESSENKOV, V. G.: Trans. I. A. U. 8, 702 (1954). — FESSENKOV, V. G., and D. A. ROSHKOVSKI: Astr. J. (U.S.S.R.) 29, 381 (1952). — FOWLER, W. A.: Procéssus nucléaires, p. 88, 1954. — FREUNDLICH, E. F.: Proc. phys. Soc. A 67, 192 (1954).

GOLDSWORTHY, F. A.: 3rd Symposium, p. 1062, 1958.

HÄRM, R., and M. SCHWARZSCHILD: Astrophysic. J. 121, 445 (1955). — HERBIG, G. H.: Non-stable Stars, p. 3, 1957. — HILTNER, W. A.: Astrophysic. J. Suppl. 2, 389 (1956). — HOUTERMANS, F.: Naturwissenschaften 44, 157 (1957). — HOYLE, F.: Monthly Notices 108, 372 (1948); Astrophysic. J. Suppl. 1, 121 (1954); Gas Dynamics, p. 181, 1955. — HOYLE, F., and R. A. LYTTLETON: Proc. Cambridge Phil. Soc. 35, 592 (1939). — HUBBLE, E., and M. L. HUMASON: Astrophysic. J. 74, 43 (1931). — HULST, H. C. V. D.: Rech. Astr. Obs. Utrecht, 12, part 2 (1949); Particules solides, p. 393, 1955. — HULST, H. C. V. D., C. A. MULLER and J. H.

OORT: Bull. Astr. Netherl. **12**, 117 (1954). — HULST, H. C. V. D., E. RAIMOND and H. V. WOERDEN: Bull. Astr. Netherl. **14**, 1 (1957).

JAGER, C. DE, and L. NEVEN: Les molécules, p. 357, 1957. — JOHNSON, H. L., and W. W. MORGAN: Astrophysic. J. **117**, 313 (1953). — JOHNSON, H. L., and A. R. SANDAGE: Astrophysic. J. **121**, 616 (1955). — JORDAN, P.: Astr. Nachr. **276**, 193 (1948).

KAHN, F. D.: Bull. Astr. Netherl. **12**, 187 (1954); Gas Dynamics, p. 60, 1955a; Particules solides, p. 578, 1955b; 3rd Symposium, p. 1058, 1958; Astrophysic. J. **129**, 205 (1959). — KAHN, F. D. and L. WOLTJER: Submitted to Astrophysic. J. (1959). — KEENAN, P. C., and G. KELLER: Astrophysic. J. **117**, 241 (1953). — KELLER, G., and R. E. MEYEROTT: Astrophysic. J. **122**, 32 (1955). — KERR, F. J., J. V. HINDMAN and MARTHA STARR-CARPENTER: Nature (Lond.) **180**, 677 (1957). — KIPPER, A. Y.: Tartu Astr. Obs. Publ. **32**, 63 (1952). — KLEIN, O., G. BESKOW and L. TREFFENBERG: Ark. Mat. Fys. 33 B, No. 1 (1946). — KOPAL, Z.: Non-stable Stars, p. 123, 1957. — KOPYLOW, I. M.: Publ. Obs. Crim. **11**, 81 (1954). — KURCHATOV, I. V.: J. Nucl. Energy **4**, 193 (1957). — KUSHWAHA, R. S.: Astrophysic. J. **125**, 242 (1957).

LEDOUX, P.: Procéssus nucléaires, p. 200, 1954. — LINDBLAD, B.: Ark. Mat. Fys. **19 A**, No. 21 (1925).

MCCREA, W. H.: Observatory **75**, 206 (1955).; Monthly Notices **117**, 562 (1957). — MCVITTIE, G. C.: Astronomical J. **61**, 451 (1956). — MAYALL, N. U.: Publ. Obs. Univ. Michigan **10**, 19 (1950). — MAYER, M. G., and E. TELLER: Physic. Rev. **76**, 1226 (1949). — MENON, T. K.: 3rd Symposium, p. 1075, 1958; see also Proc. Inst. Rad. Eng. **46**, 230. — MESTEL, L.: Monthly Notices **113**, 716 (1953); see also Astrophysic. J. **126**, 550 (1957). — MESTEL, L., and L. SPITZER: Monthly Notices **116**, 503 (1956). — MEURERS, J.: Veröff. Sternw. Bonn, No. 45 (1957). — MILLS, B. Y., A. G. LITTLE and K. V. SHERIDAN: Nature (Lond.) **117**, 178 (1956). — MILNE, E. A.: Monthly Notices **83**, 118 (1923). — MINKOWSKI, R.: Publ. Astr. Soc. Pacific **61**, 151 (1949). — MORGAN, H. R., and J. H. OORT: Bull. Astr. Netherl. **11**, 379 (1951). — MORGAN, W. W., A. E. WHITFORD and A. D. CODE: Astrophysic. J. **118**, 318 (1953). — MÜNCH, G.: Astrophysic. J. **125**, 42 (1957); 3rd. Symposium, p. 1035, 1958.

NEYMAN, J., ELIZABETH L. SCOTT and C. D. SHANE: Astrophysic. J., Suppl. **1**, 269 (1954).

ÖPIK, E. J.: Monthly Notices **111**, 178 (1951); Irish Astr. J. **2**, 219 (1953). — OORT, J. H.: Publ. Kapt. Astr. Lab. Groningen 40, (1926); Bull. Astr. Netherl. **4**, 269 (1928); Lectures on galactic structure, at Leiden, 1954a; Bull. Astr. Netherl. **12**, 177 (1954b); Gas Dynamics, p. 147, 1955. — OORT, J. H., and H. C. V. D. HULST: Bull. Astr. Netherl. **10**, 187 (1946).

PETRIE, R. M.: Monthly Notices **118**, 80 (1958). — POTTASCH, S.: Bull. Astr. Netherl. **13**, 77 (1956); Bull. Astr. Netherl. **14**, 29 (1958).

RAIMOND, E.: Bull. Astr. Netherl. **12**, 269 (1957). — RING, J.: Observatory **75**, 249 (1955). — ROMAN, N.: Astrophysic. J. **112**, 554 (1950). — ROUSKOL, E. L.: Particules solides. p. 650, 1955. — RUBEN, G., and A. MASSEVICH: Astr. J. (U. S. S. R.) **34**, 724 (1957).

SALPETER, E. E.: Procéssus nucléaires, p. 116, 1954. — SANDAGE, A. R.: Stellar Populations, p. 41, 1958a; Astrophysic. J. **127**, 513 (1958b). — SANDAGE, A. R., and M. SCHWARZSCHILD: Astrophysic. J. **116**, 463 (1958). — SAVEDOFF, M. P.: Gas Dynamics, p. 218, 1955; Astrophysic. J. **124**, 533 (1956). — SCHATZMAN, E.: Ann. Astrophys. **17**, 300 (1954a); Ann. Astrophys. **17**, 382 (1954b); at a seminar on OB associations, at Manchester, 1956. — SCHLÜTER, A.: Gas Dynamics, p. 144, 1955. — SCHMIDT, M.: Bull. Astr. Netherl. **13**, 15 (1956); Bull. Astr. Netherl. **13**,

247 (1957). — Schwarzschild, M.: Astronomical J. **57**, 57 (1952); Structure and Evolution of Stars. Princeton U. P., 1958. — Schwarzschild, M., and Barbara Schwarzschild: Astrophysic. J. **112**, 248 (1950). — Schwarzschild, M., and L. Spitzer: Observatory **73**, 77 (1953). — Schwarzschild, M., L. Spitzer and R. Wildt: Astrophysic. J. **114**, 398 (1951). — Seaton, M. J.: Monthly Notices **111**, 368 (1951); Particules solides, p. 462, 1955a; Ann. Astrophys. **18**, 188 (1955b).— Shajn, G. A., and V. F. Hase: Publ. Obs. Crim. **8**, 3 (1952). — Slettebak, A.: Astrophysic. J. **110**, 498 (1949); Astrophysic. J. **119**, 146 (1954); Astrophysic. J. **121**, 653 (1955). — Spitzer, L.: Astrophysic. J. **94**, 332 (1941); J. Wash. Acad. Sci. **41**, 309 (1951); Astrophysic. J. **120**, 1 (1954). — Spitzer, L., and J. L. Greenstein: Astrophysic. J. **114**, 407 (1951). — Spitzer, L., and M. Schwarzschild: Astrophysic. J. **114**, 385 (1951); Astrophysic. J. **118**, 106 (1953). — Strömgren, B.: Astrophysic. J. **89**, 526 (1939); Astrophysic. J. **108**, 242 (1948). — Struve, O.: Sky and Telescope **13**, 181 (1954). — Sweet, P. A.: Monthly Notices **110**, 548 (1950).

Tayler, R. J.: Astrophysic. J. **120**, 332 (1954); Proc. phys. Soc. B **70**, 1049 (1957). — Taylor, G. I.: Proc. roy. Soc. A **186**, 273 (1946).

Unsöld, A.: Z. Astrophysik **1**, 138; **2**, 209 (1931); Physik der Sternatmosphären (2nd. ed.). Springer 1955; in the George Darwin Lecture at the R.A.S., 1957; see also Monthly Notices **118**, 3 (1958).

Vyssotsky, A. N.: Publ. Astr. Soc. Pacific **69**, 109 (1957).

Walker, M. F.: Astrophysic. J., Suppl. **2**, 365 (1956); Astrophysic. J. **125**, 636 (1957). — Westerhout, G.: Bull. Astr. Netherl. **13**, 201 (1957). — Whipple, F. L.: Astrophysic. J. **104**, 1 (1946). — Woerden, H. v., G. W. Rougoor and J. H. Oort: C. R. Acad. Sci. (Paris) **244**, 1691 (1957). — Woltjer, L.: Bull. Astr. Netherl. **14**, 39 (1958). — Woolley, R. v. d. R., and O. J. Eggen: Observatory **78**, 149 (1958).

Zeipel, H. v.: Monthly Notices **84**, 665, 684, 702 (1924).

Die Entstehung von Sternen durch Kondensation diffuser Materie

Von

R. EBERT, S. v. HOERNER und St. TEMESVARY

Mit 24 Abbildungen

Inhaltsverzeichnis

Einleitung 185
A. Altersbestimmung von Sternen und Sterngruppen 187
    I. Die Entwicklung der Sterne (TEMESVÁRY) 187
        a) Grundzüge der Theorie der Sternentwicklung 187
            1. Energieerzeugung 189
            2. Homologietransformationen 193
            3. Konvektionszonen 201
            4. Entartung 213
            5. Durchmischung 216
        b) Gegenwärtiger Stand der Modelltheorie 218
            1. Sterne vom Typ der Sonne und kleinere Massen 219
            2. Schalenquellenmodelle 220
            3. Massive Sterne 222
            4. Sterne besonderer chemischer Zusammensetzung 224
        c) Vergleich und Anwendung 224
            1. Kritische Zusammenfassung 224
            2. Vergleich mit der Beobachtung 227
        Liste der in A.I. verwendeten Symbole 230
    II. Kontraktionsalter (v. HOERNER) 232
        a) Theorie 232
        b) Vergleich mit der Beobachtung 235
        c) Altersbestimmung 237
    III. Dynamische Altersangaben (v. HOERNER) 238
        a) Massenabschätzung offener Haufen aus der Leuchtkraftfunktion . 238
        b) Auflösung durch äußere Kräfte 241
        c) Auflösung durch inneren Energieaustausch 242
        d) Expandierende Assoziationen 245
        e) Einzelne schnelle Sterne 247
    IV. Zusammenfassung der Altersangaben 248
B. Voruntersuchungen zur Sternentstehung (v. HOERNER) 249

I. Geschwindigkeit und Spektraltyp 250
   a) Die Streuung der räumlichen Geschwindigkeiten 251
   b) z-Komponente und Schichtdicke 258
   c) Die Dicke der sternerzeugenden Schicht 261
II. Die zeitliche Rate der Sternentstehung. 261
   a) Aus der Leuchtkraftfunktion 261
   b) Theoretischer Ansatz 263
   c) Die zeitliche Entstehungsrate offener Sternhaufen 265
III. Sternentstehung einzeln oder in Haufen? 270
   a) Offene Haufen 270
   b) Assoziationen 271
IV. Sterne und interstellare Materie (zusammen mit TEMESVÁRY) . . . 273
   a) Die räumliche Verteilung von Sternen frühen Typs und interstellarer Materie 275
   b) Die Massenabgabe von Sternen an das interstellare Medium . . 277
   c) Die Erzeugung schwererer Elemente in Sternen 279
   d) Bilanz der weißen Zwerge und des Heliums 282
V. Zusammenfassung 288
C. Physikalische Theorien zur Sternentstehung (EBERT) 290
I. Einleitung 290
II. Sternverjüngung durch Aufsammlung 293
III. Sternentstehung durch Kondensation 296
   a) Temperaturen des interstellaren Gases 296
   b) Jeanssches Kriterium und ähnliche Instabilitätsbedingungen . . 297
   c) Konzentration von Staub durch Strahlungsdruck 300
   d) Zerfall in Untersysteme 301
   e) Entstehung von Sternen in der Umgebung von O-Sternen . . . 304
IV. Magnetfelder und Drehimpuls 308
   a) Einfluß von Magnetfeldern auf die Kontraktion 308
   b) Drehimpulsabführung durch Turbulenz und bei rotierender Zentralmasse 310
   c) Drehimpulsabführung durch Magnetfelder bei der Sternentstehung 311
V. Zusammenfassung 319
Schlußbemerkung 320
Literaturverzeichnis 321

Einleitung

Die Frage nach der Entstehung der Sterne ist gegenwärtig eines der interessantesten Probleme der Astrophysik, und wir möchten als erstes sowohl der „Gesellschaft Deutscher Naturforscher und Ärzte" als auch den Initiatoren der Aufgabenstellung für die Anregung zu einer umfassenden Behandlung dieses Themas danken.

Einleitung

Die Anzahl der jährlich erscheinenden Arbeiten, die sich — direkt oder indirekt — mit der Entstehung der Sterne befassen, wächst ständig. Daher kann zur Zeit von einer abschließenden Meinungsbildung noch nicht die Rede sein; die vorliegende Arbeit hat somit mehr die Aufgabe einer vorläufigen Zwischenbilanz, die das bisherige Material sichten, einiges Neue hinzufügen und zu weiteren Arbeiten anregen möchte.

Zunächst ist uns die Aufgabe gestellt, eine zusammenfassende und kritische Übersicht über die Gründe zu geben, die zur Annahme des definierten Alters bestimmter Sterngruppen und der gegenwärtigen Entstehung von Sternen führen. Da die Theorie der Sternentwicklung die wichtigste Methode der Altersbestimmung ist und zugleich auch die Verschiedenheit der Farbenhelligkeits-Diagramme verschiedener Sterngruppen zu erklären vermag, haben wir dieser Theorie und den Sternmodellen, auf denen sie basiert, eine ausführliche Behandlung im ersten Kapitel gewidmet, um das bereits Gesicherte von den noch offenen Problemen trennen zu können. Es folgen zwei davon unabhängige Methoden der Altersbestimmung: kontrahierende Sterne und dynamische Altersangaben. Als Test auf die Verläßlichkeit der Altersbestimmungen vergleichen wir schließlich die Ergebnisse der verschiedenen Methoden miteinander; wir erhalten das Resultat, daß an der Existenz junger und jüngster Sterne, und somit an der gegenwärtigen Entstehung von Sternen, kein Zweifel möglich ist.

Im Hinblick darauf, daß die Theorie der Sternentstehung erst im Werden begriffen ist, haben wir uns dazu entschlossen, eine Reihe von „Voruntersuchungen zur Sternentstehung" in einem gesonderten Kapitel vorzulegen. Wir möchten damit für weitere Bearbeitungen dieses Fragenkreises einige Unterlagen und Rahmenvorstellungen bereitstellen, wobei wir verschiedene, in der eigentlichen Aufgabenstellung nicht erwähnte Fragen in die Darstellung mit einbezogen haben. Außer der räumlichen Verteilung und den Unterschieden der Geschwindigkeits-Streuungen von Sternen verschiedenen Spektraltyps haben wir noch die zeitliche Rate der Sternentstehung eingehend untersucht. Die Frage, ob das Gros der Sterne einzeln oder in Haufen entsteht, wurde angeschnitten, und im Anschluß an eine Diskussion der Anreicherung schwerer Elemente durch Sternentwicklung wurden verschiedene Materie- und Massenbilanzen durchgeführt.

Der gegenwärtige Stand des Problems der Sternentstehung läßt sich dadurch kennzeichnen, daß zwar eine größere Anzahl von Arbeiten zur Behandlung einzelner Phasen und Teilfragen vorliegt, jedoch noch keine „durchgehende" Theorie der Sternentstehung von einer normalen Wolke über die verschiedenen Phasen der Kontraktion bis zum fertigen Stern. Vor allem fehlt noch die Berechnung derjenigen Phase, die beim Einsetzen der Gravitations-Instabilität beginnt und bis zu der von

HENYEY u. Mitarb. berechneten letzten Phase der Kontraktion reicht. Eine numerische Behandlung dieses „Zusammenfallens" ist in Vorbereitung und soll später durchgeführt werden.

Nach einer Darstellung der Aufsammlung interstellarer Materie durch Sterne diskutieren wir im dritten Kapitel ausführlich die vorhandenen Ansätze zu einer physikalischen Theorie der Sternentstehung durch Kondensation interstellarer Materie, wobei es sich zumeist um die erste Phase der Kontraktion, bis zum Einsetzen der Gravitations-Instabilität, handelt. Einen gesonderten Abschnitt widmen wir dem Einfluß der interstellaren Magnetfelder sowie einem der Hauptprobleme der Kontraktion, dem Abtransport des Drehimpulses.

A. Altersbestimmung von Sternen und Sterngruppen

A. I. Die Entwicklung der Sterne

A. I. a) Grundzüge der Theorie der Sternentwicklung

Die Angabe des Alters eines Sterns setzt einige theoretische, durch Beobachtungen belegbare Kenntnisse seiner Entwicklung von der Entstehung bis zum Untergang, bzw. bis zum Eintritt eines praktisch unveränderlichen Endzustandes voraus. Da man unter einem Stern einen selbstleuchtenden Körper versteht, liegt es nahe, zunächst nach den Quellen der ausgestrahlten Energie zu fragen. Als solche kommen, soviel wir wissen, zwei verschiedene Energieformen in Betracht: 1. Gravitationsenergie, die der Stern durch Kontraktion gewinnt, und 2. Kernenergie, die bei der Umwandlung chemischer Elemente durch Kernreaktionen im Innern des Sterns freigesetzt wird. Die zuletzt genannte Energiequelle setzt neben dem Vorhandensein von geeigneten Reaktionspartnern gewisse Mindestwerte der Temperatur und Dichte im Sterninnern voraus, die um viele Größenordnungen über den in dem interstellaren Medium gegebenen Werten liegen. Daher gliedert sich der Lebensweg eines Sterns ganz natürlich in drei Abschnitte, von denen der mittlere dadurch gekennzeichnet ist, daß der Stern seinen Energiehaushalt vorwiegend aus Kernenergie bestreitet, während die Gravitationsenergie als Energiequelle nur während der ersten Phase, der Kontraktion des Protosterns bis zum Einsetzen von Kernreaktionen, und in der Endphase beim Abklingen der letzten möglichen Kernumwandlungen, eine Rolle spielt. Dieser grundsätzliche Sachverhalt ist schon seit langem bekannt und läßt sich durch die folgende grobe Abschätzung [vgl. A. S. EDDINGTON (1928)] begründen:

Von der bei einer homologen Kontraktion eines Sterns der Masse M aus dem Zustand einer unendlichen Verdünnung bis auf einen Radius R

gewonnenen Gravitationsenergie stand der Bruchteil

$$\int_0^t L\,dt = \frac{(3\gamma-4)\beta}{(5-n)(\gamma-1)}\frac{GM^2}{R} \tag{1}$$

zur Ausstrahlung zur Verfügung. Das gibt mit $\gamma = \frac{c_p}{c_v} = \frac{5}{3}$; $\beta = 1$ und dem Polytropenindex $n = 3$ als Kontraktionsalter

$$t_K = 10^{7.37} \left(\frac{M}{M_\odot}\right)^2 \left(\frac{R_\odot\, L_\odot}{R\, \tilde L}\right) \text{Jahre}^1. \tag{2}$$

Da die Sonne nach geologischen Befunden schon seit einigen Milliarden Jahren größenordnungsmäßig mit der gegenwärtigen Leuchtkraft strahlt, müßte sie — wenn die Kontraktion ihre einzige Energiequelle wäre — jünger sein als die Erde, deren Rinde schon seit $4{,}5 \cdot 10^9$ Jahren erkaltet ist, wie das Häufigkeitsverhältnis der verschiedenen Bleiisotope lehrt. Da die Kontraktionsphase demnach nur einen kleinen Bruchteil des Sternlebens ausmacht, werden die allermeisten beobachteten Sterne sich in der mittleren Phase befinden, in der die Kernreaktionen die Quelle der ausgestrahlten Energie sind.

Die Gesamtentwicklung eines Sterns ist — von der Wechselwirkung mit dem interstellaren Medium abgesehen — wahrscheinlich durch vier Anfangsgrößen vollständig bestimmt:

1. seine Masse, 2. seine chemische Zusammensetzung, 3. seinen Gesamtdrehimpuls, und 4. seine Magnetfelder.

Zwei davon, nämlich die Masse und der Gesamtdrehimpuls, sind einfache Größen, für die die Angabe einer Zahl in der entsprechenden Dimension genügt. Für die Magnetfelder muß neben ihrer Stärke auch ihre Form, d. h. ihre Zusammenhangsverhältnisse, angegeben werden. In der chemischen Zusammensetzung bestimmen die Massenanteile des Wasserstoffs X und des Heliums Y im wesentlichen das effektive Molekulargewicht. Die Häufigkeit des Kohlenstoffs und Stickstoffs sind für die Ergiebigkeit des CN-Zyklus der Energieerzeugung wichtig. Die Elemente der Sauerstoffgruppe haben Bedeutung für die Absorptionsprozesse im Sterninnern und damit für den Strahlungstransport der Energie. Die Häufigkeit der leicht ionisierbaren Metallatome bestimmt den Elektronendruck an der Oberfläche des Sterns, der seinerseits sowohl für den Ionisationsgrad wie für die Absorption durch das H⁻-Ion bestimmend ist. Für die Frage der Erzeugung schwererer Elemente durch Kernreaktionen in einem späteren Entwicklungszustand des Sterns wird speziell die Häufigkeit des Eisens wichtig sein.

Nach dem als Russell-Vogtsches-Theorem bekannten Eindeutigkeitssatz der Theorie des Sternaufbaus [H. VOGT (1957), Kap. II] ist durch

---

[1] Eine Liste mit der Erklärung aller verwendeten Symbole befindet sich am Ende dieses Kapitels.

die Masse und die chemische Zusammensetzung eines Sterns sein Zustand und seine weitere Entwicklung eindeutig bestimmt, vorausgesetzt, daß in der Energieerzeugung die Kontraktionsenergie keine Rolle spielt, daß der Stern keiner Rotation oder äußeren deformierenden Kräften unterworfen ist, daß er keine Magnetfelder besitzt, und daß er keine Wechselwirkung mit dem interstellaren Medium hat. Der Erweiterung dieses Theorems auf alle vier oben genannten bestimmenden Größen stehen im Augenblick noch die mathematischen Schwierigkeiten einer hinreichend exakten hydromagnetodynamischen Theorie der Kontraktion und der Rotation entgegen. Wir können daher nur vermuten, daß der Aufbau und die Entwicklung eines Sterns durch diese vier Größen bestimmt sind. Wegen dieser Unsicherheit werden wir die Kontraktion vom Protostern bis zum Einsetzen der Kernreaktionen — d. h. den ersten Abschnitt der Sternentwicklung — gesondert unter A.II. behandeln und den Einfluß der Rotation und der Magnetfelder nur in ihrer Auswirkung auf die Homogenität der chemischen Zusammensetzung der Sternmaterie im Abschnitt a 5 ,,Durchmischung'' berücksichtigen.

1. Energieerzeugung

Für die Energieerzeugung von Bedeutung sind die folgenden drei bekannten Reaktionen:

Bei Temperaturen zwischen 10 und $20 \cdot 10^6$ Grad die Umwandlung des Wasserstoffs in Helium [W. A. FOWLER (1954)] durch die pp-Reaktion mit

$$\varepsilon_{pp} = \text{const.} \, X^2 \varrho \, T^4 \tag{3}$$

$$K_{pp} = 10^{18.80} \, \text{erg/g H} \rightarrow \text{He}$$

und den CN-Zyklus mit

$$\varepsilon_{CN} = \text{const.} \, X X_{CN} \varrho \, T^{18} \tag{4}$$

$$K_{CN} = 10^{18.78} \, \text{erg/g H} \rightarrow \text{He}$$

und bei $T \approx 2 \cdot 10^8$ Grad, wenn aller Wasserstoff ,,verbrannt'' ist, die Umwandlung des Heliums in Kohlenstoff durch den Salpeterzyklus [E. E. SALPETER (1952)] mit

$$\varepsilon_S = \text{const.} \, Y^3 \varrho^2 \, T^{18} \tag{5}$$

$$K_S = 10^{17.77} \, \text{erg/g He} \rightarrow \text{C}.$$

Die pp-Reaktion und der CN-Zyklus geben für etwa $T = 15 \cdot 10^6$ Grad das gleiche ε, wenn man für die Kohlenstoff-Stickstoff-Häufigkeit $X_{CN} = 0.005 \, X$ annimmt [W. A. FOWLER (1956)]. Bei niedrigen Temperaturen wird die Energieerzeugung durch die erste Reaktion bestimmt, bei höheren Temperaturen durch die zweite. Der entscheidende Unterschied der beiden Reaktionen liegt in der sehr viel höheren Temperatur-empfindlichkeit des CN-Zyklus, die in Sternen, deren Mittelpunktstemperatur $T_c \geq 15 \cdot 10^6$ Grad ist, eine sehr viel stärkere Konzentration der Energiequellen und damit einen konvektiven Kern zur Folge hat (vgl. Abschn. 2).

Für das weitere Schicksal eines Sterns ist es daher eine entscheidende Frage, welche der beiden Reaktionen seine Energieerzeugung und damit seinen Aufbau bestimmt.

Um einen groben Überblick über die Verteilung der Sterne auf die beiden Möglichkeiten zu erhalten, wollen wir zwei Modelle für chemisch homogene Sterne zugrunde legen, und zwar das Eddingtonsche Standardmodell (die Polytrope $n = 3$) für die Sterne der pp-Reaktion mit

$$T_c = 19.7 \cdot 10^6 \left(\frac{M}{M_\odot}\right)\left(\frac{R_\odot}{R}\right) \mu \qquad (6)$$

und das Cowlingsche Punktquellmodell für die Sterne des CN-Zyklus mit konvektivem Kern

$$T_c = 20.8 \cdot 10^6 \left(\frac{M}{M_\odot}\right)\left(\frac{R_\odot}{R}\right) \mu \qquad (7)$$

$$M_K = 0.145 \, M, \quad r_K = 0.169 \, R$$

In beiden Fällen ist der Anteil des Strahlungsdrucks am Gesamtdruck vernachlässigt. M_K ist die Masse des konvektiven Kerns.

Mit den Werten für M und R für die Sterne der Hauptreihe nach der Tabelle bei ALLEN (1955), § 99 erhält man $\frac{T_c}{\mu}$ (vgl. Tab. 1) und in einem M-μ-Diagramm die sich ein wenig überlagernden Gebiete der beiden Reaktionen (vgl. Abb. 1), wobei die Grenzlinie etwa von $\mu = \frac{1}{2}$ ür B5 Stern e bis $\mu = 1$ für M 2-Sterne verläuft. Da $\mu \leq 2$ ist, kann, wie man aus der Tab. 1 sieht, mit den beiden genannten Modellen der Fall $X = 0$, $T_c \approx 10^{8.3}$ Grad, in dem der Salpeter-Zyklus laufen würde, nicht erfaßt werden.

Da unsere eigentliche Frage der Zeitskala gilt, müssen wir wissen, welcher Anteil der Sternmasse an der Energieerzeugung teilnimmt. Dazu differenzieren wir Gl. (3) logarithmisch und erhalten, indem wir auch weiterhin durch den ganzen Stern eine konstante chemische Zusammensetzung annehmen

$$d \ln \varepsilon_{pp} = d \ln \varrho + 4 \, d \ln T$$

und mit der für eine Polytrope geltenden Beziehung bei Vernachlässigung des Strahlungsdrucks und konstantem Molekulargewicht

$$d \ln P = d \ln \varrho + d \ln T = \left(1 + \frac{1}{n}\right) d \ln \varrho$$

$$\frac{P}{P_c} = \left(\frac{\varepsilon}{\varepsilon_c}\right)^{\frac{1+n}{4+n}}. \qquad (8)$$

Für $n = 3$ entnimmt man der Emdenschen Tafel [c. f. A. S. Eddigton (1927), s. 105] für $\frac{\varepsilon}{\varepsilon_c} = 10^{-1} \to \frac{P}{P_c} = u^{n+1} = 0.2681$, $u = \frac{T}{T_c} = 0,72$, d. h. die Temperatur wäre an dieser Stelle z. B. von $T = 15.0 \cdot 10^6$ Grad auf $10.9 \cdot 10^6$ Grad abgefallen, und $q = \frac{M_r}{M} = 0.31$, $x = \frac{r}{R} = 0.22$.

Die gesuchte Zeitskala ist gegeben durch

$$t = \frac{q\,M\,X\,K}{L} \tag{9}$$

d. h. für $q_{pp} = 0.31$ und $K_{pp} = 10^{18.80}$ erg/g H

$$t_{pp} = 10^{18.0} \left(\frac{M}{M_\odot}\right)\left(\frac{L_\odot}{L}\right) X \; [\text{sec}]. \tag{10}$$

Für den CN-Zyklus liefert die Differentiation von Gl. (4), wenn man berücksichtigt, daß innerhalb des konvektiven Kerns die Druck-Dichte-Beziehung durch den Polytropenindex $n = \frac{3}{2}$ gegeben ist,

$$\frac{d\ln\varepsilon}{d\ln P} = \frac{18 + n}{1 + n} = 7.8. \tag{11}$$

Das heißt aber, die Energieerzeugung fällt schon auf etwa $\frac{1}{8}$ homogener Schichtdicke ab und verläuft also ganz innerhalb des konvektiven Kerns. Da dieser durch die Konvektion durchmischt ist und daher eine homogene chemische Zusammensetzung besitzt, ist $q_{CN} = \frac{M_K}{M} = 0.145$. Das gibt mit

$$K_{CN} = 10^{18.78} \text{ erg/g H}$$

$$t_{CN} = 10^{17.65} \left(\frac{M}{M_\odot}\right)\left(\frac{L_\odot}{L}\right) X \; [\text{sec}]. \tag{12}$$

Die beiden Zeitskalen finden sich zusammen mit der Kontraktionszeitskala t_K nach Gl. (2) in der Tab. 1. Außerdem sind in der Abb. 1 die Grenzen eingezeichnet, innerhalb deren t_{pp} bzw. t_{CN} größer als das Weltalter von $5 \cdot 10^9$ Jahren = $10^{17.2}$ sec und kleiner als 10^7 Jahre ist. Dabei wurde in dem für vollständig ionisierte Materie geltenden Ausdruck für das Molekulargewicht

$$\mu = \frac{4}{2 + 6X + Y} \tag{13}$$

$X + Y \approx 1$ angenommen.

Das Diagramm teilt die Sterne schematisch in drei Gruppen ein: Eine erste Gruppe mit Sternen großer Masse, deren Entwicklung so rasch verläuft, daß wir nur verhältnismäßig wenige Exemplare vorfinden werden, selbst wenn die Rate der Sternentstehung über die Massen gleichverteilt wäre. Eine zweite Gruppe umfaßt die Sterne, die für ihren Lebensweg weniger als ein Weltalter brauchen. Wir werden sie nach Maßgabe ihrer Masse, ihres Alters und ihrer anfänglichen chemischen Zusammensetzung über ihre Entwicklungslinie verteilt finden. Während die Sterne der ersten Gruppe ihre Energie sicher nur aus der CN-Reaktion gewinnen, kann in der zweiten Gruppe unter Umständen auch die pp-Reaktion eine Rolle spielen, wenn der Wasserstoffgehalt nicht zu niedrig ist. Schließlich haben wir eine dritte Gruppe, deren Entwicklung so langsam verläuft, daß sie noch Sterne enthalten wird, die ebenso alt sein können wie das ganze System, neben anderen Sternen, die unter Umständen ganz jung sind. Sofern ihr Wasserstoffgehalt nicht schon zu niedrig ist, werden die meisten Sterne dieser Gruppe ihre Energie aus der pp-Reaktion gewinnen.

Tabelle 1. *Die Mittelpunktstemperatur T_c und die Zeitskalen t_{pp} und t_{CN} der Wasserstoffumwandlung für die Sterne der Hauptreihe nach der Masse-Leuchtkraft-Beziehung von* ALLEN (1955). *Die Werte der zweiten Spalte gelten für das Eddingtonmodell, für das Cowling-Modell sind sie um 0,12 größer. Die letzte Spalte gibt das Kontraktionsalter nach Gl.* (2)

$\log \dfrac{M}{M_\odot}$	$\log\left(\dfrac{T_c}{\mu_e}\right)$ [Grad]	$\log\left(\dfrac{t_{pp}}{X}\right)$ [sec]	$\log\left(\dfrac{t_{CN}}{X}\right)$ [sec]	$\log t_K$ [sec]
−1.0	6.78	19.50		15.85
−0.8	6.92	19.30		15.79
−0.6	7.05	19.30		15.92
−0.4	7.18	19.00		15.75
−0.2	7.26	18.60		15.43
0.0	7.30	18.00	17.65	14.87
+0.2	7.32	17.48	17.13	14.37
0.4	7.38		16.65	13.95
0.6	7.42		16.15	13.49
0.8	7.51		15.75	13.18
1.0	7.60		15.35	12.87
1.2	7.66		15.05	12.63
1.4	7.72		14.65	12.29

Die Existenz der zur ersten Gruppe gehörenden O-Sterne ist zwar an sich schon ein Beweis dafür, daß laufend Sterne entstehen müssen,

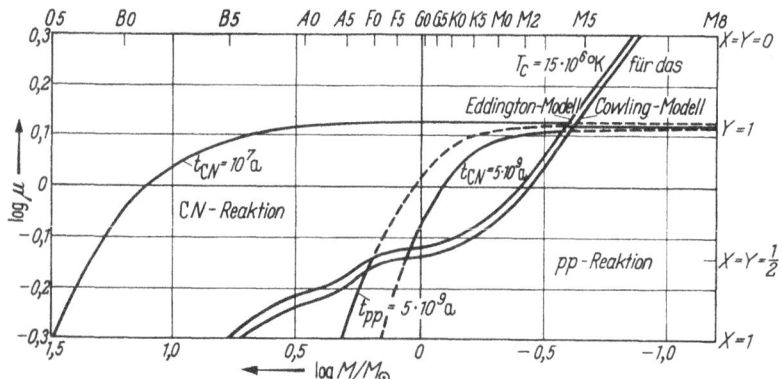

Abb. 1. Grenze zwischen der CN-Reaktion (Cowlingmodell) und der pp-Reaktion (Eddingtonmodell) bei homogener chemischer Zusammensetzung für die Sterne der Hauptreihe. Für die Zeitskalen t_{CN} und t_{pp} wurde $X + Y = 1$ angenommen. Benutzte Werte für L, M, R nach ALLEN (1955) § 99 2. Tabelle, Spektraltypen nach 1. Tabelle

jedoch haben diese Sterne einen zu exzeptionellen Charakter, als daß wir unsere Argumente ganz auf sie stützen könnten. Sie vollenden ihre Entwicklung gleichsam an dem Tage, an dem sie geboren werden, wie der kleine Euphorion, und erlauben uns nicht, sie mit gewöhnlichen Sternen zu vergleichen. Dagegen stellt die zweite Gruppe den eigentlichen Querschnitt durch das Sternenleben dar. Wenn ihre Mitglieder auch verschiedenes Alter haben werden, so gehören sie doch alle einer Generation an und sind gleichsam Zeitgenossen, die sich nur entsprechend ihrer Anfangsbedingungen ganz verschieden entwickeln können. Anders in der dritten Gruppe, in der sich Sterne aller Generationen nebeneinander finden. Hier wird es besondere Mühe bereiten, die Angehörigen verschiedener Epochen zu trennen.

Die Entwicklung eines Sterns wird vor allem in einer Änderung seiner chemischen Zusammensetzung durch die Umwandlung des Wasserstoffs in Helium bestehen. Dabei wird sich, sofern der Stern nicht durch Konvektion oder durch rotationsbedingte Zirkulationen völlig durchmischt wird, die anfängliche chemische Zusammensetzung, die man wohl als homogen voraussetzen darf, in eine inhomogene Zusammensetzung verwandeln. Insofern stellt unser Diagramm (Abb. 1) nur die „Nativität" eines Sterns dar, d.h. die aus den Anfangsbedingungen (M und μ) unmittelbar zu folgernden Zukunftsaussichten. Leider hat die Abb. 1 nur qualitative Bedeutung, nicht so sehr wegen der Unsicherheit der verwendeten empirischen Werte für M, R und L, sondern wegen der Benutzung fester Modelle. Für ein Punktquellenmodell, in dem der Einfluß des Guillotinefaktors $\tau \sim \varrho^\alpha$ mit $\alpha = 0{,}25$ mit berücksichtigt wird [M. SCHWARZSCHILD (1946)], würde z. B. die Linie $T_c = 15 \cdot 10^6$ Grad um 0,1 in $\log \mu$ tiefer verlaufen. Die Abbildung könnte zu der irrigen Meinung verführen, als ob ein zwischen den Grenzlinien liegender Punkt je nach Umständen mal zu dem einen, mal zu dem anderen Modell gehören könnte. In Wirklichkeit kann man für jedes Wertepaar (M,μ), $\mu(r) = $ konst. vorausgesetzt, den Aufbau des Sterns integrieren und erhält nach dem Eindeutigkeitssatz nur eine einzige Lösung. So ließe sich die Linie, auf der beide Reaktionen zu gleichen Teilen zur Leuchtkraft des Sterns beitragen, genau berechnen, ohne daß man empirische Werte für M, R und L verwenden müßte. Solche Rechnungen sind bis jetzt noch nicht durchgeführt, aber in absehbarer Zeit zu erwarten.

2. Homologietransformationen

Bei der Abschätzung der Zeitskalen haben wir vorausgesetzt, daß sich die Leuchtkraft eines Sterns während seiner Entwicklung nicht wesentlich ändert. Das wird aber kaum zutreffen, da sich der Aufbau des Sterns durch die Umwandlung des Wasserstoffs in Helium in der energieerzeugenden Zone ändern wird. Je nachdem, ob die Leuchtkraft mit dem infolge der Kernumwandlungen zunehmenden mittleren Molekulargewicht zu- oder abnimmt, werden die Zeitskalen t_{pp} und t_{CN} eine obere oder eine untere Grenze für die wirkliche Entwicklungszeit der Sterne darstellen. Über diese Auswirkungen der durch die Energieerzeugung bedingten Änderungen in der chemischen Zusammensetzung kann man einige Aufschlüsse erhalten, wenn man die schwächere Voraussetzung macht, daß die Sterne während ihrer Entwicklung nur eine *homologe* Änderung ihres Aufbaues erfahren, d. h. die relative Dichteverteilung in ihrem Innern erhalten bleibt.

Aus dieser Definition der Homologie

folgt unmittelbar
$$\varrho = f(x) \frac{3 M}{4 \pi R^3} \; ; \quad x = \frac{r}{R} \tag{14}$$

$$\varrho \sim \frac{M}{R^3} \tag{15}$$

und aus der Definition des Massenelementes

$$M_r = 4 \pi \int_0^r \varrho \, r^2 dr = 3 M \int_0^x f(x) \, x^2 dx \tag{16}$$

$$M_r \sim M . \tag{17}$$

Ferner folgt aus der hydrostatischen Gleichgewichtsbedingung

$$\frac{dP}{dr} = - \frac{G M_r}{r^2} \varrho \tag{18}$$

$$P = \frac{9 G}{4 \pi} \cdot \frac{M^2}{R^4} \int_x^1 \frac{f(x)}{x^2} \left[\int_0^x f(x) \, x^2 dx \right] dx \tag{19}$$

$$P \sim \frac{M^2}{R^4} \tag{20}$$

und aus der Zustandsgleichung

$$P = \frac{\mathfrak{R} \varrho T}{\mu} + \frac{a T^4}{3} = \frac{\mathfrak{R} \varrho T}{\mu \beta} = \frac{a T^4}{3 (1 - \beta)} \tag{21}$$

$$T = \frac{P}{\varrho} \cdot \frac{\mu \beta}{\mathfrak{R}} \tag{22}$$

$$T \sim \frac{M}{R} \mu \beta \tag{23}$$

und

$$\frac{1 - \beta}{\beta^4} = \frac{a \mu^4 P^3}{3 \mathfrak{R}^4 \varrho^4} \tag{24}$$

$$\frac{1 - \beta}{\beta^4} \sim M^2 \mu^4 . \tag{25}$$

Nun ist der Aufbau eines Sterns noch durch zwei weitere Gleichungen erst festgelegt, nämlich die Gleichung für das Energieströmungsgleichgewicht und die Energietransportgleichung. Die erstere lautet:

$$L_r = 4\pi \int_0^r \varepsilon \varrho\, r^2 dr \qquad (26)$$

mit

$$\varepsilon = \varepsilon_0\, \varepsilon_1\, \varrho\, T^m, \qquad (27)$$

wobei ε_0 eine Konstante ist, während für die pp-Reaktion

$$\varepsilon_1 = X^2;\quad m \approx 4$$

und für die CN-Reaktion

$$\varepsilon_1 = X X_{CN};\quad m \approx 18$$

gilt. Die Gleichung für den Strahlungstransport

$$-\frac{L_r}{4\pi r^2} = -\frac{c}{\varkappa \varrho}\,\frac{d}{dr}\left(\frac{a T^4}{3}\right) \qquad (28)$$

geht mit Hilfe von (18) und (21) über in die sogenannte Masse-Leuchtkraft-Beziehung

$$L_r = 16\pi c G\, \frac{M_r}{\varkappa}\, \frac{1-\beta}{4-3\beta}\left[1 + \beta\left(\frac{d\ln\mu}{d\ln P} - \frac{d\ln\varrho}{d\ln P}\right)\right] \qquad (29)$$

mit

$$\varkappa = \varkappa_0\, \varkappa_1\, \frac{\varrho}{T^{3.5}}, \qquad (30)$$

wobei $\varkappa_0$ eine Konstante ist und für [vgl. L. H. ALLER (1956a)],

$$(1 - X - Y) > 0.02: \quad \varkappa_1 = (1 + X)(1 - X - Y)\,\frac{\bar g}{\tau}$$

mit

$$\tau \sim \varrho^\alpha,$$

während für

$$(1 - X - Y) \leq 0.02: \quad \varkappa_1 = (1 + X)(X + 4Y)\,\frac{\bar g}{\tau}$$

ist. Bei reiner Streuung gilt $\varkappa = 0.19(1 + X)$. In Gl. (29) ist $\frac{d\ln\varrho}{d\ln P}$, wie man aus Gl. (14) und (19) ersieht, lediglich eine Funktion von x.

Wenn wir sowohl den Anteil des Strahlungsdruckes am Gesamtdruck wie auch die Variation des Molekulargewichts im Innern des Sterns vernachlässigen, d. h. $\beta = 1$ sowie $\frac{d\ln\mu}{d\ln P} = 0$ annehmen, so erhalten wir aus der Energieerzeugungsgleichung (26) sowie aus der Masse-Leuchtkraft-Beziehung (29) mit Hilfe der Homologiebeziehungen (15), (17), (20), (23) und (25) je eine Beziehung für die Leuchtkraft

$$L \sim M^{m+2}\, \mu^m\, \varepsilon_1\, R^{-(m+3)} \qquad (31)$$

und

$$L \sim M^{5.5}\, \mu^{7.5}\, \varkappa_1^{-1}\, R^{-1/2}. \qquad (32)$$

Hiermit lassen sich sowohl die Leuchtkraft wie der Radius homologer Sterne als Funktion der Masse und der chemischen Zusammensetzung ausdrücken.

$$L^{2m+5} \sim M^{10m+31} \mu^{14m+45} \varkappa_1^{-2m-6} \varepsilon_1^{-1} \tag{33}$$

$$R^{2m+5} \sim M^{2m-7} \mu^{2m-15} \varkappa_1^2 \varepsilon_1^2 \tag{34}$$

und über ihre Definitionsgleichung

$$L = 4\pi R^2 \frac{ac}{4} T_e^4 \tag{35}$$

auch die effektive Temperatur

$$T_e^{4\,(2m+5)} \sim M^{6m+45} \mu^{10m+75} \varkappa_1^{-2m-10} \varepsilon_1^{-5}. \tag{36}$$

Schließlich folgt aus Gl. (26) und (29)

$$L^{2m+15} \sim [T_e^{10m+31} \mu^{-(2m+15)} \varkappa_1^{m+2} \varepsilon_1^{11/2}]^{4/3}. \tag{37}$$

Nun wollen wir sehen, was diese Transformationen leisten? Dazu haben wir zunächst in Abb. 2a und 3 die Leuchtkraft und den Radius als Funktion der Masse sowie die Leuchtkraft als Funktion der effektiven Temperatur nach den bei ALLEN (1955), l. c. §§ 98, 99 gegebenen Tabellen aufgetragen

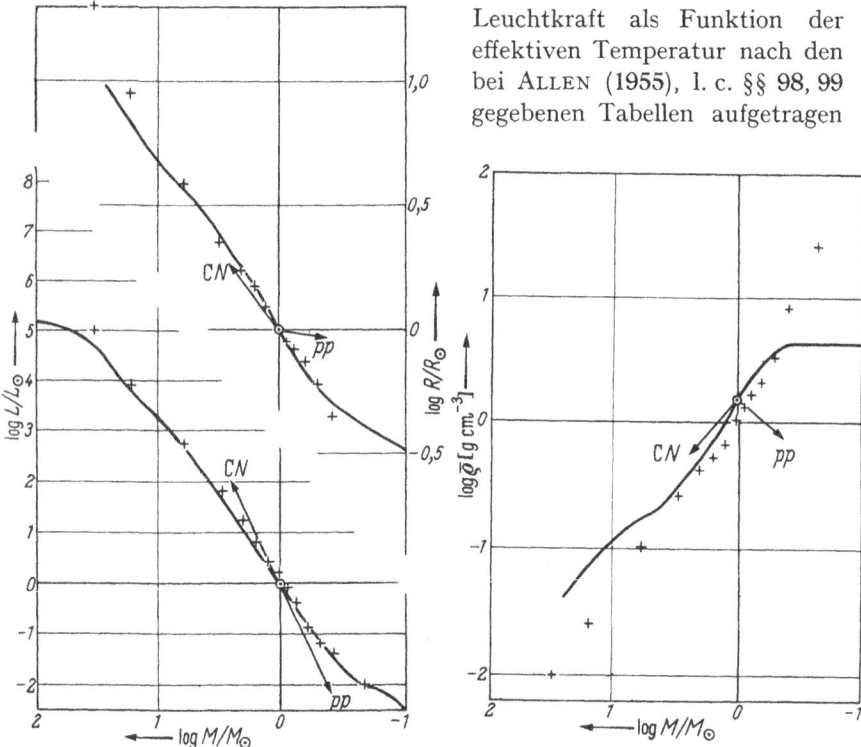

Abb. 2a u. b. Leuchtkraft, Radius und mittlere Dichte der Hauptreihensterne als Funktion der Masse nach den empirischen Werten bei ALLEN (1955) für die $M-L-R$-Beziehung (———) l. c. § 99 2. Tabelle, und für die einzelnen Spektraltypen (+) l. c. § 99 1. Tabelle. Die eingezeichneten Gradienten ergeben sich aus der Homologietransformation bei konstanter chemischer Zusammensetzung

und die Gradienten eingezeichnet, die sich aus (33), (34) und (37) für homologe Sterne mit derselben chemischen Zusammensetzung ($\mu, \varkappa_1, \varepsilon_1$) ergeben müßten. Die Übereinstimmung ist nicht allzu gut. Immerhin glaubte man daraus schließen zu dürfen, daß die Sterne zwischen A0 und F5, die ja zur CN-Reaktion gehören, etwa die gleiche chemische Zusammensetzung haben [c. f. B. STRÖMGREN (1952)]. (Dabei wurde allerdings $m = 20$ gesetzt.) Sicher gilt das nicht für die

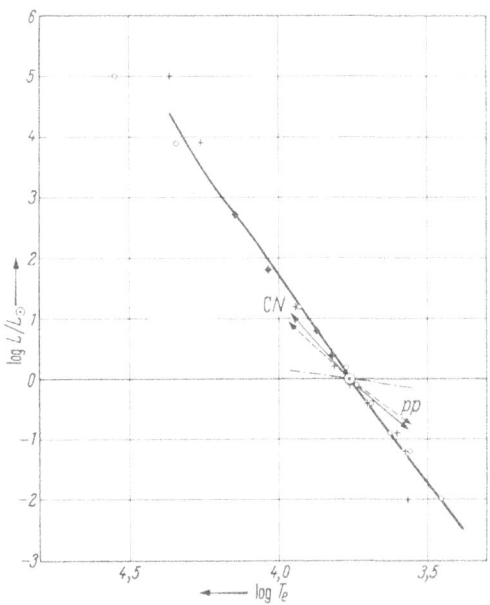

Abb. 3. H-R-Diagramm der Hauptreihensterne nach Werten von ALLEN (1955). (————) § 99, 2.Tabelle. (+) § 99, 1 Tabelle. (○) §§ 98 und 99, 1. Tabelle, mit den Homologiebeziehungen für ———— konstante chemische Zusammensetzung, - - - - - - konstante Masse, - - - - - konstante Masse und konstante chemische Zusammensetzung der äußeren Teile des Sternes

Sterne der pp-Reaktion, wie man deutlich an der Kurve für R sieht, und noch deutlicher an der mittleren Dichte, wenn man mit Gl. (34)

$$\bar{\varrho}^{2m+5} \sim M^{-2(2m-13)} \mu^{-3(2m-15)} \varkappa_1^{-6} \varepsilon_1^{-6} \tag{38}$$

bildet (vgl. Abb. 2b). Der Befund, daß vor allem die im Mittel wohl älteren Sterne der pp-Reaktion keine konstante chemische Zusammensetzung haben, entspricht im Grunde genau unserer Erwartung, da die Sternentwicklung ja wesentlich durch die Umwandlung von Wasserstoff in Helium, also durch die Änderung der chemischen Zusammensetzung bestimmt sein dürfte.

Indem wir aus Gl. (33) und (36) μ statt M eliminieren und $\varkappa_1$ sowie ε_1 vernachlässigen, da die Exponenten von μ viel größer sind, erhalten wir

$$L^{2m+15} \sim [T_e^{14m+45} M^{2m+15}]^{4/5}. \tag{39}$$

Die Richtung, in der ein Stern fester Masse M wandert, wenn sich μ ändert, ist ebenfalls in Abb. 3 eingezeichnet. Aus Gl. (33) und (36) sieht man, daß für abnehmenden Wasserstoffgehalt X sowohl die Leuchtkraft wie auch die effektive Temperatur zunehmen. Dabei fällt der Stern jedoch unter die Hauptreihe.

Nun haben wir bisher, wider besseres Wissen, eine homogene chemische Zusammensetzung innerhalb des Sterns angenommen. Statt dessen müssen wir in Gl. (31) ein immer größer werdendes Molekulargewicht μ_i entsprechend dem abnehmenden X_i in ε_1 einsetzen, in Gl. (32) dagegen das den äußeren Teilen des Sterns entsprechend konstant bleibende $\mu_a \equiv \mu$ und das entsprechende X in $\varkappa_1$. Wenn wir in den anstelle von Gl. (33) und (36) sich dann ergebenden Beziehungen $\varepsilon_1 \mu_i^m$ eliminieren, erhalten wir

$$L \sim T_e^{4/5} M^{11/5} \mu^6 \varkappa_1^{-1/5}, \tag{40}$$

d. h. ein Stern, dessen Masse und chemische Zusammensetzung der äußeren Teile unverändert bleiben, entwickelt sich noch stärker von der Hauptreihe weg. Dabei müßten wir allerdings in Gl. (32) eigentlich $\frac{d \ln \mu}{d \ln P} > 0$ mit berücksichtigen, wodurch der Anstieg in L wieder etwas stärker würde.

Damit sind die Möglichkeiten der Homologietransformationen aber noch nicht erschöpft. Wir schreiben statt Gl. (31) und (32)

$$L = A\, M^{m+2}\, \mu_i^m\, \varepsilon_1\, R^{-(m+3)} \tag{41}$$

$$L = B\, M^{5.5}\, \mu^{7.5}\, \varkappa_1^{-1}\, R^{-1/2}, \tag{42}$$

wobei A und B Größen sind, die außer von der relativen Dichteverteilung vor allem von den Konstanten ε_0 und $\varkappa_0$ abhängen. Wenn wir nun von einem Stern L, M und R kennen und auf Grund eines bestimmten Modells auch A und B, dann haben wir zwei Gleichungen, um zwei noch unbekannte Größen zu bestimmen. Von dieser Methode ist Gebrauch gemacht worden, um unter der Annahme homogener chemischer Zusammensetzung ($\mu_i = \mu$) die Größen X und Y zu bestimmen, aus denen sich μ, $\varkappa_1$ und ε_1 zusammensetzen [M. SCHWARZSCHILD (1946) und J. EPSTEIN (1950)]. Das übereinstimmende Ergebnis war, daß der Gehalt an schweren Elementen allgemein relativ klein ist $(1-X-Y) \approx$ ≈ 0.02, während sich für das Verhältnis X/Y verschiedene Werte ergaben, da die Methode doch sehr empfindlich von A und B abhängt.

Uns dagegen wäre daran gelegen, auf diese Weise das Verhältnis X/X_i zu bestimmen, das zusammen mit der Masse und der Leuchtkraft eines Sterns ein geeignetes Maß für sein Alter abgeben würde.

Wenn die Homologieannahme zutrifft, daß Sterne mit dem gleichen Energieerzeugungsmechanismus annähernd homolog aufgebaut sind, dann müßte es genügen, für die beiden Mechanismen je einen typischen

Stern mit inhomogener chemischer Zusammensetzung zu integrieren, und so für jeden der beiden Typen die Konstanten A und B zu bestimmen. Um dies zu erproben, nehmen wir für die Sonne bestimmte Werte von X und X_i an und legen so A und B fest.

Zunächst müssen wir aber noch die in Gl. (41) und (42) von der chemischen Zusammensetzung abhängenden Größen als Funktionen von X und X_i bestimmen. Indem wir $X + Y = 1$ setzen und $\varkappa_1 = (1+X)(X + 4Y) = (1 + X)(4 - 3X)$ annehmen, erhalten wir

$$f(X) = \frac{\varkappa_1}{\mu^{7.5}} = (1 + X)(4 - 3X)[3 + 5X]^{7.5} \cdot 2^{-15} \qquad (43)$$

$$f(X_i) = \frac{1}{\varepsilon_1 \mu_i^4} = \left[\frac{(3 + 5X_i)^2}{X_i}\right]^2 \cdot 2^{-8}, \quad m = 4,$$

bzw. $\qquad\qquad\qquad\qquad\qquad\qquad\qquad\qquad\qquad\qquad\qquad\qquad\qquad\qquad$ (44)

$$f(X_i) = \frac{1}{\varepsilon_1 \mu_i^{18}} = \frac{(3 + 5X_i)^{18}}{X_i} \cdot 2^{-36}, \quad m = 18.$$

In Abb. 4 sind die Kurven $\log f(X)$ und $\log f(X_i)$ über X aufgetragen.

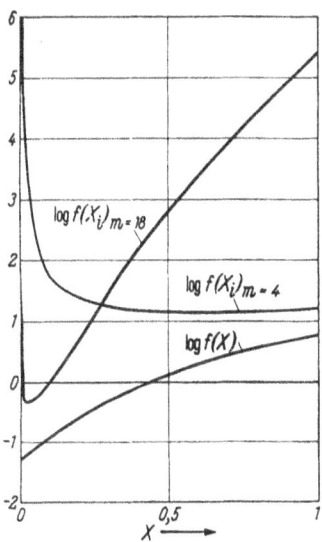

Abb. 4
Die chemischen Faktoren in den Homologietransformationen (41), (42) nach den Gl. (43) u. (44)

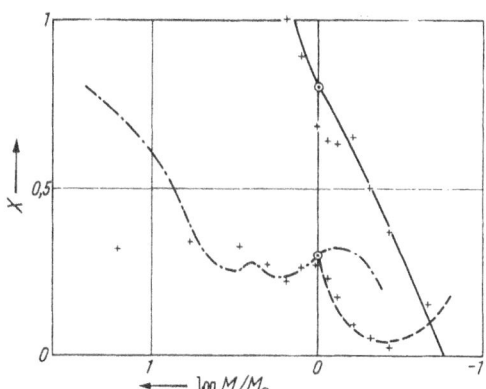

Abb. 5. Wasserstoffgehalt der äußeren Teile der Hauptreihensterne X (———) und der inneren Teile X_i (— — —) für $m = 4$ und (-·-·-·-) für $m = 18$, (+) wie in Abb. 2, bei festen Werten der Modellgrößen $A = A_\odot$ und $B = B_\odot$

Ferner haben wir aus den empirischen Werten für L, M, R nach ALLEN (1955) (l. c. § 99) die Größen

$$\left(\frac{L}{L_\odot}\right)^{-1}\left(\frac{M}{M_\odot}\right)^{m+2}\left(\frac{R}{R_\odot}\right)^{-m+3} = \frac{f(X_i)/A}{f(X_i)_\odot/A_\odot} \qquad (45)$$

und

$$\left(\frac{L}{L_\odot}\right)^{-1}\left(\frac{M}{M_\odot}\right)^{5.5}\left(\frac{R}{R_\odot}\right)^{-0.5} = \frac{f(X)/B}{f(X)_\odot/B_\odot} \qquad (46)$$

berechnet. Indem wir nun auf $X_\odot = 0.80$ und $X_{i\odot} = 0.30$ normieren und $A = A_\odot$, $B = B_\odot$ annehmen, erhalten wir mit $\log f(X) = 0.59$ und $\log f(X_i) = 1.25$ für $m = 4$ bzw. $\log f(X_i) = 1.44$ für $m = 18$ für die chemische Zusammensetzung der Hauptreihensterne das in Abb. 5 wiedergegebene Resultat. Leider ist dies für unsere Zwecke nicht brauchbar, denn um systematische Altersunterschiede feststellen zu können, müßte man wenigstens für einzelne Sterngruppen eine einigermaßen gleichmäßige chemische Zusammensetzung der äußeren Teile — also annähernd konstante Werte für X — erhalten. Da nicht einzusehen ist, warum der Wasserstoffgehalt der äußeren Teile des Sternes nach den späten Typen zu systematisch abnehmen sollte, war offensichtlich die Annahme $B = $ const nicht berechtigt.

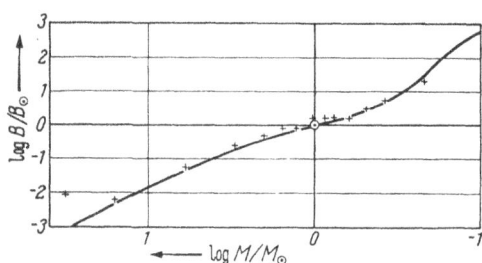

Abb. 6. Variation der Modellgröße B längs der Hauptreihe bei konstantem Wasserstoffgehalt X der äußeren Teile der Sterne

Wir kommen zu dem Schluß, daß eine detailliertere Analyse der Sterne in bezug auf ihre chemische Zusammensetzung und damit auf ihr Alter mit Hilfe der Homologiebetrachtungen nicht möglich ist, da sie nicht nach ähnlichen Modellen aufgebaut sind. Um dies noch besser zu veranschaulichen, haben wir in Abb. 6 den log des Verhältnisses $B/B_\odot$, das sich aus Gl. (46) unter Annahme $X = $ const ergibt, aufgetragen. Da B mit abnehmender Masse nach den späten Spektraltypen zu wächst, könnte die systematische Änderung des Sternaufbaues längs der Hauptreihe unter Umständen auf einem Effekt beruhen, der bewirkt, daß für gegebene Masse und chemische Zusammensetzung die Leuchtkraft entsprechend Gl. (42) nach den späten Typen zu wächst. Es liegt die Vermutung nahe, daß die mit abnehmender Masse und effektiver Temperatur zunehmende Dicke der Konvektionszonen im Stern hierfür verantwortlich ist, da sich bekanntlich mit zunehmender Ausdehnung der konvektiven Gebiete vor allem im tieferen Innern des Sternes eine höhere Gesamtleuchtkraft ergibt als im Falle des Strahlungsgleichgewichtes [L. BIERMANN (1949)].

Zusammenfassend können wir feststellen, daß sich auf Grund der Homologiebetrachtungen keine präziseren Angaben über das Alter der Sterne machen lassen, da neben der Unterscheidung der beiden energieerzeugenden Reaktionen offenbar noch weitere physikalische Effekte in Betracht gezogen werden müssen, die ein so einfaches Verfahren nicht zulassen. Bevor wir uns in den folgenden Abschnitten deren Betrachtung zuwenden, seien zum Gebrauch in späteren Kapiteln in Tab. 2 noch einmal die Werte für das Kontraktionsalter und die Zeitskalen der Wasserstoffverbrennung für die Hauptreihensterne in Abhängigkeit vom Spektraltyp nach ALLEN (1955) (l.c. § 99, 1. Tabelle) zusammengestellt.

3. Konvektionszonen

In Gebieten im Stern, in denen Konvektion herrscht, tritt zu dem Energietransport durch Strahlung

$$F_R = -\frac{4\,ac\,T^3}{3\varkappa\varrho}\frac{dT}{dr} \quad (47)$$

noch der Energietransport durch Konvektion

$$F_K = \frac{1}{2} c_p \varrho\, l v \left(\left|\frac{dT}{dr}\right| - \left|\frac{dT}{dr}\right|_{ad}\right) \quad (48)$$

Tabelle 2. *Das Kontraktionsalter t_K und die Zeitskalen der Wasserstoffumwandlung t_{CN} (für das Cowlingmodell) und t_{pp} (für das Eddingtonmodell) nach Gl. (2), (10) und (11) mit den empirischen Werten von* ALLEN (1955) *in Jahren*

Sp	$M/M_\odot$	$t_K\big/\dfrac{2}{5-n}$	t_{CN}/X	t_{pp}/X
O5	32	$1.2 \cdot 10^4$	$4.5 \cdot 10^6$	
B0	16	$8.3 \cdot 10^4$	$2.8 \cdot 10^7$	
B5	5.9	$4.2 \cdot 10^5$	$1.7 \cdot 10^8$	
A0	3.0	$1.5 \cdot 10^6$	$6.6 \cdot 10^8$	$1.5 \cdot 10^9$
A5	2.0	$3.4 \cdot 10^6$	$1.8 \cdot 10^9$	$4.0 \cdot 10^9$
F0	1.55	$6.0 \cdot 10^6$	$3.5 \cdot 10^9$	$7.8 \cdot 10^9$
F5	1.25	$1.2 \cdot 10^7$	$7.1 \cdot 10^9$	$1.6 \cdot 10^{10}$
G0	1.02	$1.5 \cdot 10^7$	$9.1 \cdot 10^9$	$2.0 \cdot 10^{10}$
G1	1.00	$2.3 \cdot 10^7$	$1.4 \cdot 10^{10}$	$3.2 \cdot 10^{10}$
G5	0.87	$2.5 \cdot 10^7$	$1.6 \cdot 10^{10}$	$3.5 \cdot 10^{10}$
K0	0.76	$4.1 \cdot 10^7$	$2.7 \cdot 10^{10}$	$6.0 \cdot 10^{10}$
K5	0.62	$9.6 \cdot 10^7$		$6.9 \cdot 10^{10}$
M0	0.49	$1.2 \cdot 10^8$		$1.1 \cdot 10^{11}$
M2	0.38	$1.9 \cdot 10^8$		$1.3 \cdot 10^{11}$
M5	0.22	$4.5 \cdot 10^8$		$3.1 \cdot 10^{11}$

hinzu. Dabei ist l der „Mischungsweg" und v die Konvektionsgeschwindigkeit

$$v^2 = \frac{1}{8} l^2 \frac{g}{T}\left(\left|\frac{dT}{dr}\right| - \left|\frac{dT}{dr}\right|_{ad}\right) \quad (49)$$

[vgl. BIERMANN (1949)]. Indem wir das Zeichen V für den doppellogarithmischen Temperaturgradienten einführen,

$$V = \frac{d\ln T}{d\ln P}\,;\quad \Delta V = V_K - V_{ad}$$

erhalten wir mit der hydrostatischen Grundgleichung

$$\frac{dP}{dr} = -g\varrho\,,\quad g = \frac{G M_r}{r^2}$$

die obigen Gleichungen in der folgenden Form

$$F_R = \frac{4\,cg(1-\beta)}{\varkappa} V_K \quad (50)$$

$$F_K = \frac{l}{2} c_p g \frac{\varrho^2 T}{P} v\, \Delta V \quad (51)$$

$$v^2 = \frac{1}{8} l^2 g^2 \frac{\varrho}{P}\, \Delta V \quad (52)$$

und, indem wir v eliminieren, für den Gesamtenergiefluß

$$\frac{L_r}{4\pi r^2} = F_R + F_K = \frac{4\,cg\,(1-\beta)}{\varkappa} V_K + \frac{l^2\,g^2\,c_p\,\varrho^{5/2}\,T}{2\sqrt{8}\ P^{3/2}} (V_K - V_{ad})^{3/2}$$
$$= \frac{4\,cg\,(1-\beta)}{\varkappa} V_R .\qquad(53)$$

Die Konvektion tritt auf, wenn der für reinen Strahlungstransport benötigte Gradient V_R größer ist als der adiabatische Gradient V_{ad}. Der sich dabei einstellende Gradient V_K liegt dann zwischen V_R und V_{ad}, d. h. es gilt

$$V_R > V_K > V_{ad} .\qquad(54)$$

So erhält man als Bedingung für das Auftreten von Konvektion aus $V_R > V_{ad}$

$$\frac{\varkappa}{1-\beta} \frac{L_r}{M_r} > 16\,\pi\,c\,G\,V_{ad} = 10^{5.0}\,V_{ad} .\qquad(55)$$

Der adiabatische Gradient hat im allgemeinen den Wert $V_{ad} = 0.40$. In Gebieten, in denen sich der Ionisationsgrad der Materie ändert, kann sein Wert bis auf die Größenordnung 0.10 abnehmen, weil durch die Ionisation die Zahl der Freiheitsgrade erhöht wird. Ebenso erniedrigt er sich etwas mit zunehmendem Anteil des Strahlungsdruckes am Gesamtdruck

$$V_{ad} \approx \frac{2}{5} \cdot \frac{1+3\,(1-\beta)}{1+6\,(1-\beta)} .\qquad(56)$$

Es gibt demnach drei verschiedene Ursachen für das Auftreten von Konvektion: Es kann das Verhältnis $\varkappa/(1-\beta)$ oder das Verhältnis L_r/M_r zu groß werden, und schließlich kann V_{ad} infolge von Ionisation oder zunehmendem Anteil des Strahlungsdruckes abnehmen.

Wir haben es im Stern mit zweierlei Konvektionszonen zu tun, einer äußeren, der sog. Wasserstoffkonvektionszone (WKZ), und einer inneren, dem konvektiven Kern. Wir wollen uns zunächst der ersteren zuwenden, die für den Aufbau und die Entwicklung, vor allem der späteren Typen, von Bedeutung ist. Als Ursache der WKZ sieht man im allgemeinen die Erniedrigung des adiabatischen Gradienten durch die Ionisation des Wasserstoffs an. Das ist nur bedingt richtig, wie zunächst zu zeigen wäre. Dazu verfolgen wir den Temperatur-Druckverlauf von außen nach innen durch die Atmosphäre eines Sternes mit nicht zu hoher effektiver Temperatur (vgl. Abb. 7). Zunächst behält die Temperatur bei wachsendem Druck den konstanten Wert etwa der Grenztemperatur. Dadurch nimmt mit zunehmendem Druck $(1-\beta)$ ab. Die Absorption beruht vorwiegend auf der des H^--Ions. $\varkappa$ ist daher dem Elektronendruck proportional, der seinerseits proportional P zunimmt. Voraussetzung für die H^--Absorption ist allerdings, daß einige freie Elektronen zur Bildung des H^- zur Verfügung stehen. Diese stammen, solange die

Wasserstoffionisation noch nicht eingesetzt hat, vorwiegend von den leicht ionisierbaren Metallen, deren Häufigkeit an dieser Stelle eine entscheidende Rolle spielt. Mit zunehmendem $\varkappa/(1-\beta)$ wächst auch der Temperaturgradient und die Temperatur beginnt langsam zu steigen. Sowie bei hinreichend hoher Temperatur die Ionisation des Wasserstoffs effektiv zu werden beginnt, wächst der Elektronendruck und mit ihm $\varkappa$ und V_R so rasch an, daß sich der Temperaturverlauf aufsteilt und die Konvektionsbedingung [Gl. (55)] erfüllt wird. An dieser Stelle hat in

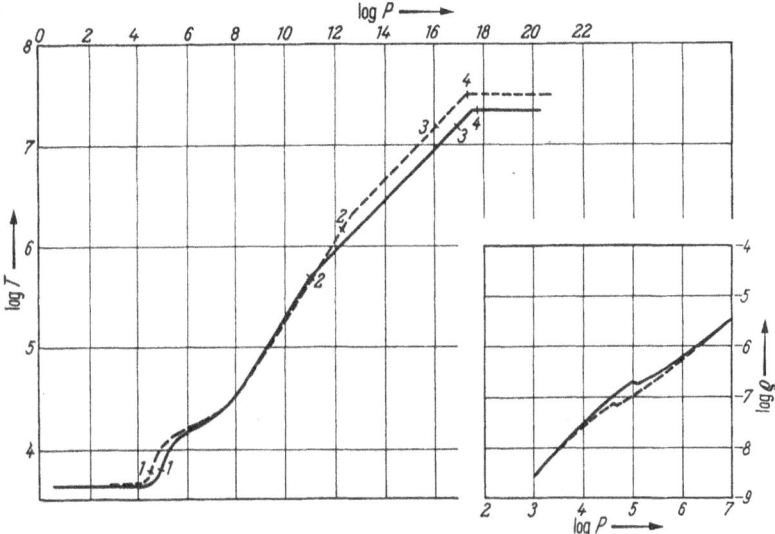

Abb. 7. Druck-Temperaturschichtung in zwei Schalenquellenmodellen $1.2\,M_\odot$. $T_e = 5300°$ nach KIPPENHAHN u. a. (1958). ——— $l = H$; $L = 20\,L_\odot$. − − − − $l = 2H$; $L = 160\,L_\odot$. Es bedeuten: 1 Äußere Grenze der WKZ, 2 Innere Grenze der WKZ, 3 $\varepsilon_{CN} = \varepsilon_{pp}$, 4 Isothermie

der Regel V_{ad} noch seinen normalen Wert 0,4 und nimmt erst im weiteren Verlauf mit zunehmender Wasserstoffionisation ab. Der Eintritt der konvektiven Instabilität an der äußeren Grenze der WKZ wird demnach durch den Faktor $\varkappa/(1-\beta)$ in der Konvektionsbedingung bewirkt. Der nun folgende Temperaturverlauf wird durch das Zusammenwirken des konvektiven und des Strahlungstransportes der Energie bestimmt. Auch wenn die Konvektion bereits den überwiegenden Teil des Energiestromes transportiert, wird sich immer noch ein nur aus Gl. (53) bestimmbarer Wert des Temperaturgradienten V_K zwischen V_R und V_{ad} einstellen. Erst in größerer Tiefe, und im allgemeinen erst, nachdem die Ionisation des Wasserstoffes nahezu abgeschlossen ist, kann der Fall eintreten, daß sich der Temperaturgradient dem adiabatischen Wert so weit nähert, d. h. $\Delta V \ll V_{ad}$ wird, daß praktisch eine adiabatische, oder besser gesagt, isentrope Temperaturschichtung herrscht. Mit dem

Abschluß der Wasserstoffionisation, wenn V_{ad} wieder seinen Normalwert erreicht hat, ist aber im allgemeinen die Schichtung noch nicht wieder stabil. Vielmehr muß $\varkappa/(1-\beta)$ erst wieder klein genug werden. $(1-\beta)$ wächst, wenn $V_K > 0.25$. Das Absorptionsgesetz hat im Sterninnern die Form $\varkappa \sim \varrho/T^{3.5}$, das gibt für $\beta \approx 1$ mit $\varrho \sim P/T$

$$\varkappa' = 1 - 4.5\, V_K \tag{57}$$

' bedeutet doppellogarithmische Differentiation nach P. $\varkappa/(1-\beta)$ nimmt demnach für $V_K \approx V_{ad} \approx 0.4$ stetig ab, bis $\varkappa$ den für reine Elektronenstreuung geltenden konstanten Wert erreicht. Dann bleibt nur noch die Zunahme von $(1-\beta)$ wirksam. Der Eintritt der Stabilität hängt mithin davon ab, wie rasch V_R wieder abnimmt. Aus

$$V_R = \frac{L_r \varkappa}{16\pi c G (1-\beta) M_r} \tag{58}$$

erhält man für $L_r = L = $ const. durch doppellogarithmische Differentiation für $\varkappa \sim \varrho/T^{3.5}$

$$V'_R = 2 - 8.5\, V_K - M'_r \tag{59}$$

und für $\varkappa = $ const.

$$V'_R = 1 - 4\, V_K - M'_r. \tag{60}$$

Die Annäherung des Strahlungsgradienten an den adiabatischen wird, wie man sieht, im allgemeinen nicht sehr flach verlaufen, wie gelegentlich behauptet worden ist. Nur wenn die WKZ bis in solche Tiefe reicht, daß die Änderung von M_r merkbar wird, kann die Annäherung langsamer werden, da M'_r wesentlich negativ ist. Der Eintritt der Stabilität an der inneren Grenze der WKZ hängt somit von dem Absorptionsgesetz und der Massenverteilung im Stern ab.

Da (53) eine Gleichung dritten Grades für $\sqrt{\varDelta V}$ darstellt und die explizite Berechnung von $V_K = \varDelta V + V_{ad}$ daher einige Mühe macht, hat man die Berechnung des Aufbaues der WKZ dadurch zu vereinfachen gesucht, daß man sie im wesentlichen als adiabatisch geschichtet annahm. Dann liegt das Problem darin, die zugehörige, konstant bleibende Entropie zu wissen, deren Wert an sich durch den Temperatur-Druckverlauf in den nichtadiabatischen äußeren Teilen der WKZ bestimmt wird. In den Fällen, in denen die Gl. (53) nicht explizit gelöst wurde, nahm man entweder den Entropiewert an der Stelle, an der die Instabilität einsetzt, und rechnete die ganze WKZ isentrop, vgl. z. B. D. E. OSTERBROCK (1953), oder man berücksichtigte den nichtadiabatischen Teil der WKZ pauschal, indem man nach Eintritt der Instabilität zunächst noch mit dem für reinen Strahlungstransport geltenden Gradienten weiterrechnete — mit der Begründung, daß im nichtadiabatischen Teil die Konvektion nur unwesentlich zum Energietransport beiträgt — bis die Temperatur soweit gestiegen ist, daß das Produkt aus der

Dichte der thermischen Energie plus der Ionisationsenergie multipliziert mit der Schallgeschwindigkeit den Gesamtenergiefluß ergibt. Von dieser Stelle an wurde dann isentrop weitergerechnet, vgl. F. HOYLE und M. SCHWARZSCHILD (1955). Dabei wurde angenommen, daß in dem nichtadiabatischen Teil die Konvektion bei hinreichend starker Instabilität praktisch mit Schallgeschwindigkeit verläuft.

Die hier gemachten Voraussetzungen treffen aber in zwei wesentlichen Punkten nicht ganz zu: 1. Selbst wenn der konvektive Energiestrom nahezu den ganzen benötigten Energietransport leistet, braucht die Schichtung noch nicht adiabatisch zu sein, und 2. die Konvektionsgeschwindigkeit kann die Schallgeschwindigkeit nicht erreichen und bleibt insbesondere im adiabatischen Fall um einige Größenordnungen unter der Schallgeschwindigkeit.

Um die erste Behauptung zu verdeutlichen, definieren wir als K das Verhältnis des konvektiven Energiestromes zum Gesamtenergiefluß

$$K = \frac{F_K}{L_r/4\pi r^2} = \frac{F_K}{\sigma T_e^4}, \tag{61}$$

dann folgt aus Gl. (53)

$$\Delta V = (1-K) V_R - V_{ad}, \tag{62}$$

d. h. wegen $\Delta V > 0$ muß $K < 1$ bleiben, was trivial ist, da immer ein Teil der Energie durch Strahlung transportiert wird. Da andererseits V_R sehr groß sein kann, braucht, auch für einen relativ kleinen Wert von $(1-K)$, noch keine Adiabasie zu herrschen, d. h. $\Delta V \ll V_{ad}$ zu sein. Dies wird noch deutlicher, wenn man mit Gl. (51) und (52) K selbst bildet [vgl. UNSÖLD (1955), Gl. (55, 43) und (55,45)] und dabei die Schallgeschwindigkeit

$$v_s = \sqrt{\gamma \frac{P}{\varrho}} \tag{63}$$

und die homogene Schichtdicke

$$H = \frac{dr}{d\ln P} = \frac{P}{g\varrho} \tag{64}$$

einführt und zur Vereinfachung $\gamma = \frac{c_p}{c_v} = \frac{5}{3}$ annimmt. Dann wird

$$K = \sqrt{\frac{5}{6}} \left(\frac{l}{H}\right)^2 \frac{\left(\frac{3}{2}P\right) v_s}{(\sigma T_e^4) c} (\Delta V)^{3/2} \tag{65}$$

und man sieht, daß es allein von dem Druck- und Temperaturverlauf abhängt, ob und wann mit wachsendem K der Überschußgradient ΔV so klein wird, daß die Schichtung praktisch adiabatisch ist. So wird z. B. nach VITENSE (1953) in der Sonne die Schichtung bei $\log P = 5.1$ instabil. Ab $\log P = 5.5$ transportiert die Konvektion bereits nahezu den ganzen Energiestrom, aber erst ab log $P = 8.0$ wird die Schichtung adiabatisch.

Zum Beweis der zweiten Behauptung bilden wir mit Gl. (52), (63) und (64) das Verhältnis der Konvektionsgeschwindigkeit zur Schallgeschwindigkeit

$$\frac{v}{v_s} = \frac{l}{H} \sqrt{\frac{\Delta V}{8\gamma}} . \tag{66}$$

Wegen $1 \approx l/H < 2$ und $\gamma > 1$ müßte $\Delta V \gtrsim 2$ sein, damit die Schallgeschwindigkeit erreicht würde. Das heißt aber, die beiden Annahmen einer nahezu adiabatischen Schichtung $\Delta V \ll V_{ad}$ und einer mit Schallgeschwindigkeit erfolgenden Konvektion schließen einander aus. Weiter würde $\Delta V \gtrsim 2$ auch $V_K > 2$ bedeuten. Durch Differentiation der Zustandsgleichung erhält man aber

$$V = \frac{1 + \frac{1}{2} x(1-x) - \beta \frac{d \ln \varrho}{d \ln P}}{4 - 3\beta + \frac{1}{2} \beta x(1-x) \left[\frac{5}{2} + \frac{\chi}{kT} + 4 \frac{1-\beta}{\beta}\right]} . \tag{67}$$

Das heißt, selbst bei Berücksichtigung der Ionisation des Wasserstoffes und des Strahlungsdruckes müßte $\varrho' < -1$ sein, also eine starke Dichteinversion herrschen, damit die Konvektionsgeschwindigkeit sich der Schallgeschwindigkeit nähern kann. Außerdem muß man berücksichtigen, daß für $V > 1$ die eigentliche homogene Schichtdicke, in der wir den Mischungsweg messen sollten, nicht mehr H ist, sondern H/V, weil dies dann den kleineren Wert hat. Das bedeutet aber, wenn wir $l = \alpha H$ mit $\alpha \leq 2$ und $\gamma \approx (1 - V_{ad})^{-1}$ annehmen, daß

$$v < \frac{\alpha}{2\sqrt{2}} (1 - V_{ad}) v_s < 0.7 v_s \tag{68}$$

bleiben muß.

Die unter Umständen entscheidende Bedeutung der nichtadiabatischen Teile der WKZ für die aus dem Sternaufbau zu ziehenden Schlüsse über die Sternentwicklung und die damit verbundene Notwendigkeit, die WKZ explizit zu berechnen, ist erst in letzter Zeit mit voller Deutlichkeit erkannt worden [E. BÖHM-VITENSE (1957) und (1958); R. KIPPENHAHN, ST. TEMESVÁRY, L. BIERMANN (1958)], vgl. auch den Abschnitt b2 ,,Schalenquellenmodelle". Dabei ist zu bemerken, daß wir den Effekt einer horizontalen Abstrahlung von den heißeren Elementen während der Dauer ihres Aufsteigens noch nicht berücksichtigt haben, durch den der Grad der Abweichung von einer adiabatischen Schichtung naturgemäß noch erhöht wird [E. VITENSE (1953) und E. BÖHM-VITENSE l. c.].

Der Einfluß der nichtadiabatischen Teile der WKZ auf den Aufbau des Sternes läßt sich, wie schon gesagt, nur auf Grund expliziter numerischer Rechnungen hinreichend genau abschätzen. Wir zitieren daher mit freundlicher Genehmigung der Verfasser aus R. KIPPENHAHN u. a. (1958).

„In grober Näherung kann man für die von uns untersuchten Bereiche im $L-T_e$-Diagramm sagen:

a) Verringert man auf irgendeine Weise die Effektivität der Konvektion im oberen (stets nichtadiabatischen) Bereich der WKZ, so gehören zum inneren Teil der WKZ bei gleichem Druck höhere Temperaturen. Vergrößert man die Effektivität, so erhält man einen entsprechend flacheren Anstieg der Temperatur an der oberen Grenze der WKZ.

b) Beim gleichem M, L, R liefert eine effektivere Konvektion an der oberen Grenze der WKZ eine „dickere" WKZ als eine weniger effektive Konvektion. Dabei verstehen wir unter der „Dicke" die Druckdifferenz zwischen äußerer und innerer Grenze der WKZ.

c) Eine Erhöhung der Effektivität der Konvektion im nichtadiabatischen Bereich der WKZ liefert Modelle, die bei gleicher effektiver Temperatur größere Leuchtkräfte und eine „dickere" WKZ haben.

d) Die unter a), b), c) beschriebenen Effekte sind bei Sternen am Anfang der Sequenz, also bei $T_e \approx 7000°$ von geringerer Auswirkung auf den inneren Aufbau als bei niedrigeren effektiven Temperaturen.

Die Behauptung a) sieht man leicht ein, wenn man beachtet, daß der sich an der oberen Grenze der WKZ einstellende Temperaturgradient zwischen dem adiabatischen und dem Strahlungsgradienten liegt. Ist die Effektivität größer, dann liegt er (unter sonst gleichen Verhältnissen) näher bei dem (niedrigeren) adiabatischen Gradienten als bei geringerer Effektivität. Die Behauptung b) folgt nun mit Hilfe von a). Verringert man die Effektivität, so erhält man einen steileren Temperaturanstieg, und das Maximum der Absorption wird rascher erreicht. Der Absorptionskoeffizient und mit ihm der Strahlungsgradient fallen rascher wieder ab, und die Stabilität an der inneren Grenze der WKZ wird rascher erreicht. Bei der Behauptung d) ist zu berücksichtigen, daß für höhere effektive Temperaturen der photosphärische Wert des Absorptionskoeffizienten dem Maximalwert der Absorption schon näher ist, und daher die Dicke der WKZ und damit ihr Einfluß auf den inneren Aufbau geringer ist als bei kühleren Modellen."

Unter Effektivität der Konvektion wird dabei ihr Einfluß auf den Aufbau des Sternes über den Temperaturgradienten verstanden. Ein geeignetes Maß dafür wäre z. B. $(V_R-V_K)/(V_K-V_{ad})$. Ein einfaches Mittel, innerhalb der Modellrechnung den Grad der Effektivität zu variieren, liegt in einer Änderung der Länge des Mischungsweges l in Einheiten der homogenen Schichtdicke H. Wir haben der zitierten Arbeit die Abb. 7 entnommen, die den Druck-Temperaturverlauf für einen Stern mit $1.2\,M_\odot$-Masse und einer effektiven Temperatur von $5300°$ wiedergibt, wobei einmal $l = H$ und zum andern Mal $l = 2H$ angenommen wurde. Die chemische Zusammensetzung der äußeren Teile des Sternes ist in beiden Fällen dieselbe. Da es sich um ein Schalenquellenmodell handelt, nimmt der Anteil des Heliumkernes an der Gesamtmasse des Sternes bei dieser Erhöhung von l von 22% auf 28% zu. Die Dicke der WKZ wächst um 1.7 Zehnerpotenzen im Druck, und die Leuchtkraft nimmt um einen Faktor 8 zu. Damit glauben wir, den in Abb. 6 erhaltenen Anstieg des Modellfaktors B von der Sonne zu den späteren Typen hin durch eine Zunahme der Dicke und der Effektivität der WKZ wenigstens qualitativ erklären zu können.

Wir kommen nun zu der inneren Konvektionszone, dem konvektiven Kern. Die Frage, in welchen Sternen er vorhanden ist, hängt, wie wir im Abschnitt 1a gesehen haben, von der Höhe der Mittelpunktstemperatur ab, die darüber entscheidet, in welchem Ausmaße die CN-Reaktion an der Energieerzeugung beteiligt ist. Da die Sterne der CN-Reaktion zu den „schnellebigeren" Objekten gehören, wird die Frage sinnvoll, welche Entwicklung der konvektive Kern infolge der Umwandlung des Wasserstoffes erfährt. Zunächst werden wir nach unseren Ergebnissen aus a1 und a2 erwarten dürfen, daß mit abnehmendem Wasserstoffgehalt X_i die Mittelpunktstemperatur des Sternes steigen wird. Wegen des hohen Exponenten im Energieerzeugungsgesetz wird selbst bei einer geringfügigen Temperaturerhöhung die Abnahme von X überkompensiert werden und eine merkliche Zunahme der Leuchtkraft erfolgen. Obwohl die Homologiebetrachtungen bei zunehmender Inhomogenität der chemischen Zusammensetzung ihre Gültigkeit verlieren, bleiben diese Annahmen doch qualitativ richtig und haben sich durch Modellrechnungen bestätigen lassen (vgl. A.I.b2 und b3), denn durch die Verschmelzung von je vier Wasserstoffkernen zu einem Heliumkern wird die Zahl der Impulsträger vermindert, so daß, um den Zentraldruck größenordnungsmäßig aufrecht zu erhalten, eine entsprechende leichte Kontraktion der inneren Teile des Sternes sowohl die Zahl der Teilchen pro Volumeneinheit, wie auch ihre Energie, d. h. die Temperatur, etwas erhöhen muß. Bei der Kontraktion des Kernes wird etwas Gravitationsenergie gewonnen, die der Stern zur Expansion der äußeren Teile verwendet.

An der Oberfläche des durchmischten konvektiven Kernes müßte nach dieser Vorstellung ein Dichtesprung $\varrho_i/\varrho_a = \mu_i/\mu_a$ herrschen, da Druck und Temperatur trotz des Unterschiedes im Molekulargewicht auf beiden Seiten den gleichen Wert haben. Modelle, die eine solche Diskontinuität berücksichtigen und als Folge der zunehmenden Inhomogenität der chemischen Zusammensetzung des Sternes eine Schrumpfung des konvektiven Kernes der Masse und dem Radius nach, sowie eine Expansion der äußeren Teile des Sternes ergeben, sind von SCHÖNBERG, CHANDRASEKHAR (1942), LEDOUX (1947) und HARRISON (1944) und (1947) gerechnet worden [c. f. auch STRÖMGREN (1952)].

Wenn der konvektive Kern jedoch mit zunehmendem Molekulargewicht μ_i der Masse nach wirklich abnimmt, wird gar keine Diskontinuität in μ und damit auch kein Dichtesprung auftreten, sondern sich eine stabil geschichtete Übergangszone bilden, in der das Molekulargewicht kontinuierlich von seinem ungeänderten äußeren Wert auf den innerhalb des konvektiven Kernes gegebenen Wert übergeht. Zum Verständnis der im Abschnitt b3 behandelten Modellrechnungen, die diesen eben erwähnten Umstand berücksichtigen, wollen wir eine kurze mehr qualitative Betrachtung anstellen[1].

Solange der Einfluß des relativen Strahlungsdruckes auf den adiabatischen Gradienten nach Gl. (56) unerheblich ist, wird der konvektive Kern eine homologe Änderung erfahren, und es wird nach Gl. (15), (20) und (23) gelten

$$\varrho_c \sim \frac{M_n}{R_n^3}\ ;\quad P_c \sim \frac{M_n^2}{R_n^4}\ ;\quad T_c \sim \frac{M_n}{R_n}\mu\beta\ .$$

Dabei sind aber M_n und R_n nicht die Masse und der Radius des Sternes selbst, sondern Masse und Radius jener Polytropen $n = 3/2$, deren innersten Teil der konvektive Kern darstellt. Würde diese Polytrope nun eine homologe Kontraktion erleiden, also nur R_n abnehmen und M_n konstant bleiben, so würde neben ϱ_c und T_c vor allem der Mittelpunktsdruck P_c stark zunehmen. Da dies nach dem oben Gesagten aber erst am Ende der Entwicklung beim Einsetzen der eigentlichen Kontraktion des gesamten Sternes zu erwarten ist, müssen wir annehmen, daß auch die Masse M_n mit wachsendem μ abnimmt. Tatsächlich sind M_n und R_n die beiden Parameter, die erst durch die an den konvektiven Kern anzupassende äußere Hülle bestimmt werden. Indem wir

$$\varrho_c \sim \mu^\delta \quad \text{und} \quad T_c \sim \mu^\tau$$

definieren, wälzen wir die ohne numerische Integration der äußeren Hülle zunächst bestehende Unkenntnis von M_n und R_n auf die Exponenten δ und τ ab und erhalten für $\beta \approx 1$

$$M_n \sim \mu^{-\frac{\delta + 3(1-\tau)}{2}} \quad \text{und} \quad R_n \sim \mu^{-\frac{1+\delta-\tau}{2}}.$$

Entsprechend der schon gemachten Annahme, daß der Stern keinen nennenswerten Gewinn an Gravitationsenergie erzielt, und daher der Mittelpunktsdruck zunächst höchstens abnimmt, aber nicht zunimmt, wollen wir voraussetzen, daß

$$0 < \tau + \delta \leq 1$$

gilt. Dabei muß bemerkt werden, daß δ und τ natürlich keineswegs konstante Werte darzustellen brauchen, sondern sich vielmehr im Laufe der Entwicklung ändern können.

---

[1] Dieser Abschnitt wurde bei der Korrektur vor der Drucklegung eingefügt.

Um nun zunächst die Änderung der Leuchtkraft mit μ abzuschätzen, nehmen wir an, daß L durch den Mittelpunktswert der Energieerzeugung $\varepsilon_c \sim X \varrho_c T_c^m$ und eine entsprechende charakteristische Masse $M_\varepsilon \sim \varrho_c (r_\varepsilon)^3$ gegeben sei, wobei r_ε die charakteristische Länge nach Gl. (11) ist

$$r_\varepsilon = \left|\frac{dr}{d\ln\varepsilon}\right| = \frac{H_c}{7.8} \quad \text{für} \quad m = 18 \quad \text{und} \quad n = \frac{3}{2}$$

und die homogene Schichtdicke H_c, d. i. die Strecke auf der der Druck auf den e-ten Teil des Mittelpunktswertes abfällt, R_n proportional ist. Damit erhalten wir

$$L \sim \varepsilon_c M_\varepsilon \sim X \varrho_c^2 T_c^m (r_\varepsilon)^3 \sim X \mu^{(m+3/2)\tau + 1/2\delta - 3/2}$$

und mit

$$\mu = \frac{4}{3 - Z + 5X}$$

$$L \sim \frac{X}{(3 - Z + 5X)^{(m+3/2)\tau + 1/2\delta - 3/2}}. \tag{69}$$

Die Differentiation von (69) nach X ergibt, daß die Leuchtkraft mit abnehmendem Wasserstoffgehalt zunimmt, solange

$$X \gtreqless \frac{2(3-Z)}{5[(2m+3)\tau + \delta - 5]}$$

ist, z. B. mit $\tau \approx \delta = 1/2$, $m = 18$, $Z = 0$ für $X \gtreqless 0.08$.

Andererseits muß wegen $X \lesseqgtr 1$ und $\delta > 0$

$$\tau > \frac{31 - 2Z - 5\delta}{5(2m+3)}$$

sein, damit ein solches Maximum in der Leuchtkraft existieren kann.

Bei der Ableitung von (69) war stillschweigend vorausgesetzt worden, daß die Energieerzeugung ganz innerhalb des konvektiven Kernes erfolgt. Man erhält Gl. (69) aber auch, indem man Gl. (26) für die Polytrope $n = 3/2$ integriert [vgl. EDDINGTON (1927), S. 101ff. und WRUBEL (1958), Gl. (38.18)]

$$L_K \sim \frac{T_c^{m+3/2} \varrho_c^{1/2}}{\mu^{3/2}} \int_0^{z_K} u^{3+m} z^2 \, dz.$$

Dabei sind u und z die Emden-Variabeln.

Wegen $T = u \cdot T_c$, $\varrho = u^n \varrho_c$ und $M_r = \left(-z^2 \frac{du}{dz}\right) \cdot M_n$ gilt für die Masse des konvektiven Kernes

$$M_K \sim \left(-z^2 \frac{du}{dz}\right)_K M_n,$$

d. h.

$$M_K \sim \left(-z^2 \frac{du}{dz}\right)_K \mu^{3/2\tau - 1/2\delta - 3/2}. \tag{70}$$

Da der Exponent von μ negativ ist, solange $\tau - \delta/3 < 1$ bleibt, wird M_K mit wachsendem μ abnehmen, sofern $\left(-z^2 \frac{du}{dz}\right)_K$ nicht zunimmt.

Diese Lage der Oberfläche des konvektiven Kernes innerhalb der polytropen Verteilung ist über die Stabilitätsbedingung (55) ebenfalls durch δ und τ gegeben, denn aus $\left[\frac{\varkappa L \nu}{(1-\beta) M_r}\right]_K = \text{const.}$ folgt für $\varkappa \sim (1+X) \frac{\varrho}{T^{3.5}}$

$$\frac{\int_0^{z_K} u^{3+m} z^2 \, dz}{u_K^{3.5} \left(-z^2 \frac{du}{dz}\right)_K} \sim \frac{(3-Z+5X)^{(m-6.5)\tau+3\delta-1}}{X(1+X)}$$

und im Falle reiner Elektronenstreuung für $\varkappa \sim (1+X)$

$$\frac{\int_0^{z_K} u^{3+m} z^2 \, dz}{u_K^{1.5} \left(-z^2 \frac{du}{dz}\right)_K} \sim \frac{(3-Z+5X)^{(m-3)\tau+2\delta-1}}{X(1+X)}.$$

Gemäß der Stabilitätsbedingung muß diese Größe nach der instabilen Seite hin, d. h. nach innen, zunehmen. Wenn sie auch mit der Zeit, also mit abnehmenden Wasserstoffgehalt zunimmt, so bedeutet das, daß sich die Stabilitätsgrenze nach kleinerem z und damit auch nach kleinerem $\left(-z^2 \frac{du}{dz}\right)$ zu verschiebt. Das ist, wie sich durch Differentiation nach X ergibt, der Fall, wenn

$$\frac{(1+2X)(3-Z+5X)}{5X(1+X)} > \begin{cases} (m-6.5)\tau + 3\delta - 1 \\ (m-3)\tau + 2\delta - 1 \end{cases} \text{bzw.}$$

gilt. Für hinreichend kleines X ist dies sicher erfüllt, für $X = 1$ aber nur so lange

$$\tau < \frac{34 - 3Z - 30\delta}{10(m-6.5)} \quad \text{für} \quad \varkappa \sim (1+X) \frac{\varrho}{T^{3.5}}$$

bzw.

$$\tau < \frac{34 - 3Z - 20\delta}{10(m-3)} \quad \text{für} \quad \varkappa \sim (1+X).$$

Indem wir diese Bedingung mit der oben für die Zunahme der Leuchtkraft erhaltenen unteren Grenze von τ kombinieren, erhalten wir als Bedingung für δ mit $Z = 0$

$$\delta < \frac{6m + 505}{50m + 155} \quad \text{bzw. für reine Streuung} \quad \delta < \frac{m + 48}{5m + 15}.$$

Das ergibt z. B. für $m = 18$ $\delta < 0.58$ und damit $0.14 < \tau < 0.30$ bzw. für reine Streuung $\delta < 0.63$ und $0.14 < \tau < 0.23$.

Wie man sieht, verschärft sich unsere eingangs aufgestellte Forderung für den Anfang der Entwicklung bei $X \approx 1$ auf $\delta + \tau < 0.88$ bzw. 0.86.

Dieses Resultat steht in guter Übereinstimmung mit den im Abschnitt b3 behandelten Modellrechnungen von KUSHWAHA (1957) und HENYEY u. a. (1958).

Auffällig ist, daß τ innerhalb verhältnismäßig enger Grenzen eingeschränkt bleiben muß, damit die beginnende Umwandlung des Wasserstoffes eine gleichzeitige Zunahme der Leuchtkraft und Schrumpfung des konvektiven Kernes zur Folge hat.

Neben der inzwischen ebenfalls durch Modellrechnungen erwiesenen Schrumpfung des konvektiven Kernes (vgl. A.I.b3) besteht noch die Möglichkeit, daß er quasi von innen her abnimmt, dadurch, daß die konvektive Durchmischung abstirbt und ein isothermer ganz aus Helium bestehender Kern entsteht, der schließlich von einer Schalenquelle der Wasserstoffumwandlung umgeben ist. Die Entstehung eines solchen isothermen Kernes wird gern damit begründet, daß die Zeitskala der konvektiven Durchmischung bei abnehmendem Wasserstoffgehalt zu lang würde verglichen mit der Zeitskala der Wasserstoffumwandlung selbst.

Diese Meinung bedarf einer Berichtigung. Aus Gl. (51) und (52) erhält man mit $\dfrac{\Re}{\mu c_p} = \dfrac{2}{5}$ und $\beta \approx 1$

$$v^3 = 10^{-1} \frac{l}{H} \cdot \frac{F_K}{\varrho} \tag{71}$$

und, indem man den Energiestrom in der Mittelpunktsregion $r \leq H$ zu

$$F_K \approx \frac{4\pi \int_0^H \varrho \varepsilon r^2 dr}{4\pi H^2} \approx \frac{H}{3} \varrho_c \varepsilon_c$$

abschätzt, für die Zeitskala τ_K der konvektiven Durchmischung

$$\tau_K^3 = \frac{l^3}{v^3} \approx 30 \frac{l^2}{\varepsilon_c}. \tag{72}$$

Die Zeitskala der Wasserstoffumwandlung ist

$$\tau_\varepsilon = \frac{X}{\dot{X}} = \frac{X M_K K_{CN}}{L} \tag{73}$$

und wegen $\varepsilon_c > L/M_K$ das Verhältnis der beiden Zeitskalen

$$\left(\frac{\tau_K}{\tau_\varepsilon}\right)^3 = \frac{30\, l^2 L^2}{M_K^2 K_{CN}^3 X^3} \tag{74}$$

oder mit $l \approx r_K = 0.169\, R$; $M_K = 0.145\, M$

$$\frac{\tau_K}{\tau_\varepsilon} \leq 10^{-10.8} \left(\frac{L}{L_\odot} \frac{M_\odot}{M} \frac{R}{R_\odot}\right)^{1/3} X^{-1}. \tag{75}$$

Das Verhältnis ist so klein, daß danach die Durchmischung des konvektiven Kernes bei abnehmendem X erst im allerletzten Moment aufhören kann, selbst wenn man noch berücksichtigen würde, daß in Wirklichkeit

$r_K < H_c$ ist, da für das Cowling-Modell der Druck an der Oberfläche des Kernes noch etwa die Hälfte des Mittelpunktdruckes beträgt.

Wir werden im nächsten Abschnitt zu zeigen versuchen, daß die Entstehung eines isothermen Kernes darauf beruht, daß durch die bei Entartung sehr hohe Wärmeleitfähigkeit des Elektronengases der zum Energietransport benötigte Temperaturgradient unter den adiabatischen Wert sinkt und dadurch mit der Konvektion auch die Durchmischung aufhört.

Die tatsächliche Existenz eines nichtdurchmischten Heliumkernes setzt dann allerdings noch voraus, daß auch in und außerhalb der Schalenquelle keine Konvektion mehr herrscht, wodurch nach Gl. (55) eine Mindestmasse für den isothermen Kern definiert ist. SCHÖNBERG und CHANDRASEKHAR (1942) erhielten für diese mit den Werten der Sonne für ein Verhältnis des inneren zum äußeren Molekulargewicht $\mu_i/\mu = 2.2$ den Wert $M_K > 0.065\,M$.

4. Entartung

Ebenso wie die Konvektion ist die bei einem hinreichend großen Verhältnis $\varrho/T^{3/2}$ auftretende Entartung des Elektronengases ein für das Schicksal eines Sternes sehr entscheidendes Phänomen, dessen Berücksichtigung eine ebenfalls nicht unerhebliche Komplikation für die Theorie des Sternaufbaues bedeutet. Denn, so wie für die Konvektion die Energietransportgleichung eine kompliziertere Form annimmt, wird bei eintretender Entartung der Druck in der Zustandsgleichung eine komplizierte Funktion der Dichte und der Temperatur, die gerade an den entscheidenden Stellen explizit numerisch berechnet werden muß.

Wir beginnen mit der im vorigen Abschnitt schon berührten Frage, inwieweit durch die Entartung des Elektronengases die Konvektion unterbunden werden kann. Als Maß für die Entartung dient der Parameter $\psi = \eta/kT$, wobei η die Gibbssche freie Energie pro Elektron ist. In den Tafeln von MCDOUGALL und STONER (1938) sind Funktionen von ψ tabuliert, mit deren Hilfe sich der Druck und die Dichte angeben lassen. Für schwache Entartung, d. h. für eine gerade bemerkbar werdende Abweichung von der idealen Zustandsgleichung, hat der Entartungsparameter Werte zwischen $\psi = -5$ und -3. Für $\psi < -1$ läßt sich die folgende Approximation benutzen [R. KIPPENHAHN u. a. (1958)]

$$\psi \approx 2.30 \log \frac{\varrho}{\mu T^{3/2}} + 18.60 \tag{76}$$

Wir haben in der Abb. 1 entsprechenden Abb. 8, die den Mittelpunktswerten $\psi_c = -5; -3$ und -1 entsprechenden Linien für das Cowling-Modell bzw. für das Eddington-Modell eingezeichnet. Man sieht, daß man in den meisten uns interessierenden Sternen oberhalb der Sonne schon zu Beginn ihrer Entwicklung mit einer schwachen Entartung rechnen kann.

Einen auch bei schwacher Entartung korrekten Ausdruck für die Wärmeleitfähigkeit hat LEE (1950) erhalten. Er gibt ein dem Wärmeleitungskoeffizienten entsprechendes $\varkappa_c$ an, das in die gewohnte Gl. (47)

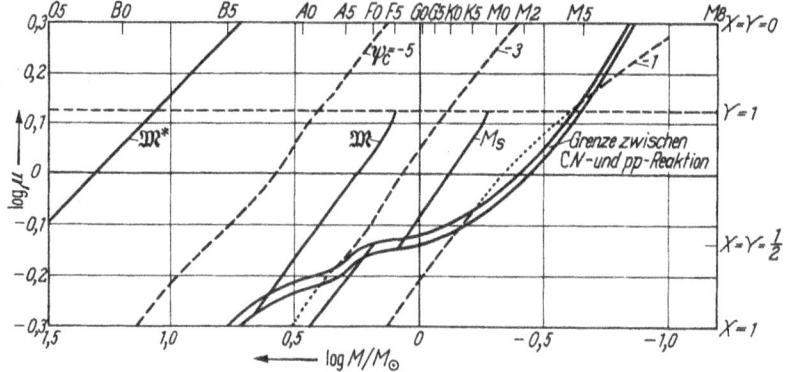

Abb. 8. Der Grad beginnender Entartung ψ_c und die Grenzen möglicher Entartung im Zentrum der Hauptreihensterne: $\mathfrak{M}^*$ für vollständige Paarentartung nach TEMESVÁRY (1954), $\mathfrak{M}$ für vollständige relativistische Elektronenentartung nach CHANDRASEKHAR (1938), M_S Grenze zwischen relativistischer und nichtrelativistischer Elektronenentartung

eingesetzt den Energietransport durch Leitung ergibt. Die Größe $\varkappa_c T$ ist dann eine reine Funktion von ψ. Wir geben in Abb. 8a das Leesche Resultat für eine chemische Zusammensetzung mit 10% Massenanteil schwerer Elemente. $\varkappa_c$ nimmt mit zunehmender Entartung rasch ab. Trotzdem ist für $\psi < -1$ der Grad der Entartung wohl noch nicht ganz ausreichend, um den Temperaturgradienten im konvektiven Kern des Cowling-Modelles unter den adiabatischen Wert zu drücken. Dabei hängt das Ergebnis sehr stark von der Wahl des Modelles ab. Ob und bei welchem Grad von Entartung die konvektive Instabilität aufgehoben wird, werden erst Modellrechnungen zeigen können, bei denen $\varkappa_c$ mitberücksichtigt wird, was bis jetzt nur für

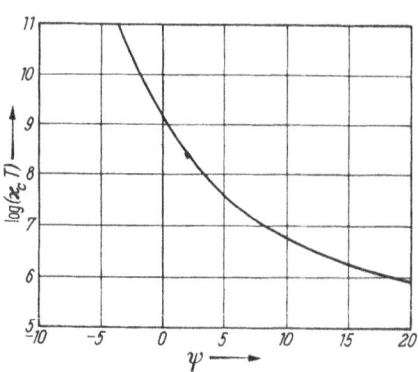

Abb. 8a. Der der Elektronenleitfähigkeit entsprechende „Absorptionskoeffizient" $\varkappa_e$ in Abhängigkeit von dem Entartungsparameter ψ nach LEE (1950)

weiße Zwerge geschehen ist. So bleibt uns zur Stütze unserer Vermutung vorläufig nur die zuerst von GAMOW (1945) gefundene Tatsache, daß der isotherme Kern teilweise entartet sein muß, da andernfalls nach einem Resultat von SCHÖNBERG und CHANDRASEKHAR (1942) für $M_K > 0.10\,M$ keine stabile Konfiguration existieren würde.

Sowohl wegen der Bedeutung, die die Schalenquellenmodelle mit teilweise entartetem isothermen Kern inzwischen erlangt haben, als auch im Hinblick auf die Frage nach dem Endzustand eines Sternes müssen wir noch den Fall der vollständigen Elektronenentartung betrachten. Hierbei ist der Druck des Elektronengases gegeben durch [vgl. S. CHANDRASEKHAR, Stellar Structure (1939)]

$$p_e = \text{const.} f(x) \tag{77}$$

mit

$$x = \text{const.} \left(\frac{\varrho}{\mu_e\,m_H}\right)^{1/3} \tag{78}$$

und

$$f(x) = (2\,x^3 - 3\,x)\sqrt{x^2 + 1} + 3\,\mathfrak{Ar\,Sin}\,x. \tag{79}$$

Das letztere ist eine Hyperbelfunktion, deren Scheitel bei $\varrho/\mu_e = 10^{6.0}$ gcm^{-3} liegt. μ_e ist das Molekulargewicht pro Elektron. Die asymptotischen Näherungen geben

$$p_e \sim \left(\frac{\varrho}{\mu_e}\right)^{5/3} \tag{80}$$

für die nichtrelativistische Elektronenentartung und

$$p_e \sim \left(\frac{\varrho}{\mu_e}\right)^{4/3} \tag{81}$$

für den relativistischen Fall. Setzt man für den Gesamtdruck im Zentrum des Sternes

$$P_c = \chi\,\frac{G M^2}{4\pi R^4} \tag{82}$$

an, wobei χ nur von der Dichteverteilung im Stern abhängt, und bezeichnet man mit $\xi = \varrho_c/\varrho_m$ das Verhältnis der Zentraldichte zur mittleren Dichte, dann erhält man für die Grenze zwischen nichtrelativistischer und relativistischer Entartung die Masse [G. W. WARREN (1944)]

$$M_S\,\mu_e^2 = 1.55\,M_\odot\,\frac{\xi^2}{\chi^{3/2}}. \tag{83}$$

In der gleichen Weise erhält man nach CHANDRASEKHAR (1939) eine Grenzmasse, oberhalb deren selbst im Zentrum eines Sternes auch bei Berücksichtigung des Strahlungsdruckes keine vollständige Entartung mehr eintreten kann

$$\mathfrak{M}\,\mu_e^2 = 3.66\,M_\odot\,\frac{\xi^2}{\chi^{3/2}}. \tag{84}$$

Beide Grenzen haben wir in Abb. 8 eingezeichnet.

Die Masse $\mathfrak{M}$ ist auf jeden Fall eine unübersteigbare Grenze für die Entstehung eines vollständig entarteten Sternes wie z. B. eines Weißen

Zwerges. Der früher aus ihr gezogene Schluß, daß ein Stern mit größerer Masse, der seine Energiequellen erschöpft hat und sich kontrahieren muß, zu einem überdichten Neutronenstern kollabieren würde, weil der Druck des entarteten Gases dem Gravitationsdruck nicht mehr das Gleichgewicht halten könne, ist hinfällig geworden, nachdem TEMESVÁRY (1952) gezeigt hat, daß an dem kritischen Punkt die Mittelpunktstemperatur des Sternes bereits so hoch ist, daß Elektron-Positron-Paare entstehen, deren Druck das Gleichgewicht aufrecht erhält. Das heißt, bei weiterer Kontraktion des Sternes tritt eine andere Entartung der Zustandsgleichung in dem Sinne auf, daß im Grenzfall der vollständigen Paarentartung nach KOPPE (1948) der Druck nur noch von der Temperatur abhängt

$$p_e = 2 \cdot \frac{7}{8} \cdot \frac{a T^4}{3}. \tag{85}$$

Die zugehörige Masse für vollständige Paarentartung

$$\mathfrak{M}^* \mu^2 = 15.18 \, M_\odot \frac{\xi^2}{\chi^{3/2}} \tag{86}$$

ist ebenfalls in Abb. 8 eingezeichnet worden.

Für Sterne, die zum größten Teil aus vollständig entarteter Materie bestehen wie die Weißen Zwerge, ergibt sich zwangsläufig aus Gl. (80) und (82) eine Relation zwischen der Masse, der mittleren Dichte und dem Molekulargewicht pro Elektron, die dazu benutzt werden kann, aus den beobachteten Werten für die Masse und den Radius den Wasserstoffgehalt zu bestimmen. LEE (1950) fand so mit einer etwas differenzierteren Methode, daß der Wasserstoffgehalt der Weißen Zwerge kleiner als $1/2\%$ sein muß. Für Sirius B fand er, daß $X = 10^{-3}$ ausreicht, um die beobachtete Leuchtkraft zu erklären. Dabei wird angenommen, daß die Energieerzeugung in der Hülle des sonst ganz aus Helium bestehenden Sternes erfolgt. Die zugehörige Lebensdauer würde $3 \cdot 10^{10}$ Jahre betragen.

5. Durchmischung

Alle unsere bisherigen Überlegungen — und wie wir sehen werden, fast alle modelltheoretischen Rechnungen — beruhen auf der Annahme, daß die Materie der Sterne nicht durchmischt wird, außer in den Konvektionszonen. Nach dem bekannten von Zeipelschen Theorem hat jedoch die Rotation eines Sternes meridionale Zirkulationen zur Folge, die auch die im Strahlungsgleichgewicht befindlichen Teile des Sternes durchmischen werden, wenn die Zirkulationsgeschwindigkeit groß genug ist. Die Ermittlung dieser Geschwindigkeit bereitet einige Schwierigkeiten: Ist ω die Winkelgeschwindigkeit der Rotation eine Funktion des Ortes im Stern, dann bedeutet ein bestimmtes Rotationsgesetz

ω (r, ϑ) in der Regel eine ganz bestimmte Störung des thermodynamischen Gleichgewichtes, das durch eine entsprechende meridionale Zirkulation wieder in Ordnung gebracht werden soll. Ist die Geschwindigkeit dieser Zirkulation groß genug, so kann sie unter Umständen die Verteilung ω (r, ϑ) selbst wieder beeinflussen. Indem man von diesen Rückwirkungen absieht und die Bewegungsgleichung linearisiert, läßt sich nach SWEET (1950) die Zirkulationsgeschwindigkeit abschätzen. Wir geben in Tab. 3 das Verhältnis der Zeitskala der Wasserstoffumwandlung zu der nach der Sweetschen Theorie ermittelten Zeitskala t_m der meridionalen Zirkulation. Danach müßten wir für die frühen Typen mit Durchmischung rechnen. Da in der Kernregion des Sternes durch die Energieerzeugung das Molekulargewicht zunimmt und eine solche Zirkulation das schwere Material herausschaffen müßte, entsteht durch die so bedingte Asymmetrie in der Verteilung von μ eine neue thermodynamische Störung, die, wie MESTEL (1953) und (1957) gezeigt hat, der ursprünglichen rotationsbedingten Zirkulation entgegenwirkt. In diesem Ergebnis der Mestelschen Untersuchungen liegt der Grund dafür, daß man z. Z. in der allgemeinen Theorie der Sternentwicklung das Auftreten einer rotationsbedingten Durchmischung glaubt verneinen zu dürfen. MESTEL selbst bemerkt dazu, daß der Umstand, daß Objekte gleichen Alters, wie etwa die Sterne eines offenen oder Kugelhaufens, eine lineare Verteilung im H-R-Diagramm zeigen, darauf schließen läßt, daß neben der Masse kein zweiter Parameter, wie etwa die Rotation, bedeutsam sein könne. Dabei müßte man freilich sicher sein, daß der spezifische Drehimpuls von der Entstehung der Sterne her nicht mit der Masse korreliert ist.

Da alle diese Abschätzungen nur auf einer genäherten Theorie der Sternrotation beruhen, bleibt ein gewisses Gefühl der Unsicherheit. Es ist das Verdienst von KIPPENHAHN (1958) gezeigt zu haben, wie weit die unbedingte Anwendbarkeit einer solchen linearisierten Theorie erster Ordnung reicht, nämlich etwa so weit, wie das Verhältnis χ/α der Zeitskala der Reibung zur Umlaufperiode der meridionalen Zirkulation < 1 bleibt. Dabei ist $\chi = \overline{\omega}^2 / 2\pi G \bar{\varrho}$ und $\alpha = \bar{\nu} \dfrac{GM^2}{LR^3}$ mit der kinematischen Zähigkeit ν (vgl. Tab. 3). Man sieht, daß die hinreichende Bedingung für die Anwendbarkeit der linearisierten Theorie erster Ordnung gerade für die interessierenden Fälle nicht erfüllt ist. Andererseits zeigt die Kippenhahnsche Untersuchung, daß je nach dem sich einstellenden Rotationsgesetz die Zirkulation auch in mehrere Strömungssysteme zerfallen kann, was unter Umständen wieder eine Verringerung der Durchmischung des Sternes zur Folge haben könnte.

Die kaum zu bezweifelnde Existenz stellarer Magnetfelder macht andererseits alle Schlüsse noch unsicherer, da sich deren möglicher

Einfluß auf die Verteilung des Drehimpulses im Stern bisher nur in groben Zügen überblicken läßt. Insgesamt kann man von der weiteren Entwicklung der hydromagnetodynamischen Theorie noch Resultate erwarten, die unsere bisherigen Vorstellungen recht beträchtlich werden modifizieren können. Wir werden im Teil C noch auf die Frage des Drehimpulsverlustes, den der Stern außer durch den Strahlungsstrom und die Korpuskularstrahlung auch durch die evtl. Wechselwirkung der Magnetfelder mit dem umgebenden Medium erleidet, zurückkommen. Hier sei nur ein spezieller Punkt noch kurz berührt. Eine solche äußere Abbremsung kann unter Umständen das Auftreten einer Rotationsinstabilität zur Folge haben, die zwar, weil davon vorwiegend die äußeren Teile des Sternes betroffen werden, für die Frage der Durchmischung keine besondere Bedeutung haben wird, aber möglicherweise die Effektivität der gerade bei den schneller rotierenden frühen Typen zunehmend dünner werdenden WKZ etwas erhöht.

Tabelle 3. *Verhältnis der Zeitskala der Wasserstoffumwandlung zu der der Eddington-Sweetschen-Zirkulationen nach* STRÖMGREN (1952) *sowie der der Reibung zu der der Zirkulationen nach* KIPPENHAHN (1958)

Sp	$\bar{v}_{Rot.}$	t_{CN}/t_m	$\log (\chi/\alpha)$
Oe—Be	350 km/sec	5	2.86
O—B	94 km/sec	0.4	1.50
A	112 km/sec	1.4	3.56
F0—F2	51 km/sec	0.4	3.04
F5—F8	20 km/sec	0.1	2.12
dG	0	0	$-\infty$

A.I.b) Gegenwärtiger Stand der Modelltheorie

In den vorigen Abschnitten haben wir gezeigt, daß eine Reihe von Fragen nur mit numerischen Methoden zu lösen sein wird. Auf dem Gebiet der Berechnung von Sternmodellen ist in den letzten Jahren ein außerordentliches Maß an Arbeit geleistet worden, das nicht zuletzt der Entwicklung der elektronischen Rechenmaschinen zu danken ist. Eine vollständige Würdigung dieses ganzen Gebietes würde im Rahmen unseres Themas einen zu breiten Raum einnehmen. Darum wollen wir in den folgenden Abschnitten nur anhand einiger, für die einzelnen Sternmassen typischer Modelle die für unser Problem wichtigen Fragen diskutieren. Ein guter Teil der in einzelnen Fällen zu übenden Kritik ist bereits in den vorigen Abschnitten enthalten und braucht hier nicht wiederholt zu werden. Allgemein möchten wir der Erwartung Ausdruck geben, daß mit zunehmender Verwendung programmgesteuerter Rechenmaschinen die bisher noch vorherrschende Methode, Sternmodelle zonenweise aus quasianalytischen Lösungen zusammenzusetzen, von der einer durchgehenden, möglichst expliziten Integration abgelöst werden wird, da hierbei das Ineinandergreifen der verschiedenen physikalischen Bedingungen und Phänomene viel offenkundiger zutage tritt.

1. Sterne vom Typ der Sonne und kleinere Massen

Für Rote Zwerge $M < 0.9 M_\odot$ hat OSTERBROCK (1953) eine Reihe von Modellen gerechnet, bei denen die WKZ mit berücksichtigt wurde. Im Gegensatz zu früheren, ohne äußere Konvektionszone gerechneten Modellen ergab sich nun für die pp-Reaktion eine mit der beobachteten Leuchtkraft verträgliche normale chemische Zusammensetzung. Der Gehalt an Elementen schwerer als Helium beträgt nur einige Prozent. Wegen der niedrigen effektiven Temperatur ($< 5000°$) und höheren atmosphärischen Dichte dürfte der Einfluß der nichtadiabatischen Teile der WKZ vernachlässigbar sein. Die geometrische Dicke der WKZ beträgt etwa 30% des Sternradius. Lediglich für den kühlsten Stern, Kr60A mit $T_e = 2940°$ ergab sich ein sehr niedriger Wasserstoffgehalt. Neben der Unsicherheit der bolometrischen Korrektur wird als Grund beginnende Entartung in Erwägung gezogen.

Von den Versuchen, ein Modell der Sonne zu erhalten, ist der interessanteste ein von SCHWARZSCHILD u. Mitarb. (1957b) gerechnetes Modell, das außer der WKZ auch die Änderung der chemischen Zusammensetzung durch die Energieerzeugung während der bisherigen Lebenszeit der Sonne berücksichtigt. Wenn man die Masse der Sonne, ihre ursprüngliche homogene chemische Zusammensetzung, die gleich der heutigen der äußeren Teile ist, und ihr Alter vorgibt, für das SCHWARZSCHILD $5 \cdot 10^9$ Jahre annimmt, dann ist nach dem Eindeutigkeitssatz der heutige Zustand der Sonne bereits festgelegt. Da außer der Masse und der äußeren chemischen Zusammensetzung noch der Radius und die Leuchtkraft als beobachtbare Größen zur Verfügung stehen, würden sich grundsätzlich noch zwei Größen adjustieren lassen. Das können Ausgangsgrößen sein, z. B. das Alter oder der H- und He-Gehalt. Es können aber auch andere Parameter sein, deren Wert noch problematisch ist, z. B. die Konstante im Energieerzeugungsgesetz oder die Länge des Mischungsweges. SCHWARZSCHILD hat nach seiner Methode, bei der die Entropie der WKZ als freier Parameter eingeht, praktisch nur noch einen Parameter frei und dafür den Heliumgehalt gewählt. Er erhielt für $Z = 2\%$, $Y = 20\%$. In diesem Modell wurde der nichtadiabatische Teil der WKZ nicht berücksichtigt. Indem SCHWARZSCHILD den von ihm erhaltenen Wert für die Entropie im adiabatischen Teil der WKZ mit den Werten vergleicht, die VITENSE (1953) für $l = H$ und $l = 2H$ erhielt, und findet, daß sein Entropiewert ebenso wie die Dicke seiner WKZ mit etwa 125000 km zwischen den beiden Werten von VITENSE (66 bzw. 160000 km) liegt, schließt er, daß $l \approx 1.5 H$ sein sollte. Der Vitenseschen Rechnung fehlt zwar die von der Vollständigkeit des Modelles, d. h. von der Erfüllung der inneren Randbedingungen herrührende Information, dafür ist sie aber auch nicht mit den Unsicherheiten der Absorptions- und Energieerzeugungsgesetze im Innern behaftet. SCHWARZSCHILDS Entropiewert für die WKZ ist dagegen zu einem guten Teil von der inneren Grenze der WKZ her bestimmt. So erscheint diese Kombination als ein vielleicht ganz angemessenes Verfahren zur Bestimmung der problematischen Größe von l, nur unterscheidet sich der äußere Heliumgehalt in den beiden Rechnungen um einen Faktor 2.

Hervorzuheben ist noch, daß nach SCHWARZSCHILDS Rechnung die Helligkeit der Sonne während der $5 \cdot 10^9$ Jahre ihres angenommenen Alters mit der zunehmenden Inhomogenität ihrer chemischen Zusammensetzung um eine halbe Größenklasse heller geworden ist. Danach müßte vor etwa zwei Milliarden Jahren, zur Zeit des Praekambriums die Oberflächentemperatur der Erde in der Nähe des Gefrierpunktes gelegen haben. Allgemein entspricht dieses Resultat nicht nur der in A.I.a begründeten theoretischen Erwartung, sondern würde, wenn es sich noch durch erdgeschichtliche Konsequenzen bestätigen läßt, bemerkenswerterweise das erste, direkt erhaltene, quantitative Datum aus der Geschichte der Sonne darstellen und um so bemerkenswerter sein, als es sich bei dieser kosmogonisch offensichtlich um ein Objekt an der Grenze der Datierbarkeit handelt.

2. Schalenquellenmodelle

Die von HOYLE und SCHWARZSCHILD (1955) und HASSELGROVE und HOYLE (1956) durchgeführten Rechnungen zur Entwicklung von Schalenquellenmodellen mit teilentartetem isothermen Kern, die für festgehaltene Sternmasse eine Serie quasistationärer Modelle mit größer werdendem Heliumkern als Entwicklungslinie einer festen Masse im H-R-Diagramm ergeben, machen nicht nur die Existenz von Riesensternen mit nicht viel mehr als Sonnenmasse verständlich, sondern konnten vor allem erstaunlich gut das recht komplizierte H-R-Diagramm der Kugelhaufen M 3 und M 92 erklären. Charakteristisch für dieses ist, daß nach dem Rechtsabmarsch von der Hauptreihe, den alle Sternhaufen mitmachen, bei einer effektiven Temperatur von etwa 5500° ein plötzlicher Anstieg zu höheren Leuchtkräften erfolgt. Die Autoren glauben diesen Anstieg durch den eben bei dieser Oberflächentemperatur wirksam werdenden Einfluß der WKZ erklären zu können. Dieses Resultat ist neuerdings sowohl durch eine Untersuchung von E. BÖHM-VITENSE (1958) über die WKZ solcher Modelle mit stark verdünnter Atmosphäre wie auch durch eine Wiederholung der Hoyle-Schwarzschildschen Rechnung [KIPPENHAHN u. a. (1958)] in Frage gestellt worden. In der letzteren Arbeit wurden die gleichen Werte für die Masse (1.2 $M_\odot$) und die chemische Zusammensetzung benutzt und auch sonst die gleichen physikalischen Bedingungen angenommen wie in der Originalarbeit. Nur in zwei Punkten unterscheidet sich die zweite Rechnung von der ersten. Einmal erfolgte die Integration von einer vorgegebenen Grenztemperatur an kontinuierlich bis zu dem Fit an der Oberfläche des Kernes, die durch die doppelte Bedingung der eingetretenen Isothermie und einer hinreichenden Entartung definiert ist; und zum anderen wurde dabei die nichtadiabatische Schichtung der WKZ explizit gerechnet. Das Resultat zeigt Abb. 9. In der Nähe der Hauptreihe ist die Übereinstimmung zunächst recht befriedigend, wenn man berücksichtigt, daß die Originalrechnung (.) hier für 1.1 $M_\odot$ ausgeführt wurde. Anstatt aber nach oben abzubiegen, folgt die neue Entwicklungslinie mehr den früher für Riesen der Population I erhaltenen Resultaten (+). Die Autoren weisen nach, daß die Abweichung der Resultate auf die in A.I.a 3 behandelten Unterschiede in der Berücksichtigung des nichtadiabatischen Teiles der WKZ zurückzuführen ist, der für diese Modelle mit sehr viel dünnerer Atmosphäre einen bedeutend größeren Einfluß hat als für die Hauptreihensterne der

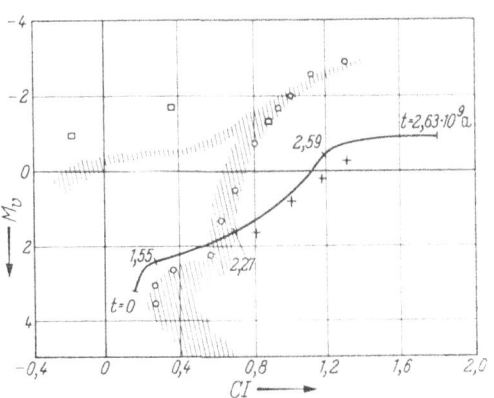

Abb. 9. Entwicklungssequenz eines Schalenquellenmodelles mit 1.2 $M_\odot$ der Population II nach KIPPENHAHN, TEMESVÁRY, BIERMANN (1958) mit den Resultaten von HOYLE und SCHWARZSCHILD (1955) für ● 1.1 $M_\odot$, ○ 1.2 $M_\odot$, □ 1.2 $M_\odot$ der Population II, c. f. l. c. Fig. 5; + 1.1 $M_\odot$ der Population I l. c. Fig. 4

gleichen effektiven Temperatur. Dabei ist die horizontale Abstrahlung der Turbulenzelemente noch nicht berücksichtigt, die nach BÖHM-VITENSE (l. c.) die Effektivität der WKZ noch weiter herabsetzen soll.

Für unser Problem ergibt sich die folgende Konsequenz: Während nach der Hoyle-Schwarzschildschen Rechnung an der Spitze der Entwicklungslinie die Temperatur in dem Kern des Sternes so hoch wird, daß die Heliumverbrennung einsetzt und sich die Entwicklung eines doppelten Schalenquellenmodelles nach links, zur Hauptreihe zurück wahrscheinlich machen ließ, verschwindet die neue Entwicklungssequenz bei einer effektiven Temperatur von 2800° praktisch in einem der Beobachtung nicht mehr zugänglichen Bereich, ohne daß die Zentraltemperatur, mit $T_c = 44 \cdot 10^6$ Grad, den für die Heliumumwandlung nötigen Wert erreicht hätte. Der Massenanteil des Heliumkernes ist mit etwa 50% der gleiche wie am Umkehrpunkt der ersten Rechnung. Die Zeitskala ist infolge der niedrigeren Leuchtkraft etwas länger.

Da die erhaltenen Lösungen immer empfindlicher von allen den Einfluß der WKZ auf den Aufbau des Sternes bestimmenden Größen, wie z. B. dem Mischungsweg abhängen, je weiter man im H-R-Diagramm von etwa 5500° aus nach rechts geht, wird die Klärung des weiteren Schicksals dieser Sterne noch einige Mühe bereiten.

3. Massive Sterne

Wir fragen nun, für welchen Bereich der Sternmasse das zuletzt behandelte Schalenquellenmodell von aktueller Bedeutung sein wird. Zunächst ist leicht einzusehen, daß dies für kleinere Massen nicht der Fall sein dürfte. Zwar würde nach hinreichend langer Zeit wohl in jedem Stern unterhalb der Sonnenmasse nur noch ein isothermer Kern vorhanden sein, wie das in guter Annäherung bei den Weißen Zwergen der Fall ist; es ist heute nur noch nicht soweit, denn die gegenwärtige Sonne enthält in ihrem Zentrum noch immer 30% Wasserstoff, obwohl als ihr Alter das Weltalter angenommen worden ist [SCHWARZSCHILD (1957)]. Nach den größeren Massen zu ist, wie die Chandrasekharsche Grenzmasse $\mathfrak{M}$ zeigt, eine Begrenzung des Bereiches der Schalenquellenmodelle dadurch gegeben, daß keine hinreichende Entartung mehr auftritt. Dann wird, wie wir in A.I.a3 gesehen haben, mit abnehmendem X_c der konvektive Kern schrumpfen und dabei ein Gebiet mit kontinuierlich nach innen abnehmendem Wasserstoffgehalt zurücklassen. Es war daher eine Frage an die Modelltheorie, welchen Verlauf diese Entwicklung weiterhin nehmen wird. Solche Entwicklungssequenzen sind u. a. von TAYLER (1951b) und (1956) und zuletzt von KUSHWAHA (1957) gerechnet worden. Wir ziehen aus den Resultaten der letzteren Arbeit die folgenden für unser Problem interessanten Schlüsse:

1. Der Entartungsparameter der chemisch homogenen Anfangsmodelle ist selbst bei der kleinsten der gerechneten Massen (2.5 $M_\odot$) mit $\psi_c < -2$ zu klein, als daß der konvektive Kern von innen her absterben könnte.

2. Im Laufe der Entwicklung nimmt ψ stetig ab. Das heißt, eine Schalenquelle kann nur entstehen, wenn die Durchmischung im Zentrum des konvektiven Kernes von Anfang an unterbunden wird.

3. Mit der Leuchtkraft wächst zunächst auch der Radius des Sternes, und die Entwicklung verläuft im H-R-Diagramm nach rechts oben. Wenn jedoch der konvektive Kern soweit geschrumpft ist, daß die außerhalb des Kernes erzeugte Energie einen merkbaren Beitrag zur Gesamtleuchtkraft liefert, erreicht der Sternradius ein Maximum, und die Entwicklung verläuft bei weiter zunehmender Leuchtkraft im H-R-Diagramm nach links oben auf die Hauptreihe zurück. Es will uns scheinen, daß dieser Umkehrpunkt um so näher an der Hauptreihe liegt, je größer die Masse des Sternes ist. Gerechnet wurde er nur für den größten Massenwert (10 $M_\odot$).

4. Die Ursache für diese Richtungsumkehr der Entwicklungslinie und den damit verbundenen Wiederanstieg der effektiven Temperatur sieht Kushwaha in dem durch die rasche Abnahme des Wasserstoffgehaltes im konvektiven Kern, der an der Umkehrstelle von 10% auf 2% absinkt, bedingten Anstieg der Mittelpunktstemperatur. Das heißt aber unserer Meinung nach nichts anderes, als daß wir, nachdem die eigentliche Kernregion „ausgebrannt" ist, einen echt kontrahierenden Stern vor uns haben, wie nicht anders zu erwarten war. Bemerkenswert ist noch, daß der anfängliche Massenanteil des konvektiven Kernes und damit die bei Einsetzen der Kontraktion verbleibende Region gleichmäßiger verteilter Energiequellen für einen Stern von $10\,M_\odot$ 24% beträgt.

Anmerkung bei der Korrektur vor der Drucklegung

Inzwischen ist uns durch die freundliche Vermittlung von Herrn Prof. Böhm, Kiel, dem dafür herzlich gedankt sei, der Vorabdruck einer Arbeit von Henyey, Le Levier und Levee (1958) bekannt geworden, die die Entwicklung von Sternen mit $1.5 - 20\,M_\odot$ umfaßt. Bis auf Unterschiede, die sich auf den mit $X_0 = 0.68$ niedriger angenommenen ursprünglichen Wasserstoffgehalt zurückführen lassen, stimmen die Ergebnisse mit denen Kushwahas ($X_0 = 0.90$) gut überein. Die Autoren stellen fest, daß während der Umwandlung des im konvektiven Kern enthaltenen Wasserstoffes die Mittelpunktsdichte um etwa 50% und die Mittelpunktstemperatur um etwa 25% zunimmt, und daß diese prozentuale Änderung über den ganzen untersuchten Bereich von Sternmassen erstaunlich konstant ist. Gemäß unserer im Abschnitt a 3 gegebenen Definition entspricht dies einem $\delta = 0.54$ und einem $\tau = 0.30$. Die gleiche Feststellung läßt sich auch aus Kushwahas Resultaten ablesen.

Eine notwendige Korrektur erfuhr die Ausdehnung dieser Ergebnisse und unserer daran geknüpften Betrachtungen auf noch größere Sternmassen durch eine kürzlich erschienene Arbeit von Schwarzschild und Härm (1958), die die Entwicklung von Sternen mit $30 - 200\,M_\odot$ behandelt. Dabei wurde der Strahlungsdruck mit berücksichtigt und insbesondere sein Einfluß auf den adiabatischen Gradienten in Rechnung gesetzt. Wie man aus Gl. (56) ersieht, hat eine Zunahme von $(1 - \beta)$ eine Abnahme von V_{ad} zur Folge. Damit wird aber die gleichzeitige stabilisierende Wirkung einer Zunahme von $(1 - \beta)$ in dem Term auf der linken Seite der Konvektionsbedingung (55) verringert. Als Folge davon zeigt der konvektive Kern in den von Schwarzschild und Härm gerechneten Modellen die Tendenz, der Masse nach zuzunehmen. Das würde aber, wie wir bereits in a 3 gesehen haben, eine Diskontinuität der chemischen Zusammensetzung zur Folge haben. Nun ist bei den im Innern so massereicher Sterne herrschenden Temperaturen der Absorptionskoeffizient allein durch die Elektronenstreuung gegeben und würde daher auf der stabilen Seite einen größeren Wert haben als innerhalb der Konvektionsgrenze. Da dies mit der Bedingung (55) nicht verträglich ist, bildet sich bei gleichzeitiger Schrumpfung des eigentlichen konvektiven Kernes ein der Masse nach zunehmendes Übergangsgebiet aus, in dem der Wasserstoffgehalt von seinem ursprünglichen Wert auf den des konvektiven Kernes abnimmt. Insgesamt steht dem Stern so ein größerer Vorrat an Wasserstoff zur Verfügung, wodurch seine Lebenszeit nicht unbeträchtlich heraufgesetzt wird. Außerdem verlaufen die Entwicklungslinien im H-R-Diagramm sehr viel flacher und weiter nach rechts, so daß unsere oben an Kushwahas Resultate unter 3. geknüpfte Vermutung, daß die

Umkehr um so näher bei der Hauptreihe erfolgt, je größer die Masse ist, nicht zutrifft. Als Grenzmasse, oberhalb deren dieses neue Entwicklungsschema einsetzt, ergaben sich 10 $M_\odot$.

4. Sterne besonderer chemischer Zusammensetzung

Im Zusammenhang mit der Frage, wie sich die gegenseitige Lage der Hauptreihe der Populationen I und II fixieren läßt, spielt das Problem eine Rolle, ob die Position der Subzwerge vom Typ F im H-R-Diagramm durch einen Mangel an schwereren Elementen erklärt werden kann. Zu diesem Zwecke hat REIZ (1954) eine Modellrechnung für Sterne mit vernachlässigbarem Gehalt an schweren Elementen durchgeführt. Das benutzte Verfahren beruht weitgehend auf Homologietransformationen. Das Resultat liefert Modelle, die für $Z = 0$ und $X = 0.6$ bis 0.9 etwa $1^{m}\!.5$ unter der Hauptreihe im Bereich der Spektralklassen F0—F6 liegen, und damit etwa der Lage der in Frage stehenden Subzwerge entsprechen. Obwohl die angenommenen Massenwerte klein sind (0.84—$0.3\, M_\odot$), spielt die Entartung gerade noch keine Rolle. Die den angegebenen Mittelpunktswerten für Temperatur und Dichte entsprechenden Werte des Entartungsparameters liegen bei $\psi = -0.8$. Ob der Einfluß der WKZ trotz der relativ hohen effektiven Temperatur von etwa 7000° das Ergebnis dieser Analyse, bei der es empfindlich auf die Bestimmung der Leuchtkraft ankommt, nicht doch etwas verändern kann, wird noch einer besonderen Untersuchung bedürfen.

A.I.c) Vergleich und Anwendung

1. Kritische Zusammenfassung

Unsere bisherigen Betrachtungen betrafen im allgemeinen nur die mittlere, durch die Freisetzung der Kernenergie bestimmte Phase der gesamten Sternentwicklung. Lediglich in den Abschnitten b2 und 3 konnten wir die Entwicklung der Sterne noch ein wenig in Richtung auf ihren dritten und letzten Lebensabschnitt hin verfolgen. Über diesen kann die Theorie zur Zeit noch keine zusammenhängenden und detaillierten Aussagen machen. Das hat folgende Gründe:

1. Obwohl allen Sternen gemeinsam ist, daß sie mit zunehmender Inhomogenität der chemischen Zusammensetzung die Hauptreihe unter einer Expansion der äußeren Teile bei wachsender Leuchtkraft verlassen, ist ihr weiterer Weg noch nicht im gleichen Maße gesichert. Die hinreichend massiven Sterne scheinen, sowie der Wasserstoff im geschrumpften konvektiven Kern nahezu aufgebraucht ist, unmittelbar in eine Kontraktionsphase einzutreten, wobei die Energieerzeugung in den außerhalb liegenden, noch nicht ganz erschöpften Teilen weiterläuft. Dabei ist noch nicht gewiß, ob es sich bereits um den Beginn der letzten,

endgültigen Kontraktion handelt. Bei den weniger massiven Sternen, die einen isothermen Kern mit nach außen wandernder Schalenquelle entwickeln, ist noch unklar, wie und wo sie in die Endphase ihrer Entwicklung eintreten.

2. Die Kontraktion als eigentliche Endphase ist noch nicht explizit gerechnet worden. Ältere Arbeiten, die eine homologe Kontraktion voraussetzen, können im Rahmen unseres Problemes nicht mehr als eine ausreichende Darstellung der Entwicklung angesehen werden.

3. Sowie bei der Kontraktion die Temperatur und Dichte im Sternmittelpunkt über die etwa der Salpeterreaktion entsprechenden Werte hinaus wachsen, ergeben sich eine Reihe noch nicht hinreichend gelöster kernphysikalischer und thermodynamischer Probleme, die unter anderem den Einfluß der Paarerzeugung und der unter Umständen zunehmenden Neutrinoverluste auf die Stabilität des Sternes betreffen.

4. Zur Frage eines möglicherweise unstetigen Endes der dritten Entwicklungsphase haben sich keine wesentlichen neuen Gesichtspunkte ergeben, außer denen, daß die Supernovae zur Erklärung bestimmter Abschnitte innerhalb der Theorie der Synthese schwerer Elemente benötigt werden und daß für die Erklärung der gewöhnlichen Novae nach einem Gedanken von J. BINGE (private Mitteilung) das jetzige Bild der kernchemischen Entwicklung der Sterne ausgezeichnet zu den Voraussetzungen einer älteren Theorie von BIERMANN (1939) paßt, nach der sich in kontrahierenden Sternen, die vorwiegend aus Elementen schwerer als Helium bestehen, Instabilitätszonen aufbauen können, deren plötzliches Umklappen einen Nova-Ausbruch ergeben würde.

Zusammenfassend läßt sich nach alledem bezüglich der Lebensdauer der Sterne mit mehr als einer Sonnenmasse folgendes sagen:

1. Es kann als gesichert gelten, daß die Sterne etwa 10 bis 20% ihrer Wasserstoffmasse in der Nähe der ursprünglichen, das heißt für eine homogene chemische Zusammensetzung geltenden Hauptreihe umwandeln.

2. Da mit zunehmender Inhomogenität der chemischen Zusammensetzung die Leuchtkraft zunimmt, kann die Zeitskala t_{CN} als obere Grenze für die Dauer dieser Entwicklungsphase angesehen werden, sofern sie mit dem Anfangswert der Leuchtkraft ermittelt wurde.

3. Da das der ersten Entwicklungsphase des Sternes entsprechende Kontraktionsalter wesentlich kürzer ist als die Zeitskala t_{CN}, kann diese auch als obere Grenze für die gesamte Lebensdauer des Sternes angesehen werden, vorausgesetzt, daß die Entwicklung in der dritten und letzten Phase ebenfalls wesentlich rascher verläuft als die Umwandlung der ersten 10—20% der Wasserstoffmasse.

4. Da in vielen Fällen nicht so sehr das Alter eines einzelnen Sternes interessieren wird als vielmehr das einer ganzen Gruppe von Sternen,

wie etwa das eines Sternhaufens, kann zu diesem Zweck ein von v. HOERNER (1957) angegebenes Verfahren benutzt werden, das zur Altersbestimmung offener Sternhaufen lediglich die Spektralklasse der frühesten im Haufen vorkommenden Sterne verwendet, genauer gesagt die Lage der linken Kante der nach rechts abbiegenden Hauptreihe des Haufens im HR-Diagramm. Die für diese Methode geltenden Voraussetzungen entsprechen völlig den von uns für t_{CN} gemachten, vor allem der, daß an der Stelle des Abbiegens von der Hauptreihe stets der gleiche Bruchteil q der Wasserstoffmasse verbraucht sei. Tab. 4a enthält das in dieser Weise dem Spektraltyp des sogenannten Knies entsprechende Alter.

Tabelle 4a. *Altersbestimmung aus dem Spektraltyp am Knie der Hauptreihe nach* v. HOERNER (1957)

Sp	M_v	$\mathfrak{M}$	t
O 5	−5.5	39 $\mathfrak{M}_\odot$	4.7 · 10⁶ Jahre
B 0	−4.5	20	4.6 · 10⁶ Jahre
B 5	−2.2	6.7	4.6 · 10⁷ Jahre
A 0	−0.2	3.5	3.2 · 10⁸ Jahre
A 5	+1.3	2.2	1.2 · 10⁹ Jahre
F 0	+2.3	1.7	2.7 · 10⁹ Jahre
F 3	+3.0	1.26	3.8 · 10⁹ Jahre
—	—	1.02	7.8 · 10⁹ Jahre

Beim Vergleich mit den Werten für t_{CN} nach der Tab. 2 ist zu beachten, daß t_{CN} die Lebenserwartung eines Sternes ist, der sich bei gleichem Spektraltyp auf der ursprünglichen Hauptreihe befindet, während die Werte der Tab. 4a das bei der Entwicklung bis zum Knie erreichte Alter darstellen.

Auch die nachstehenden, durch die in A.I.b besprochenen expliziten Rechnungen sich ergebenden Alter passen sich innerhalb der einem Faktor ≈ 2 entsprechenden allgemeinen Unsicherheit gut ein:

Tabelle 4b. *Alter aus den Modellrechnungen*

$M/M_\odot$	q	t [10⁹ Jahre]	Autor
10	0.15	0.034	⎫
5	0.12	0.156	⎬ KUSHWAHA (1957)
2.5	0.10	0.924	⎭
1.2	0.22	2.0 + 2.3 = 4.3	KIPPENHAHN u. a. (1958)
1.1	0.19	6.2	HOYLE-SCHWARZSCHILD (1955)

Zum Schluß müssen wir noch auf einen Punkt hinweisen, der vielleicht am schärfsten den gegenwärtigen Stand und die Art unserer Kenntnisse über das Alter der Sterne beleuchtet.

Bei allen unseren bisherigen Betrachtungen und Rechnungen haben wir stets die Masse des einzelnen Sternes als zeitlich konstant angenommen und nur die chemische Zusammensetzung sich ändern lassen. Da dabei die Leuchtkraft stetig zunimmt, konnten wir eine obere Grenze für die Lebenserwartung des Sternes angeben. Die Leuchtkraft wird aber nicht in dem gleichen Maße zunehmen, wenn die Masse des Sternes mit der Zeit abnimmt, und die Masse der Sterne mit mehr als einer

Sonnenmasse muß abnehmen, einerseits weil die für sie aus der Energieerzeugung sich ergebende Lebenserwartung zu kurz ist, als daß wir das Schicksal dieser Sterne offen lassen könnten, und andererseits weil diese Massen zu groß sind, um letzten Endes in den quasistationären Zustand eines vorwiegend aus vollständig entarteter Materie bestehenden Sternes gelangen zu können.

Dabei wird es von dem Verhältnis der zeitlichen Rate der Massenabnahme zur Rate der Kernumwandlungen abhängen, ob die Lebenserwartung des Sternes abnimmt oder zunimmt. Da eine Erhöhung der Lebenserwartung auch als „Verjüngung" aufgefaßt werden kann, sieht man, daß neben der in Teil C.II noch zu behandelnden Verjüngung durch Materieaufsammlung unter Umständen auch eine Verjüngung durch Massenverlust möglich ist.

Für den zeitlichen Verlauf dieser notwendigen Massenabnahme kann man drei einfache Modelle heranziehen: Erstens: die Abnahme der Sternmasse erfolgt kontinuierlich und ändert sich nur langsam etwa mit der Zeitskala der Sternentwicklung, oder zweitens: die Massenabnahme erfolgt diskontinuierlich in mehreren einzelnen Raten, oder drittens: sie erfolgt in einer einzigen Rate, was einer Sternkatastrophe gleichkommt. Die verschiedenen Auswirkungen auf die Zeitskala t_{CN} liegen auf der Hand.

Für alle drei Modelle lassen sich praktisch vorkommende Beispiele angeben: zu dem ersten, die Korpuskularstrahlung emittierenden Sterne sowie die infolge zu schneller Rotation oder unter der Einwirkung eines nahen Begleiters masseabsprühenden Sterne. Zu dem zweiten, die gewöhnlichen (rekurrierenden) Novae und die Hüllen abwerfenden Sterne. Zu dem dritten Modell, die Supernovae und der evtl. Zerfall in Doppelsterne.

Da wir in Abschnitt B. IV diese Massenabgabe von Sternen an das interstellare Medium noch ausführlicher behandeln werden, wollen wir hier nur vorwegnehmen, daß nach der sich dabei ergebenden Bilanz die unserem ersten Modell entsprechenden Prozesse vorherrschen dürften.

In Anbetracht der relativen Häufigkeit der Doppelsterne mit mehr als $0.5\, M_\odot$ wird die Möglichkeit der Beseitigung der Überschußmasse durch Zerfall in Mehrfachsysteme unter Umständen eine nicht zu vernachlässigende Rolle spielen. Jedoch würde die Behandlung dieses Vorganges eine bedeutende Weiterentwicklung der hydromagnetodynamischen Theorie der Sternrotation voraussetzen (vgl. A.I.a 5).

2. Vergleich mit der Beobachtung

Im vorigen Abschnitt schilderten wir den gegenwärtigen Stand der Theorie der Sternentwicklung sowie die sich hieraus ergebende Möglichkeit der Altersabschätzung. Um zunächst ein Gefühl für den Grad der Verläßlichkeit solcher Abschätzungen zu gewinnen, soll der gegenwärtige Abschnitt untersuchen, wieweit die Theorie der Sternentwicklung sich bestätigt durch Vergleiche mit Beobachtungen und mit anderweitigen Abschätzungen. Wir vergleichen theoretische und beobachtete HR-Diagramme von Sternhaufen, verweisen dann auf die Altersabschätzungen der zwei folgenden Abschnitte und bringen zum Schluß einige mehr indirekte Prüfungen.

1. Setzt man voraus, daß alle Sterne eines Sternhaufens verschiedene Masse, aber gleiches Alter haben, so läßt sich die Gestalt der Hauptreihe für jedes Alter aus der Theorie ableiten. Ein Versuch dieser Art wurde z. B. durchgeführt von N. ROMAN (1955) für den offenen Sternhaufen NGC 752, siehe Abb. 10. Das als Ausgang benutzte Cowlingmodell ist zwar nicht in der Lage, die normale Hauptreihe sehr befriedigend wiederzugeben, doch wird die durch die Sternentwicklung erfolgte Abweichung der Hauptreihe von der als Ausgang benutzten durch die Theorie gut dargestellt.

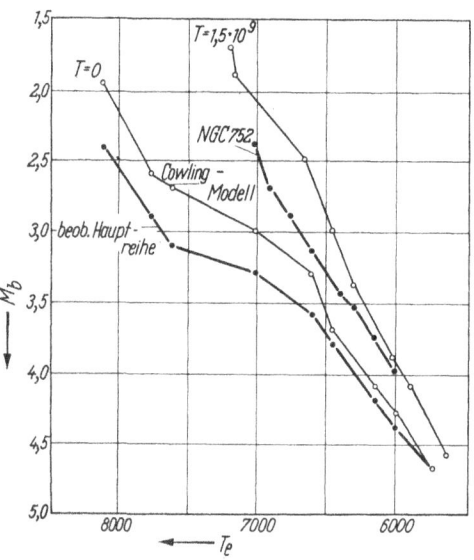

Abb. 10. Normale Hauptreihe und die von NGC 752, verglichen mit dem Cowlingmodell zu Beginn und nach $1.5 \cdot 10^9$ Jahren. Nach N. ROMAN (1955)

Ein anderer Weg wurde von LOHMANN (1957) gewählt, der aus den beobachteten Hauptreihen mehrerer Sternhaufen in einem iterativen Verfahren eine allen Haufen gemeinsame „ursprüngliche" Hauptreihe zurückrechnet, diese als Ausgang benutzt und nun Linien gleichen Alters berechnet, die er Isochronen nennt. In Abb. 11 zeigen wir diese Lohmannschen Isochronen im Vergleich mit den Beobachtungen einiger Sternhaufen nach H. L. JOHNSON (1954b). Die qualitative Übereinstimmung beider Bilder und eine quantitative Diskussion in Tab. 8 bei LOHMANN zeigen, daß sich sämtliche beobachteten Hauptreihen aus einer einzigen ursprünglichen Hauptreihe ableiten lassen. Wenn auch im einzelnen noch manche Abweichungen vorhanden sind (z. B. das weniger scharfe Abbiegen der Plejaden), so können wir das Ergebnis doch als eine gute Bestätigung für die Theorie der Sternentwicklung bezeichnen.

Selbst das recht komplizierte HR-Diagramm der Kugelhaufen läßt sich zumindest in groben Zügen durch die Ergebnisse theoretischer Rechnungen darstellen.

2. Eine weitere Kontrolle besteht darin, daß man die aus der Theorie der Sternentwicklung folgenden Altersangaben mit anderen Altersabschätzungen vergleicht. Dies wird in den Abschnitten A.II.c, A.III.a3 und A.III.b3 durchgeführt und in A.IV. diskutiert werden. Wir wollen hier nur vorwegnehmen, daß sich innerhalb der Fehlergrenzen stets eine befriedigende Übereinstimmung ergibt und nirgends ein krasser Widerspruch.

Auch gegenüber dem sog. Alter der Welt ergibt sich kein Widerspruch. Bei keinem Objekt war es bisher nötig, aus der Theorie der

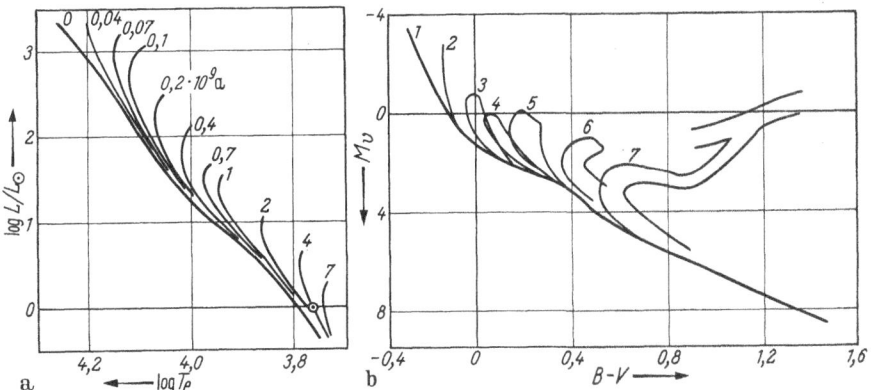

Abb. 11. *a* Linien gleichen Alters (Isochronen) nach LOHMANN (1957); *b* beobachtete Hauptreihen offener Haufen nach JOHNSON (1954b); *1* = NGC 2362, *2* = Plejaden, *3* = M 39, *4* = Coma, *5* = Praesepe, *6* = NGC 752, *7* = M 67

Sternentwicklung ein merklich höheres Alter als $6 \cdot 10^9$ Jahre anzunehmen. Für die Schnelläufer der Sonnenumgebung ergeben sich nach v. HOERNER (1957b) etwa $5.7 \cdot 10^9$ Jahre. Auch von den übrigen Sternen der Sonnenumgebung muß keinem ein höheres Alter zugeschrieben werden.

3. Schließlich erwähnen wir noch einige weitere Prüfungsmöglichkeiten.

α) Da die Sterne bei ihrer Entwicklung nach der Theorie expandieren sollten, so kann bei konstantem Drehimpuls die mit der Expansion verbundene Abnahme der Rotationsgeschwindigkeit abgeschätzt werden. Dies wurde durch SANDAGE (1955) durchgeführt für die beiden Extremfälle der starren und der reibungsfreien Rotation. Aus der Verteilung der Rotationsgeschwindigkeiten von Sternen oberhalb der Hauptreihe berechnete SANDAGE die entsprechende Verteilung für die ursprüngliche

Lage der Sterne auf der Hauptreihe. Das Ergebnis verglich er dann mit der beobachteten Verteilung der Rotationsgeschwindigkeiten von Hauptreihensternen. SANDAGE findet befriedigende Übereinstimmung; einige Abweichungen werden ausführlich diskutiert und als nicht signifikant bezeichnet.

β) Aus Sternzählungen berechnete v. HOERNER (1957a) die Masse-Leuchtkraft-Beziehung in den Kugelhaufen M3 und M92. Während für die Sterne der Sonnenumgebung im Intervall $+2 \geqq M_v \geqq -2$ die Massen um einen Faktor 2.8 ansteigen, erhält man für die Kugelhaufen im gleichen Intervall nur eine schwache Massenzunahme von höchstens einigen Prozent. Dies entspricht gerade der Theorie der Sternentwicklung, wonach alle Riesen eines Kugelhaufens von nahezu der gleichen Stelle der Hauptreihe herstammen und somit alle nahezu die gleiche Masse haben sollten. Auch die Bestimmung der Masse der RR Lyrae-Sterne von SANDAGE (1956) zu 1.25 Sonnenmassen kann mit der gleichen Begründung als Bestätigung gelten.

Liste der in Kapitel A.I. verwendeten Symbole

A Modellkonstante des Energieerzeugungsgesetzes
a Stefansche Konstante
B Modellkonstante der Masse-Leuchtkraftbeziehung
c Lichtgeschwindigkeit
F_K konvektiver Energiestrom
F_R Strahlungsstrom
G Gravitationskonstante
g Schwerebeschleunigung
H homogene Schichtdicke
K Anteil des Konvektionsstromes am Gesamtenergietransport
K_{CN} Umwandlungsrate für die CN-Reaktion
K_{pp} Umwandlungsrate für die pp-Reaktion
K_S Umwandlungsrate für die Salpetersche Reaktion
L Leuchtkraft
l Mischungsweg
M Sternmasse
M_K Masse des Kernes
M_n Masse der zum Kern gehörigen Polytropen
$\mathfrak{M}$ Chandrasekharsche Grenzmasse für Elektronenentartung
$\mathfrak{M}^*$ Grenzmasse für vollständige Paarentartung
M_S Grenzmasse zwischen relativistischer und nichtrelativistischer Entartung
m Exponent der Temperatur im Energieerzeugungsgesetz
n Polytropenindex
P Druck
P_K Druck an der Oberfläche des Kernes
p_e Druck des entarteten Elektronengases
q Massenanteil des Kernes

Liste der in Kapitel A.I. verwendeten Symbole (Fortsetzung)

R	Sternradius
R_n	Radius der zum Kern gehörigen Polytropen
$\mathfrak{R}$	Gaskonstante
r	Abstand vom Mittelpunkt des Sternes
T	Temperatur
t_K	Kontraktionsalter
t_{CN}	Zeitskala der CN-Reaktion als Energiequelle
t_{pp}	Zeitskala der pp-Reaktion als Energiequelle
v	Konvektionsgeschwindigkeit
v_s	Schallgeschwindigkeit
X	Massenanteil des Wasserstoffes
x	$= r/R$, relativer Abstand vom Zentrum
Y	Massenanteil des Heliums
Z	$= 1 - X - Y$, Massenanteil schwererer Elemente
α	$= l/H$ Länge des Mischungsweges in homogenen Schichtdicken, auch $= \dfrac{\nu G M^2}{L R^5}$
β	Verhältnis des Gasdruckes zum Gesamtdruck
γ	$= c_p/c_v$, Verhältnis der spezifischen Wärmen
δ	$= \dfrac{d \ln \varrho_c}{d \ln \mu}$
ε	Energieerzeugung pro Massen- und Zeiteinheit
ε_0	numerische Konstante ⎫ im Energieerzeugungsgesetz
ε_1	chemische Konstante ⎭
$\varkappa$	Massenabsorptionskoeffizient
$\varkappa_0$	numerische Konstante ⎫ im Absorptionsgesetz
$\varkappa_1$	chemische Konstante ⎭
$\varkappa_e$	entspricht der Elektronenleitfähigkeit
μ	mittleres Molekulargewicht
μ_e	Molekulargewicht pro Elektron
ν	kinematische Zähigkeit
ξ	$= \varrho_c/\varrho_m$, Verhältnis der zentralen zur mittleren Dichte
ϱ	Dichte
σ	$= \dfrac{ac}{4}$
τ	Guillotinefaktor, oder auch $\dfrac{d \ln T_e}{d \ln \mu}$
τ_e	Zeitskala der Wasserstoffverbrennung
τ_K	Zeitskala der Durchmischung
χ	Parameter der Massenverteilung im Stern, auch $= \dfrac{\overline{\omega}^2}{2 \pi G \overline{\varrho}}$; Ionisierungsspannung
ψ	Entartungsparameter
ω	Winkelgeschwindigkeit
∇	$= \dfrac{d \ln T}{d \ln P}$
∇_K	konvektiver Gradient
∇_R	Strahlungsgradient
∇_{ad}	adiabatischer Gradient
$\Delta \nabla$	Überschußgradient
$\odot$	Sonne

A.II. Kontraktionsalter
A.II. a) Theorie

1. Eine zweite Möglichkeit der Altersabschätzung ergibt sich aus der Betrachtung derjenigen Sterne, die so jung sind, daß sie die Hauptreihe noch nicht erreicht haben. Setzen wir voraus, daß sich die Sterne durch Kondensation interstellarer Materie bilden, so muß zunächst eine dichtere Wolke entstehen, die sich weiter kontrahiert. Ist die Temperatur genügend angestiegen, so beginnt der werdende Stern zu strahlen, und sein Weg im HR-Diagramm läßt sich verfolgen. Die abgestrahlte Energie wird aus der potentiellen Energie nachgeliefert, die bei der Kontraktion frei wird.

Die letzten Phasen dieser gravitativen Kontraktion bis zum Einsetzen der Kernreaktionen wurden in einer Arbeit von Henyey, LeLevier und Levée (1955) untersucht, von der bisher leider nur eine wenig ausführliche Mitteilung veröffentlicht wurde.

2. Die physikalischen Voraussetzungen dieser Arbeit sind folgende. Der Strahlungsdruck wird vernachlässigt. Die Absorption wird berechnet aus einer Kombination der Morseschen Ergebnisse (1940) mit Elektronenstreuung und frei-frei-Übergängen von Wasserstoff und Helium. Die Konvektion ist, falls nötig, für den Kern berücksichtigt, nicht dagegen für die äußeren Zonen. Die Energieerzeugung erfolgt aus der Kontraktion und, sobald die nötige Temperatur im Kern erreicht ist, aus dem CN-Zyklus und der pp-Reaktion.

Der jeweilige Ausgangszustand wird aus bereits vorhandenen Gleichgewichts-Konfigurationen gewonnen durch eine Homologie-Transformation auf die gewünschten Werte von Masse und Radius. Dabei ergab sich, daß die Willkür in der Wahl des Ausgangszustandes wenig Einfluß auf das Ergebnis hat: auch recht verschiedene Ausgangszustände konvergierten bereits nach wenigen Zeitschritten zu einem gemeinsamen weiteren Entwicklungswege hin.

Bei gleicher chemischer Zusammensetzung wurde für 5 verschiedene Massen gerechnet und bei gleicher Masse für 4 verschiedene Zusammensetzungen. Die Radien waren bei Beginn der Rechnung 3—5mal größer als im Hauptreihenzustand.

3. Allen 8 gerechneten Entwicklungszügen ist gemeinsam, daß die effektive Temperatur anfangs schnell ansteigt; die Helligkeit nimmt nur wenig zu, durchläuft beim Einsetzen der Kernreaktionen ein flaches Maximum und sinkt dann etwas ab. Sobald im Zentrum die Kernreaktionen voll angelaufen sind, erfolgen alle Änderungen nur noch in einer um rund 2 Zehnerpotenzen langsameren Zeitskala, der Stern ist für längere Zeit auf der Hauptreihe zur Ruhe gekommen. Die weitere Entwicklung hatten wir im Abschnitt A.I. beschrieben.

Für 25% Helium und 1% schwere Elemente münden die Entwicklungszüge für verschiedene Massen in eine Hauptreihe ein, die zwar etwas von der beobachteten Hauptreihe abweicht (bis 0.06 in log T), doch ist eine Wiederholung der Rechnungen mit neueren Absorptionsformeln angekündigt. Die Zeitdauer t_K vom Anfang der Rechnung bis zum Einmünden in die Hauptreihe zeigt Tab. 5, wobei wir noch den Faktor berechneten, um den sich der Radius des Sternes in diesem Zeitraum verkleinert. Weiterhin geben wir noch die zu den Massen $\mathfrak{M}$ im Endzustand auf der Hauptreihe gehörenden Werte der absoluten visuellen Helligkeit M_v und des Spektraltyps Sp an, nach ALLEN (1955).

4. Wieweit die zitierten Ergebnisse der Wirklichkeit entsprechen, ist schwer abzuschätzen. Es erscheint plausibel, daß vor allem die hier vernachlässigte Rotation den Verlauf der Kontraktion ganz wesentlich beeinflussen dürfte. Eine Durchführung der Rechnungen mit Rotation bringt jedoch außer einer beachtlichen rechentechnischen Komplikation noch notwendig das Problem der Magnetfelder mit sich. Eine Kontraktion dürfte auf die Dauer nur dann möglich sein, wenn die Rotation durch Magnetfelder laufend gebremst wird [LÜST und SCHLÜTER (1955)]. Und nur, falls diese Bremsung so schnell geschieht, daß die verbleibende Rotation die Kontraktion nicht merklich beeinflußt, nur dann könnten die Henyeyschen Rechnungen gerechtfertigt sein. Nur ist dann wieder die Frage, ob man für eine so starke Bremsung so starke Magnetfelder braucht, daß nun *diese* die Kontraktion merklich beeinflussen.

Zur Anwendung auf die Beobachtung ist weiterhin zu bedenken, daß über die Zeitdauer der Vorgeschichte (bis zum Einsetzen der Henyeyschen Rechnung) so gut wie nichts bekannt ist. Beide Einwände (Rotation und Vorgeschichte) fassen wir dahingehend zusammen, daß die Kontraktionszeiten t_K der Tab. 5 eine *untere Grenze* darstellen für das Alter derjenigen Sterne, die gerade eben die Hauptreihe erreichen.

Zum Vergleich mit der im Abschnitt A.I.a1 durchgeführten Abschätzung haben wir in der letzten Spalte die Kontraktionszeiten der

Tabelle 5. *Die Kontraktion junger Sterne bis zur Hauptreihe*

t_K = Dauer der von HENYEY u. Mitarb. (1955) gerechneten letzten Phasen der Kontraktion.
R_0 = Sternradius zu Beginn der Rechnung.
R_H = Sternradius am Ende der Rechnung (Hauptreihe).
M_v und Sp beziehen sich auf den Endzustand (Hauptreihe).
t_{AI} = Abschätzung aus Tab. 2.

$\mathfrak{M}$	t_K	R_0/R_H	M_v	Sp	t_{AI}
Sonnenmassen	10^6 Jahre		mag		10^6 Jahre
0.65	150	3.2	7.0	K3	80
1.00	30	3.0	4.7	G1	23
1.25	14	3.0	3.9	F5	12
1.55	8	3.3	3.0	F0	6.0
2.29	3	5.4	1.5	A3	2.6

Tab. 2 mit eingetragen, die der Polytropen $n = 3$ entsprechen. Wir sehen, daß unsere grobe Abschätzung bereits recht ähnliche Zahlen lieferte wie die Henyeyschen Rechnungen.

5. Haben alle Sterne eines Haufens verschiedene Masse, aber gleiches Alter, so muß für jeden Haufen die Hauptreihe nach unten begrenzt sein: Sterne kleinerer Masse haben, entsprechend dem Alter des Haufens, noch nicht genügend Zeit gehabt, sich bis zur Hauptreihe hin zu kontrahieren, liegen also noch rechts der Hauptreihe. Diese untere Grenze verschiebt sich mit zunehmendem Alter zu kleineren Massen.

Liegt dieser untere Abzweig der Hauptreihe in dem von HENYEY gerechneten Bereich (Tab. 5), so sollte die Länge der Hauptreihe, von den Haufensternen frühesten Spektraltyps bis zum unteren Abzweig, etwa 9.5 bolometrische Größenklassen betragen, wie v. HOERNER (1957b) durch Vergleich mit der Theorie der Sternentwicklung gezeigt hat. Dabei ist allerdings vorausgesetzt, daß für die Sterne am unteren Abzweig die Dauer t_V ihrer Vorgeschichte vernachlässigt werden kann gegenüber der Kontraktionsdauer t_K, und für die Sterne am oberen „Knie" wurde t_K gegenüber der Dauer der weiteren Entwicklung vernachlässigt. Ist die erste Voraussetzung erfüllt und die zweite nicht, so sollte die Hauptreihe länger sein als 9.5 Größenklassen, und ist allein die zweite Voraussetzung erfüllt, so sollte sie kürzer sein. Vermutlich ist die zweite Voraussetzung gut und die erste nur schlecht erfüllt; somit wäre zu erwarten, daß die Länge der Hauptreihe 9.5 bolometrische Größenklassen oder etwas weniger beträgt.

6. In Abschnitt A.I.c 1 sagten wir, daß sich aus den Entwicklungszügen (feste Masse, variable Zeit) die Gestalt der Hauptreihe eines Sternhaufens (variable Masse, feste Zeit) konstruieren läßt. Das war möglich, weil die Vorgeschichte bis zum Beginn der Rechnung vernachlässigt werden konnte. Dies ist jetzt, wie soeben gesagt, vermutlich kaum der Fall. Um trotzdem einen, wenn auch recht provisorischen, Vergleich mit der Beobachtung durchführen zu können, haben wir die Gestalt der Hauptreihe (unterhalb des Abbrechens) aus den Henyeyschen Entwicklungszügen konstruiert für die zwei Annahmen: a) Die Vorgeschichte t_V eines Sternes der Masse $\mathfrak{M}$ von seiner Entstehung bis zum Anfang der Heynyeschen Rechnungen ist für alle Sterne von gleicher Dauer; b) Die Dauer der Vorgeschichte beträgt $1/3$ derjenigen Zeit, über die sich die Henyeysche Rechnung erstreckt, plus einem konstanten Betrag:

$$\begin{align} \text{a)} \quad & t_V(\mathfrak{M}) = t_c = \text{const.} \\ \text{b)} \quad & t_V(\mathfrak{M}) = t_c + \tfrac{1}{3} t_K(\mathfrak{M}) \,. \end{align} \tag{87}$$

In Abb. 12 haben wir die beiden so konstruierten Linien eingezeichnet. Zwar hätten andere Annahmen andere Linien ergeben; es dürfte jedoch nur schwer gelingen, Linien zu erhalten, die steiler sind als a), während man leicht Linien flacher als b) bekommen kann.

Die Wahl von t_c beeinflußt übrigens nur das aus Beobachtungen abgeleitete Alter eines Haufens, nicht dagegen die Gestalt des Abzweigens.

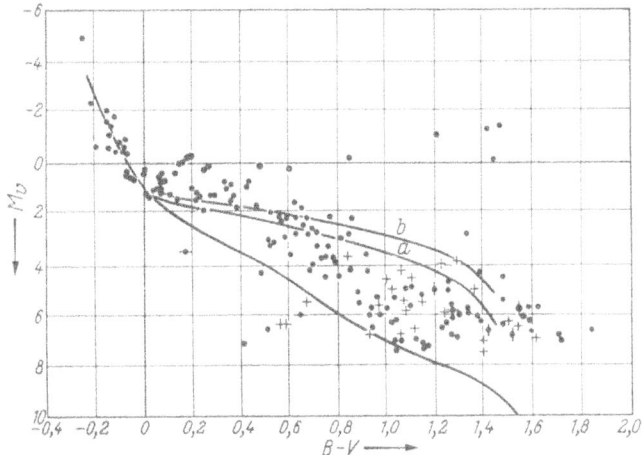

Abb. 12. Farben-Helligkeits-Diagramm von NGC 2264. ● normale Sterne, + Hα-Emission, ♦ variable Helligkeit nach WALKER (1956). Die durchgehende Linie ist die normale Lage der Hauptreihe. Noch kontrahierende Sterne sollten auf den Linien a oder b liegen, je nach Annahme (87) über ihre Vorgeschichte bis zum Beginn der Rechnungen von HENYEY u. Mitarb. (1955)

A.II. b) Vergleich mit der Beobachtung

Wir fragen zunächst danach, ob das untere Abbrechen der Hauptreihe tatsächlich beobachtet wurde in den Fällen, wo dies zu erwarten ist, und gehen dann auf die Gestalt des Abzweiges ein. Den Vergleich der sich ergebenden Kontraktionsalter einiger offener Haufen mit anderen Altersabschätzungen bringen wir in Abschnitt c).

1. Die älteren Beobachtungen reichen meist nicht bis zu genügend schwachen Sternen herunter. Durch die Henyeyschen Rechnungen sind jedoch eine Reihe von Beobachtungen an extrem jungen Haufen angeregt worden, die allerdings erst zum Teil veröffentlicht worden sind.

Die Längen der Hauptreihen von 28 offenen Haufen und Assoziationen wurden durch v. HOERNER (1957b) (Tab. 7), zusammengestellt. Erweitern wir sein Ergebnis um zwei inzwischen erschienene Arbeiten von WALKER (1957) und JOHNSON (1957a), so zeigt Tab. 6, daß die theoretischen Erwartungen sich bisher vollauf bestätigten.

In 24 Fällen, in denen die Beobachtungen nicht weit genug nach unten reichen, zeigt sich noch nichts Besonderes, während in allen 5 Fällen mit längerer Hauptreihe das vorausgesagte Abzweigen oder zumindestens eine Verbreiterung nach rechts zu beobachten ist.

2. In Abb. 12 zeigen wir das von WALKER (1956) an NGC 2264 beobachtete Farben-Helligkeits-Diagramm. Das Abbrechen der Hauptreihe bei B—V = 0.05 ist deutlich markiert und spätere Typen liegen noch rechts der Hauptreihe. Zwei Ergebnisse sind von besonderer Wichtigkeit: erstens sind eine große Anzahl dieser noch kontrahierenden Sterne T Tauri-Variablen [auch bei drei weiteren extrem jungen Haufen ist dies der Fall (WALKER, 1957)], und zweitens zeigen gerade die späteren Spektraltypen starke Rotation (von F8 ab). Leider gehen die spektroskopischen Untersuchungen nur bis $m \leq 13$, während die Variablen erst ab $m \geq 13$ einsetzen, so daß sich über die Frage einer Korrelation zwischen Rotation und Variabilität vorläufig nichts aussagen läßt.

Tabelle 6. *Unterschied ΔM_b der bolometrischen Helligkeit zwischen den Sternen frühesten Spektraltyps und der unteren Beobachtungsgrenze.* (Bei NGC 2264, 6530 und 2362 ist das obere Ende der Hauptreihe schlecht definiert)

Name	ΔM_b	Unteres Ende
NGC 2264	>15	V, A
NGC 6530	>10	V, A
Orion-Assoz.	13	V, A
h, χ Persei	11	V
NGC 2362	>10	V
Plejaden	9.5	—
Hyaden	8.0	—
Praesepe	8.0	—
21 weitere Haufen	≤6.0	—

V = starke Verbreiterung der Hauptreihe nach rechts,
A = Abzweigen der Hauptreihe nach rechts

Wir vergleichen nun die Gestalt des unteren Abzweiges mit den beiden theoretisch konstruierten Linien, die auf den Annahmen a und b von (87) basieren. Wir sagten bereits, daß Linien steiler als Linie a sich mit den Henyeyschen Rechnungen schlecht erklären lassen. Die Übereinstimmung ist also nicht besonders befriedigend, aber vermutlich geht dies auf die vernachlässigte Rotation zurück.

Weiterhin haben wir zum Vergleich nur noch die Orion-Assoziation zur Verfügung, deren Farben-Helligkeits-Diagramm JOHNSON (1957a) angibt. Die Beobachtungen reichen jedoch nur bis etwa 4 Größenklassen unter den Abzweigungspunkt, so daß die Abweichung gegen Linie *a* noch nicht so stark hervortritt. Auch sind von den schwächeren Sternen nur ein kleiner Teil untersucht worden.

WALKER hat zwar inzwischen 8 weitere junge Sternhaufen untersucht (1957), jedoch noch keine Diagramme oder Tabellen veröffentlicht. Bei NGC 6530 ist nur angegeben, daß die Hauptreihe wiederum etwa bei A0 nach rechts abzweigt, und daß 70 sehr lichtschwache Variablen gefunden wurden. Bei NGC 6611 bricht die Hauptreihe ebenfalls ab (Stelle nicht angegeben) und 13 Variablen wurden gefunden. Bei IC 5146 werden 19 Variablen erwähnt.

Über den Vergleich mit der Beobachtung läßt sich zusammenfassend sagen: Das vorausgesagte untere Abbrechen der Hauptreihe ist genau in all den Fällen beobachtet worden, wo dies nach der Theorie zu erwarten war. Die genauere Form des Abzweiges läßt sich aus der Theorie noch nicht befriedigend ableiten und die beobachtete Rotation und Variabilität der noch kontrahierenden Sterne sind in der Theorie nicht enthalten.

A.II. c) Altersbestimmung

1. Das untere Abzweigen wurde bisher nur bei drei HR-Diagrammen beobachtet. In allen drei Fällen wird angegeben, daß die A0-Sterne noch auf der Hauptreihe liegen und die späteren Typen darüber. Setzen wir den Punkt des Abzweigens bei A0.5 an, so erhält man nach ALLEN (1955) eine zugehörige Masse von $\mathfrak{M} = 2.80$ Sonnenmassen. Falls es zulässig ist, die Werte der Tab. 5 ein wenig zu extrapolieren, so erhalten wir für die Kontraktionsdauer t_K:

$$\left. \begin{array}{ll} \text{NGC 2264} & \text{WALKER} \quad (1956) \\ \text{NGC 6530} & \text{WALKER} \quad (1957) \\ \text{Ass. I Ori} & \text{JOHNSON} \quad (1957a) \end{array} \right\} t_K = 2 \cdot 10^6 \text{ Jahre}. \qquad (88)$$

Wegen der Vernachlässigung der Rotation und der Unkenntnis der Vorgeschichte ist dieser Wert als untere Grenze für das Alter der drei Objekte anzusehen.

2. v. HOERNER (1957b) hat das Alter von NGC 2264 und der Orion-Assoziation aus dem oberen „Knie" der Hauptreihe nach der Theorie der Sternentwicklung abgeschätzt. Bei NGC 2264 ist das Knie wegen geringer Sternzahl nur schlecht definiert. Ein einzelner Stern, S Mon, ist vom Typ O7, es folgt je ein Stern vom Typ B2.5, B3, B5, B6, B9 usw.; alle genannten Sterne liegen nicht oder nur wenig über der normalen Hauptreihe. Somit läßt sich nicht entscheiden, ob S Mon von gleichem Alter ist wie der übrige Haufen, oder als junger Nachzügler zu betrachten ist, was nach v. HOERNER (1957b) relativ häufig vorkommt (Bei 11 von 28 untersuchten Haufen ist ein einzelner, relativ früher Stern angegeben. In 7 Fällen liegt er deutlich abgesetzt links vom Knie, in Verlängerung der ursprünglichen Hauptreihe, sollte also jünger sein als die anderen Haufensterne.) Wir können daher nur sagen, daß das Knie links von B2.5 liegen muß; das Alter von NGC 2264 ist somit $< 15 \cdot 10^6$ Jahre, was sich zwar mit Gl. (88) verträgt, aber nicht viel besagt.

Auch in der Orion-Assoziation ist ein einzelner, früher Stern vom Typ O6, doch ist das Knie bei O9 deutlich ausgeprägt und ergibt ein Alter von $3 \cdot 10^6$ Jahren. Aus Abschnitt A.III.c wollen wir vorwegnehmen, daß für drei schnelle „Ausreißer" der Orion-Assoziation (auf Grund von Abstand und Geschwindigkeit) sich als Alter ergibt: 2.6, 2.6 und $4.8 \cdot 10^6$ Jahre. Rechnen wir also mit einem Alter der Orion-Assoziation von rund $3 \cdot 10^6$ Jahren, so paßt dieser Wert sehr gut zu der Kontraktionszeit von $2 \cdot 10^6$ Jahren.

A.III. Dynamische Altersangaben

Als dritte Möglichkeit der Altersabschätzung betrachten wir einige dynamische Vorgänge in Sternsystemen. Stark aufgelockerte Bewegungshaufen können nach BOK (1951) wegen der Gezeiten- und Scherungskräfte der Milchstraße nicht beliebig alt sein, und die dichteren galaktischen Haufen lösen sich nach SPITZER (1940) durch gegenseitigen Energieaustausch der Haufensterne auf. Da jedoch in beiden Fällen (sowie auch in späteren Kapiteln) die Masse der Haufen eine entscheidende Rolle spielt, und da über die Massen bisher wenig bekannt ist, so sehen wir uns gezwungen, zunächst vom Thema abzuweichen und eine diesbezügliche Untersuchung einzuschieben.

A.III.a) Massenabschätzung offener Haufen aus der Leuchtkraftfunktion

1. Soweit die Beobachtungen reichen und noch nicht zu dicht an der Beobachtungsgrenze liegen, steigt die LKF (Leuchtkraftfunktion) offener Haufen in Richtung der schwächeren Sterne gleichmäßig an. Wir würden viel zu kleine Massen erhalten, wenn wir sie direkt aus der Anzahl der beobachteten Sterne entnehmen wollten, da vermutlich (ähnlich wie in der Sonnenumgebung) der größte Beitrag zur Gesamtmasse von den kleineren Sternen herrührt. Dies ganze Problem ist noch wenig geklärt. Da wir jedoch Angaben über die Massen brauchen, so scheint es das beste zu sein, wenn wir eine nach dem gegenwärtigen Stand der Beobachtungen und Meinungen bestmögliche Massenabschätzung durchführen, sie dort wo möglich kontrollieren und das Ergebnis dann mit Vorsicht anwenden.

Als Arbeitshypothese werden wir die Behauptung benutzen, daß Haufensterne und Feldsterne bei ihrer Entstehung nach der gleichen LKF verteilt seien; eine Möglichkeit der Kontrolle wird sich in Tab. 8 ergeben.

2. v. D. BERGH (1957a) findet, daß sich 9 untersuchte offene Haufen zwanglos durch eine gemeinsame LKF darstellen lassen, die wir in Abb. 13 zeigen und mit der LKF der Sterne der Sonnenumgebung nach SANDAGE (1957) vergleichen. Wären Feldsterne und Haufensterne nach der gleichen LKF entstanden, so sollten oberhalb von $M \geq 4$ beide übereinstimmen, während unterhalb von $M \leq 4$ die Anzahl der Feldsterne geringer sein sollte, da sich die massiveren Sterne, je nach Alter, bereits von der Hauptreihe wieder wegentwickelt haben. (Auf die hiermit zusammenhängenden Probleme der zeitlichen Rate der Sternentstehung werden wir in Abschnitt B.II. näher eingehen.)

Daß die v. d. Berghsche LKF für die lichtschwachen Sterne nicht glatt in die LKF der Feldsterne einmündet, braucht nicht reell zu sein. Ein Blick auf Abb. 1 der v. d. Berghschen Arbeit zeigt, daß die LKF der einzelnen Haufen starke Streuungen zeigen und daß nur drei Haufen über $M = 5$ hinausreichen. Hyaden und Praesepe reichen bis $M \leq 6.5$, könnten jedoch, als relativ alte Haufen [s. v. HOERNER (1957b)], bereits am lichtschwachen Teil der LKF Abweichungen zeigen [s. V. D. BERGH (1957b)], da die weniger massiven Sterne bei zunehmender Auflösung den Haufen schneller verlassen als die massiveren Sterne. Bis $M \leq 8.0$ reichen allein die Plejaden, die jedoch bei V. D. BERGH den Vermerk tragen: "faintest magnitude group possibly incomplete".

Zum Vergleich zeichnen wir in Abb. 13 die LKF des offenen Haufens NGC 2264 nach WALKER (1956) mit ein, der bei V. D. BERGH *nicht* mit enthalten ist. Trotz großer Streuung ist der allgemeine Zug wieder der gleiche, und die LKF der Feldsterne ist durchaus eine mögliche Fortsetzung.

3. Als Grundlage für unsere Massenabschätzung (sowie auch für Abschnitt B.II.) wollen wir, im Sinne einer Arbeitshypothese, eine etwas geglättete Kombination benutzen, einerseits für $M \geq 5$ der

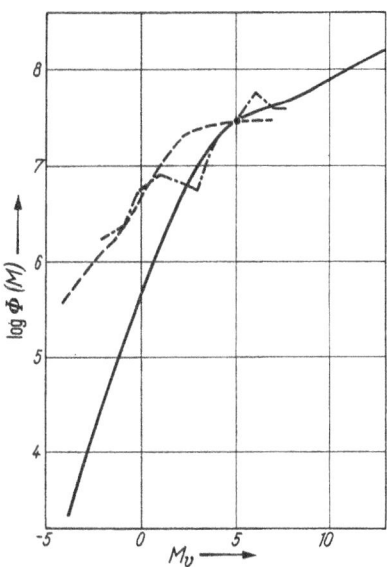

Abb. 13. Leuchtkraftfunktion. —— Feldsterne, nach SANDAGE (1957), ------ 9 offene Haufen, nach V. D. BERGH (1957a), -·-·-·- NGC 2264, nach WALKER (1956), ● Normierung

LKF der Feldsterne, andererseits für $M \leq 5$ der v. d. Berghschen LKF und derjenigen von NGC 2264. Tab. 7 zeigt die benutzten normierten Werte.

Sind in einem offenen Haufen $n(M_1, M_2)$ Sterne im Intervall $M_1 - 1/2 < M_v < M_2 + 1/2$ beobachtet worden (von den hellsten Hauptreihen-Sternen des Haufens bei M_1 bis zur Beobachtungsgrenze bei M_2), so betrug die ursprüngliche Masse $\mathfrak{M}_0$ des Haufens

$$\mathfrak{M}_0 = \frac{1000}{\sum\limits_{-\infty}^{M_2} \psi - \sum\limits_{-\infty}^{M_1-1} \psi} \cdot n(M_1, M_2) \tag{89}$$

und unter Berücksichtigung der Sternentwicklung beträgt die gegenwärtige Masse

$$\mathfrak{M} = \frac{1000 - \sum\limits_{-\infty}^{M_1-1} \mathfrak{M}\psi}{\sum\limits_{-\infty}^{M_2} \psi - \sum\limits_{-\infty}^{M_1-1} \psi} \cdot n(M_1, M_2). \tag{90}$$

Tabelle 7. *Benutzte Leuchtkraftfunktion für offene Sternhaufen, normiert auf eine Gesamtmasse von 1000 Sonnenmassen.* ψ = Anzahl Sterne im Intervall $M_v \pm {}^1/_2$. $\mathfrak{M}$ = Masse eines Sternes der vis. Helligkeit M_v

M_v	ψ	$\mathfrak{M} \cdot \psi$	$\sum_{-\infty}^{M_v} \psi$	$\sum_{-\infty}^{M_v} \mathfrak{M}\psi$	M_v	ψ	$\mathfrak{M} \cdot \psi$	$\sum_{-\infty}^{M_v} \psi$	$\sum_{-\infty}^{M_v} \mathfrak{M}\psi$
−5	0.24	6.1	0.34	8	+8	82	54	382	525
−4	0.61	11.4	0.95	19	9	101	54	483	579
−3	1.21	17.0	2.16	36	10	128	54	611	633
−2	2.37	24.4	4.5	61	11	172	64	783	697
−1	4.71	32	9.2	93	12	202	67	985	764
0	9.3	44	18.5	137	13	236	59	1221	823
1	15.5	50	34	187	14	269	50	1490	873
2	23.2	50	57	237	15	303	42	1793	915
3	31	49	88	286	16	303	34	2096	949
4	39	45	127	331	17	286	24	2382	973
5	47	44	174	375	18	252	16	2634	989
6	57	42	231	417	19	236	8	2870	997
7	69	54	300	471	20	202	3	3072	1000

4. Wir führen eine Kontrolle durch in den Fällen, wo aus der beobachteten Geschwindigkeitsstreuung und dem Virialsatz die Massen berechnet worden sind. Dies ist nur für zwei offene Haufen der Fall; zum Vergleich setzen wir auch für Kugelhaufen die gleiche (ursprüngliche) Leuchtkraftfunktion voraus, was ähnlich auch schon von SANDAGE (1954) für M3 versucht wurde. Wir benutzen die Sternzählung von TAYLER (1954a) für M92, und zum Vergleich die von SCHWARZSCHILD und BERNSTEIN (1955) korrigierte Massenbestimmung von M92 aus den von WILSON und COFFEEN (1954) gemessenen Geschwindigkeiten.

Tabelle 8. *Massenbestimmung*

	aus Leuchtkraftfunktion			aus Geschwindigkeitsstreuung			
	M_1	M_2	n	$\mathfrak{M}$	v. WIJK (1949)	BOK (1951)	MAWRIDIS (1956)	SCHWARZSCHILD und BERNSTEIN (1955)
Plejaden	−3	+5	98	554	480	400	550	—
Praesepe	0	+5	125	687	700	—	1000	—
M92	+4	+4	10540	$2.0 \cdot 10^5$	—	—	—	$(1.4 \pm 0.7) 10^5$

Die Übereinstimmung ist leidlich gut, und somit dürfte die Leuchtkraftfunktion der Tab. 7 eine brauchbare Näherung sein.

5. In Tab. 9 wenden wir die Methode auf eine Anzahl offener Haufen an, wodurch wir zu weiteren Messungen der Geschwindigkeits-Streuung Anregung geben möchten. (Bei der Angabe des Spektraltyps wurden die in Abschnitt A.II.c2 diskutierten frühen Einzelsterne nicht berücksichtigt.)

Im Folgenden wird sich ergeben, daß die älteren Haufen sich bereits im Stadium fortgeschrittener Auflösung befinden können. Ist dies der

Fall, so muß unsere Methode versagen, da sich während der Auflösung die Leuchtkraftfunktion verändern sollte, was durch v. d. BERGH (1957b) an M 67 bestätigt wurde. Für eine Mittelung beschränken wir uns daher auf Haufen mit $Sp \leq B9$; und h Per wollen wir als Extremfall auch ausschließen. Über die restlichen 13 Haufen gemittelt erhalten wir $\overline{\mathfrak{M}} = 1620$ und $\overline{\mathfrak{M}_0} = 1685$ Sonnenmassen; wegen der Unsicherheit unserer Abschätzung benutzen wir weiterhin einen abgerundeten Wert von

$$\mathfrak{M} = 1500 \text{ Sonnenmassen} \tag{91}$$

und schätzen die Unsicherheit auf einen Faktor 1.5 ein.

Tabelle 9. *Gegenwärtige Masse $\mathfrak{M}$ und ursprüngliche Masse $\mathfrak{M}_0$*
$n(M_1, M_2) =$ beobachtete Anzahl von Sternen in $M_1 - \frac{1}{2} < M_v < M_2 + \frac{1}{2}$

	Sp	M_1	M_2	n	Lit.	$\mathfrak{M}$	$\mathfrak{M}_0$
						Sonnenmassen	
h Per	B0	—7	+1	457	1	13400	13400
M 11	B8.5	—1	+2.5	308	2	4300	4620
NGC 663	B1	—4	0	64	3	3490	3530
M 29	B0	—5	—1	26	3	2810	2810
NGC 457	B2.5	—3	+1	70	3	2060	2100
NGC 2362	B1	—7	0	37	4, 5	2000	2000
M 41	B8	0	+2	53	6	990	1100
NGC 2264	B2.5	—5	+8	384	7	1010	1010
NGC 2516	B8	—1	+2	49	8	874	940
NGC 7243	B7	—1	+3	76	3	846	910
NGC 7209	B7	0	+3	70	3	798	885
M 34	B7	—1	+1	24	9	751	808
Praesepe.	A0	0	+5	125	10	686	763
Hyaden	A3	0	+6	146	10	600	666
M 39	B9	—1	+1	18	3	563	604
Plejaden.	B6	—3	+5	98	10	554	566
Urs. Maj. Strom . .	A0	0	+3	30	11	385	427

[1] SANDAGE (1957); [2] JOHNSON, SANDAGE, WAHLQUIST (1956); [3] BECKER, STOCK (1953); [4] JOHNSON (1950); [5] JOHNSON, HILTNER (1956); [6] COX (1954); [7] WALKER (1956); [8] COX (1955); [9] JOHNSON (1954a); [10] MAWRIDIS (1956); [11] GLIESE (1941).

A. III. b) Auflösung durch äußere Kräfte

1. BOK (1934) untersucht drei Arten von äußeren Einflüssen, die die Auflösung von offenen Haufen und Bewegungshaufen verursachen: die Gezeitenkraft der Milchstraße, ihre Scherung (differentielle Rotation) und der Energieaustausch zwischen Haufen- und Feldsternen.

Ist die Gezeitenwirkung der Milchstraße größer als die Eigengravitation eines Haufens, so löst er sich auf. Wie BOK (1934, 1951) gezeigt hat, hängt die Entscheidung hierüber im wesentlichen nur von der Dichte des Haufens ab, sobald alle galaktischen Daten festliegen.

Für die kritische Dichte, bei der die Gezeitenwirkung gerade gleich der Eigengravitation ist, gibt BOK an:

$$\varrho_{krit} = 6.2 \cdot 10^{-24} \text{ g/cm}^3 = 0.093 \; \mathfrak{M}_\odot/\text{pc}^3 \; . \tag{92}$$

Haufen geringerer Dichte sind instabil. Stark abgeflachte Haufen werden schon etwas eher instabil: bei einem Achsenverhältnis 1 : 3 ist die kritische Dichte doppelt so groß.

2. Auch etwas oberhalb der kritischen Dichte sind die Haufen nicht auf die Dauer stabil. BOK schätzt den Energieaustausch der Haufensterne mit den Feldsternen ab. Die resultierende Auflockerung des Haufens bringt die Dichte langsam unter den kritischen Wert. BOK findet, daß ein Haufen von anfangs 12 pc Durchmesser und einer Dichte von 2 ϱ_{krit} nach etwa $3 \cdot 10^9$ Jahren völlig aufgelöst ist.

3. Im ersten Stadium der Auflösung wird sich der Haufen vorwiegend in Richtung und Gegenrichtung des Zentrums der Milchstraße verlängern. Durch die differentielle Rotation der Milchstraße setzt später ein Scherungseffekt ein, der eine Verlängerung in Rotationsrichtung bewirkt. BOK schätzt ab, in welchem Maße instabile alte Haufen senkrecht zur Ebene der Milchstraße abgeflacht sein sollten. Aus dem Vorhandensein einiger instabiler, nicht oder kaum abgeflachter Bewegungshaufen (z. B. Ursa Major-Strom) schließt BOK, daß diese Haufen nicht älter als $3 \cdot 10^8$ Jahre sein dürften:

$$t_{Urs\,Maj} \leqq 3 \cdot 10^8 \text{ Jahre.} \tag{93}$$

Dies Ergebnis steht im Einklang mit einem Alter von $3 \cdot 10^8$ Jahren, das sich nach v. HOERNER (1957b) aus dem oberen „Knie" im HR-Diagramm und der Theorie der Sternentwicklung für den Ursa Major-Strom ergibt.

A.III. c) Auflösung durch inneren Energieaustausch

1. Durch gravitative Wechselwirkung tauschen die Sterne eines Haufens untereinander Energie aus; dadurch erhalten ab und zu einige Sterne Überentweichgeschwindigkeit und gehen dem Haufen verloren. Dieser Vorgang wurde von SPITZER (1940) untersucht und von CHANDRASEKHAR (1947) fortgeführt.

Nennen wir $\mathfrak{M}$ die Masse eines Haufens und r seinen Radius, so beträgt nach CHANDRASEKHAR (1947, Formel 5. 218)

$$T = 5.8 \cdot 10^5 \, (\mathfrak{M} \, r^3)^{1/2} \quad \begin{array}{l} T \text{ in Jahren} \\ r \text{ in Parsec} \\ \mathfrak{M} \text{ in Sonnenmassen} \end{array} \tag{94}$$

wobei wir die mittlere Sternmasse mit $1/2$ Sonnenmassen eingesetzt und einen logarithmischen Term konstant (mit $\mathfrak{M} = 1500$) gesetzt haben.

Setzt man mit SPITZER und CHANDRASEKHAR voraus, daß sich während jeder Relaxationszeit eine Maxwellsche Geschwindigkeitsverteilung einstellt, so erhält pro Relaxationszeit der Bruchteil

$$q = 0.0074 \tag{95}$$

aller Sterne Überentweichgeschwindigkeit und verläßt den Haufen.

Gegen diese Art der Behandlung des Problems lassen sich zwar einige Einwände vorbringen [s. z. B. v. HOERNER (1958)], doch dürften die Gl. (94) und (95) zumindest der Größenordnung nach richtige Ergebnisse liefern.

2. Falls die Relaxationszeit T konstant bliebe, so würde die Masse eines Haufens exponentiell abklingen mit einer Halbwertszeit von

$$\frac{T \ln 2}{q} = 93.7 \cdot T. \tag{96}$$

In Wirklichkeit nehmen jedoch $\mathfrak{M}$ und r zeitlich ab und somit auch T, die Auflösung geht also schneller. Dies wurde durch v. HOERNER (1958) anhand von drei Modellen näher untersucht.

In Modell 1 verlassen den Haufen alle Sterne, die nach Gl. (94) und (95) Überentweichgeschwindigkeit erhalten. Sie nehmen den Bruchteil

$$p = 0.0052 \tag{97}$$

der Energie des Haufens mit, wodurch sich der Haufenradius verkleinert. Die Differentialgleichung für die Auflösung des Haufens ist analytisch integrierbar; es ergibt sich, daß die Relaxationszeit linear mit der Zeit abnimmt und daß jeder Haufen nach einer Lebensdauer

$$t_H = 30 \, T_0 \tag{98}$$

nahezu restlos aufgelöst ist ($T_0=$ ursprüngliche Relaxationszeit). Eine ähnliche Untersuchung wurde von J. KING (1957) durchgeführt, der unter Vernachlässigung von Gl. (97) eine Auflösung in $42 \, T_0$ erhält.

Bei Modell 3 wird berücksichtigt, daß den Haufen bereits alle die Sterne verlassen, die weiter fliegen als die durch die Gezeitenkraft der Milchstraße gegebene Stabilitätsgrenze r_s. Die Lebensdauer eines Haufens kann sich hierdurch stark verkürzen, sie beträgt z. B. nur noch $8.4 \, T_0$, falls wir den Haufen mit $r_s = 2 \, r$ beginnen lassen.

3. Bei v. HOERNER (1958) ist die Lebensdauer als Funktion von Masse $\mathfrak{M}$ und Radius r in Form einer Tabelle angegeben. Wir möchten hier eine andere Art der Darstellung wählen, die sich mit der Auflösung nach BOK kombinieren läßt. Unsere eigentliche Frage ist, ob offene Haufen ein Alter von $5 \cdot 10^9$ Jahren erreichen können oder nicht. Wir fragen deshalb nach denjenigen Werten von $\mathfrak{M}$ und r, bei denen sich ein Haufen in einer etwas kürzeren Zeit, sagen wir in $3 \cdot 10^9$ Jahren, gerade auflöst.

Bei Modell 1 muß nach Gl. (98) die Relaxationszeit dann

$$T_0 = 3 \cdot 10^9/30 = 10^8 \text{ Jahre}$$

betragen, und aus Gl. (94) erhalten wir

$$\mathfrak{M} = \frac{3 \cdot 10^4}{r^3} \cdot \quad \begin{array}{l} \mathfrak{M} \text{ in Sonnenmassen} \\ r \text{ in Parsec} \end{array} \quad . \tag{99}$$

Ist Gl. (99) erfüllt, so löst sich der Haufen in $3 \cdot 10^9$ Jahren auf; ist die Masse größer, so dauert die Auflösung länger. Die entsprechende Linie ist in Abb. 14 eingezeichnet. Bei Modell 3 haben wir die v. Hoernersche Tabelle 5 auf $3 \cdot 10^9$ Jahre interpoliert und den resultierenden Zusammenhang zwischen $\mathfrak{M}$ und r in Abb. 14 mit eingezeichnet.

Die rechte Gerade der Abb. 14 entspricht der kritischen Dichte [Gl. (92)] nach Bok:

$$\mathfrak{M} = 0.38 \, r^3. \tag{100}$$

Der einzelne Punkt oberhalb dieser Geraden entspricht der Bokschen Abschätzung, die wir in Absatz A.III.b2 zitierten. Die gestrichelte Verbindungslinie ist frei geschätzt.

Als wichtigstes Ergebnis entnehmen wir der Abb.14: alle Haufen, deren Masse kleiner als 1000 Sonnenmassen ist, können nicht älter werden als $3 \cdot 10^9$ Jahre. Ist der Radius kleiner als etwa 5 pc, so lösen sie sich auf durch gegenseitigen Energieaustausch der Haufensterne, der Radius nimmt ab. Ist der Radius größer als etwa 10 pc, so lockern sie sich auf durch Wechselwirkung mit Feldsternen und werden durch die Gezeitenkraft und die Scherung der Milchstraße aufgelöst. Der Radius nimmt zu.

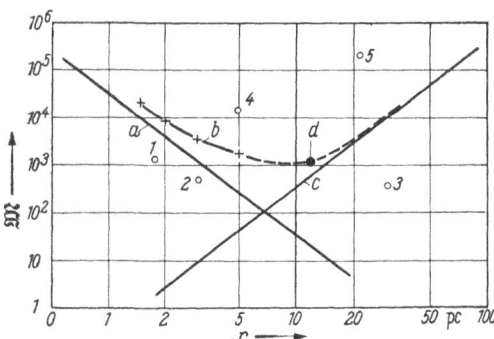

Abb. 14. Grenzwerte von Masse und Radius für die Auflösung von Sternhaufen in $3 \cdot 10^9$ Jahren. Unterhalb der Linie geht die Auflösung schneller. a Modell 1, b Modell 3, nach v. Hoerner (1958), c kritische Dichte, d siehe Abschnitt A III b 2, nach Bok (1934) und (1951). 1 Mittelwerte offener Haufen, 2 Plejaden, 3 Ursa Major Strom, 4 h-Persei, 5. Kugelhaufen

Die offenen Sternhaufen liegen fast alle unterhalb der Grenzlinie b. Setzen wir die mittlere Masse nach Gl. (91) mit 1500 Sonnenmassen an, so ergibt sich nach v. Hoerner eine mittlere Auflösungszeit von

$$\bar{t}_H = 1 \cdot 10^9 \text{ Jahren} \tag{101}$$

und für die Plejaden erhalten wir z. B. $1.2 \cdot 10^9$ Jahre. Ein extrem massereicher Haufen wie h Persei ist dagegen erst nach $1 \cdot 10^{10}$ Jahren aufgelöst, und die Kugelhaufen sind es erst nach etwa 10^{12} Jahren.

4. Diese Zahlen beziehen sich auf die Zukunft. Bezüglich der Vergangenheit läßt sich allein bei stark aufgelockerten Bewegungshaufen nach Gl. (93) ein Maximalalter angeben, nicht dagegen bei Auflösung durch gegenseitigen Energieaustausch der Haufensterne. Die Mehrzahl der offenen Haufen könnte daher, rein dynamisch, auch wesentlich älter sein als Gl. (101).

Kombinieren wir jedoch die dynamische Betrachtung mit der Theorie der Sternentwicklung, so erhalten wir nach v. HOERNER (1958) eine Möglichkeit der gegenseitigen Ergänzung und Kontrolle. Wir beginnen damit, daß alle Haufen, die noch Sterne frühen Typs enthalten, relativ jung sind. Die Mittelung, die zu Gl. (101) führte, wurde nur über relativ junge Haufen erstreckt; daraus folgt, daß Gl. (101) nicht nur die zukünftige, sondern die *gesamte* Lebensdauer offener Haufen darstellt. Wenn das richtig ist, so folgt, daß es keine (oder, je nach Streuung, nur relativ wenige) offene Haufen geben dürfte, die älter sind als 10^9 Jahre. Bei diesem Alter liegt das Knie der Hauptreihe etwa bei A 3. Bezeichnen wir als Typ eines Haufens den Spektraltyp seiner frühesten (am Knie liegenden) Sterne, so dürfte es also nur sehr wenige Haufen von späterem Typ als A 3 geben.

In Abschnitt B.II.c werden wir die beobachtete Typenhäufigkeit der offenen Haufen näher untersuchen. Hier nehmen wir nur vorweg, daß die Typen B0 bis A0 einigermaßen gleichmäßig mit durchschnittlich 8 Haufen pro Unterklasse besetzt sind; bei A 1—A 2 sinkt die Häufigkeit steil ab, und später als A 2 sind insgesamt nur 5 Haufen (von 100) registriert. Die in B.II.c diskutierte Entdeckungswahrscheinlichkeit offener Haufen nimmt zwar auch zu späteren Typen hin ab, kann jedoch das plötzliche Absinken nicht erklären. Setzt man das Abbrechen der Häufigkeit etwa bei A 1.5 an, so ergibt Tab. 4 ein zugehöriges Alter von

$$t_H = 5 \cdot 10^8 \text{ Jahren.} \tag{102}$$

Die Ergebnisse der Gl. (101) und (102) unterscheiden sich um einen Faktor 2. Die Fehlergrenze von Gl. (102) wurde im Absatz A.II.5 auf einen Faktor 1.5 abgeschätzt, und die in dem dynamischen Auflösungsalter [Gl. (101)] steckenden Unsicherheiten dürften eher noch größer sein. Wir dürfen daher sagen, daß Gl. (101) und (102) innerhalb der Fehlergrenzen übereinstimmen. Wir kommen somit zu dem Schluß, daß die Mehrzahl der offenen Sternhaufen jünger als 10^9 Jahre ist.

A. III. d) Expandierende Assoziationen

1. Aus Untersuchungen von Eigenbewegungen scheint hervorzugehen, daß einige der Assoziationen expandieren. Aus dem Gang der Geschwindigkeiten mit der Entfernung vom Zentrum läßt sich ein „Expansionsalter" ableiten. Die bisher untersuchten Fälle wurden

durch v. HOERNER (1957b) zusammengestellt und lassen sich mit dem aus der Sternentwicklung folgenden „Entwicklungsalter" vergleichen. Da die Angaben bei kleineren Geschwindigkeiten extrem unsicher werden, wiederholen wir hier nur die Fälle mit großer Geschwindigkeit (in den drei weggelassenen Fällen ist $v_{exp} \leq 2 \frac{km}{sec}$).

Tabelle 10. *Expansionsalter t_{exp} von Assoziationen, verglichen mit ihrem Entwicklungsalter t_{entw}*

Assoziation	Typ	v_{exp} km/sec	t_{exp} 10^6 Jahre	Literatur	t_{entw} 10^6 Jahre
Lacerta-Ass.	B1	8	4.2	BLAAUW und MORGAN (1953a)	6.8
II (ζ) Per-Ass.	B0.5	12	1.5	DELHAYE und BLAAUW (1953)	5.5
I_I Cep-Ass.	O6	8	4.5	MARKARJAN (1953)	0.7
			Mittel 3.4		Mittel 4.3

Die Übereinstimmung der Expansionsalter mit den Entwicklungsaltern ist nur im Mittel gut, während man im Einzelfall nur von einer Übereinstimmung der Größenordnung sprechen kann. Berücksichtigen wir jedoch, daß der Prozeß der Sternbildung sich nach BLAAUW (1956a) und v. HOERNER (1957b) über einen Zeitraum von $2 \cdot 10^6$ bis $5 \cdot 10^6$ Jahren erstreckt, so ist vielleicht nicht mehr zu erwarten.

2. Inzwischen ist an der Vorstellung der expandierenden Assoziationen einige Kritik geäußert worden. PETRIE (1957) berichtet, daß neuere Messungen der Radialgeschwindigkeiten der I (h) Per-Assoziation keinerlei Expansionseffekt ergeben haben, und HOROSHEWA (1956) kommt zu dem Schluß, daß die II Cep-Assoziation nicht expandiert, sondern als Aggregat von 5 Sternströmen aufzufassen ist, die weder in sich, noch von einem gemeinsamen Zentrum her expandieren. Betrachten wir daraufhin nochmals die in Tab. 10 zitierten Arbeiten, so scheint die Expansion der II Cep-Assoziation tatsächlich am fraglichsten zu sein. Für die Lacerta-Assoziation ergab sich zwar eine gute Korrelation zwischen μ_ε und der Deklination; tragen wir jedoch $\mu_\alpha \cos \delta$ über der Rektaszension auf, so ist fast keine Korrelation vorhanden. Am deutlichsten ist der Effekt einer Expansion wohl bei der II (ζ) Per-Assoziation zu sehen, bei der die Korrelation in jeder Richtung befriedigend sein dürfte.

3. Zusammenfassend können wir sagen, daß die Expansion der II Per-Assoziation höchstwahrscheinlich reell ist, die der Lacerta-Assoziation vermutlich auch, und die der II Cep-Assoziation vielleicht nicht. In den beiden ersten Fällen erhalten wir Expansionsalter, die von gleicher Größenordnung sind wie die aus der Theorie der Sternentwicklung folgenden Alter (sie sind zu klein um Faktoren 3.7 und 1.6).

A.III. e) Einzelne schnelle Sterne

1. Nach BLAAUW und MORGAN gibt es einige Sterne frühen Typs, die zwar in keiner Assoziation stehen, deren Eigenbewegungen es jedoch sehr wahrscheinlich machen, daß sie „Ausreißer" (runaway stars) benachbarter Assoziationen sind. Tab. 11 zeigt die bisher veröffentlichten Beispiele.

Tabelle 11. *Assoziationen mit „Ausreißern"*

Assoziation			Stern				
Name	Sp	t_{entw}	Name	Sp	v	t_* Jahre	Lit.
Orion-Assoz.	O 9	$3 \cdot 10^6$ Jahre	AE Aurigae	O 9.5	128 km/sec	$2.6 \cdot 10^6$	1
			μ Columbae	B 0 V	128 km/sec	$2.6 \cdot 10^6$	2, 3
			53 Arietis	B 2 V	80 km/sec	$4.8 \cdot 10^6$	1
I Cep-Assoz.	O 9	$3 \cdot 10^6$ Jahre	68 Cygni	O 8	45 km/sec	$5.1 \cdot 10^6$	1
Sco-Cen-Assoz.	O 9.5	$4 \cdot 10^6$ Jahre	ζ Ophiuchi	O 9.5	32 km/sec	$3.0 \cdot 10^6$	4
Lacerta-Assoz.	B 1	$6.8 \cdot 10^6$ Jahre	HD 197419	B 2 Vc	35 km/sec	$5 \cdot 10^6$	1
			HD 201910	B 5 V	35 km/sec	$5 \cdot 10^6$	1

[1] BLAAUW (1956a); [2] BLAAUW, MORGAN (1953b); [3] BLAAUW, MORGAN (1954); [4] BLAAUW (1952).

Zum Vergleich mit dem aus der Geschwindigkeit v des Sternes (und seinem Abstand von der Assoziation) folgenden Alter t_* des Sternes haben wir das [nach v. HOERNER (1957b)] aus dem bei Sp liegenden Knie folgende Entwicklungsalter der Assoziation mit angegeben. Da die Abweichungen beider Altersangaben höchstens einen Faktor 1.7 erreichen und dem Betrag nach gemittelt nur etwa 30% betragen, können wir die Übereinstimmung als durchaus befriedigend bezeichnen.

Eine Schwierigkeit sei noch erwähnt. Nach MÜNCH (1957) ist der Stern 53 Arietis vermutlich ein β Canis Majoris-Stern vom Spektraltyp B 2 IV, mit einer Periode von etwa 4 Std., in der die Radialgeschwindigkeit um etwa 5 km/sec schwankt. Da 53 Arietis deutlich über der Hauptreihe liegt, schließt MÜNCH auf ein Alter des Sternes von $50 \cdot 10^6$ Jahren, was sowohl zu dem Entwicklungsalter der Orion-Assoziation als auch zu dem Laufzeit-Alter t_* im Widerspruch stünde. Wir möchten dagegen zweierlei einwenden. Erstens: läge 53 Arietis (im ungünstigsten Fall) genau am Knie einer Isochrone, so folgt aus dem Spektraltyp B 2 nach Tab. 4 ein Alter von nur 10^7 Jahren, und ein kürzeres Alter bei jeder anderen Lage. Zweitens ist in Höhe des Knies meist eine starke Streuung zu beobachten. Bei der Orion-Assoziation reicht die Streuung von O 9 bis B 1 [v. HOERNER (1957b), Tab. 1], so daß das Hinzukommen von 53 Arietis lediglich die Streuung bis B 2 vergrößern würde.

2. Eine zweite Sorte schneller Einzelsterne wurde von MÜNCH (1956) untersucht. Es handelt sich um frühe B-Sterne in hoher galaktischer

Breite. Drei Sterne sind vom Typ B1 Ib und einer vom Typ B1 II, und die Entfernungen von der galaktischen Ebene liegen zwischen 670 und 2700 pc. Aus den beobachteten Radialgeschwindigkeiten schließt MÜNCH (bei unbekannten Eigenbewegungen) auf eine mittlere Laufzeit von etwa $2 \cdot 10^7$ Jahren von ihrem vermutlichen Start in der galaktischen Ebene bis zu ihrem jetzigen Ort. Ihr Entwicklungsalter jedoch sollte wesentlich kürzer sein. B1 I-Sterne sollten auf Isochronen liegen, deren Knie spätestens bei B0 liegt, woraus wir nach Tab. 4 ein Entwicklungsalter von höchstens $5 \cdot 10^6$ Jahren ableiten. Wir erhalten somit eine Diskrepanz von mindestens einem Faktor 4, deren Lösung noch aussteht.

A.IV. Zusammenfassung der Altersangaben

Das weitaus wichtigste Hilfsmittel der Altersbestimmung von Sternen und Sterngruppen ist durch die Theorie der Sternentwicklung gegeben, die wir anfangs eingehend diskutierten. Eine ausführliche und kritische Zusammenfassung wurde bereits zu Beginn des Abschnittes A.I.c gegeben und soll hier nicht wiederholt werden. Daher beschränken wir die gegenwärtige Diskussion auf den Vergleich der Altersangaben.

Wir hatten uns anfangs die Aufgabe gestellt, erstens die Methoden zu schildern, nach denen sich das Alter von Sternen und Sterngruppen bestimmen läßt, zweitens die Ergebnisse soweit als möglich auf ihre Verläßlichkeit zu prüfen, und drittens zu entscheiden, wieweit man zu der Annahme gezwungen ist, daß auch in der Gegenwart noch Sterne entstehen.

Drei Methoden der Altersbestimmung wurden behandelt, die voneinander völlig unabhängig sind:

1. Die Theorie der Sternentwicklung (oberes Knie der Hauptreihe).
2. Die Kontraktion junger Sterne (unteres Abbrechen der Hauptreihe).
3. Dynamische Altersangaben [a) expandierende Assoziationen, b) Ausreißer, c) Auflösung offener Haufen].

Als Kontrolle der Güte der Altersabschätzungen haben wir laufend die Ergebnisse der verschiedenen Methoden miteinander verglichen; als Zusammenfassung bringen wir in Tab. 12 alle diejenigen Fälle, für die mehrere voneinander unabhängige Abschätzungen vorliegen.

Aus der Tab. 12 lesen wir folgendes Ergebnis unserer Untersuchung ab: erstens erhalten wir stets miteinander verträgliche Angaben in all den Fällen, wo nur obere oder untere Grenzen angegeben werden können. Zweitens: bei den direkten Altersangaben beträgt die größte Abweichung einen Faktor 3.7, oder, wenn wir die in der Tabelle nicht mit enthaltenen jungen B-Sterne hoher galaktischer Breite mit einbeziehen, einen Faktor 4. Dem Betrag nach gemittelt ist die Abweichung je zweier

Angaben ein Faktor 1.6. Das bedeutet:
> Die durchgeführten Altersabschätzungen sind zuverlässig,
> der mittlere Fehler ist ein Faktor 1.6.

Weiterhin zeigt die Übereinstimmung im oberen Teil der Tab. 12, daß die Existenz relativ junger Sterngruppen kaum zu bezweifeln ist. Zusammen mit Tab. 6 von v. HOERNER (1957b) erhalten wir für Assoziationen ein Alter zwischen $0.4 \cdot 10^6$ und $7 \cdot 10^6$ Jahren. Von 19 untersuchten offenen Sternhaufen liegen 17 im Bereich von $4 \cdot 10^6$ bis 10^9 Jahren, und nur zwei sind älter als 10^9 Jahre.

Tabelle 12. *Objekte mit mehr als einer Altersbestimmung (in Jahren)*

	Entwicklung	Kontraktion	Expansion	Dynamik Ausreißer	Auflösung
Assoziationen					
Orion-Assoz.	$3 \cdot 10^6$	$\geq 2 \cdot 10^6$		$\begin{cases} 2.6 \cdot 10^6 \\ 2.6 \cdot 10^6 \\ 4.8 \cdot 10^6 \end{cases}$	
Lacerta-Assoz.	$6.8 \cdot 10^6$		$4.2 \cdot 10^6$	$\begin{cases} 5 \cdot 10^6 \\ 5 \cdot 10^6 \end{cases}$	
II (ζ) Per-Assoz.	$5.5 \cdot 10^6$		$1.5 \cdot 10^6$		
I Cep-Assoz.	$3 \cdot 10^6$			$5.1 \cdot 10^6$	
Sco-Cen-Assoz.	$4 \cdot 10^6$			$3 \cdot 10^6$	
Offene Haufen NGC 2264	$<15 \cdot 10^6$	$\geq 2 \cdot 10^6$			
Urs Maj-Strom	$3 \cdot 10^8$				$\leq 3 \cdot 10^8$
Mittlere Lebensdauer	$5 \cdot 10^8$				$1 \cdot 10^9$
Älteste Objekte M 67	$4.6 \cdot 10^9$	Verglichen mit $\frac{1}{H} = 5.4 \cdot 10^9$			
Schnelläufer	$5.7 \cdot 10^9$	(HUMASON, MAYALL, SANDAGE, 1956)			
Kugelhaufen	$5.6 \cdot 10^9$				

Eine große Anzahl der Objekte aus Tab. 12 sind somit vor einigen Millionen Jahren entstanden, was man, kosmologisch gesehen, durchaus als „Gegenwart" bezeichnen kann, und das Ergebnis dieses Kapitels lautet:
> Die gegenwärtige Entstehung von Sternen kann als gut
> gesichert bezeichnet werden.

B. Voruntersuchungen zur Sternentstehung

Nach den Ergebnissen des vorigen Kapitels müssen wir das Vorhandensein von relativ jungen Sternen und Sterngruppen als gesichert betrachten. Der Anlaß, nach einer Theorie der Sternentstehung — unter gegenwärtigen kosmischen Bedingungen — zu suchen, ist somit klar gegeben.

Wir wiesen bereits in der Einleitung darauf hin, daß eine geschlossene Theorie der Sternentstehung noch nicht vorliegt. Es ist deshalb die Aufgabe des gegenwärtigen Kapitels, eine Anzahl von einzelnen Voruntersuchungen anzustellen, die dazu verhelfen sollen, Unterlagen, Rahmen und Aufgaben einer zukünftigen Theorie soweit möglich festzulegen. In den Fällen, wo uns eine Festlegung nicht oder nicht ausreichend gelang, wollen wir zumindest auf die Wichtigkeit der angeschnittenen Fragen hinweisen und damit zu weiteren Untersuchungen anregen.

Als erstes untersuchen wir die Frage, ob die beobachtete Verschiedenheit der Raumgeschwindigkeiten von Sternen verschiedenen Spektraltyps bereits bei der Sternentstehung vorliegen muß oder als nachträglicher Alterseffekt gedeutet werden kann. Zweitens fragen wir nach der zeitlichen Rate der Sternentstehung (= Anzahl entstehender Sterne pro Zeit und Volumen, als Funktion der Zeit), um zu erfahren, ob die jungen Sterne nur wenige, gelegentliche Nachzügler sind oder ob die Sternentstehung heute genauso häufig ist wie früher.

Weiterhin ist zu fragen, ob das Gros der Sterne einzeln oder in Haufen bzw. Assoziationen entsteht, und ob die Assoziationen stabile Gebilde sein können. Schließlich stellen wir noch einige Bilanzen auf bezüglich der vorhandenen Menge von Weißen Zwergen und Helium.

B.I. Geschwindigkeit und Spektraltyp

Aus der Beobachtung und zahlreichen statistischen Auswertungen ist bekannt, daß Sterne verschiedenen Spektraltyps verschiedene Geschwindigkeits-Verteilungen besitzen. Der auffälligste Zug ist eine Zunahme der Streuung mit zunehmendem Typ, aber auch die Schwerpunkte der Geschwindigkeiten sind vom Spektraltyp abhängig, und auch die Form und Achsenrichtung der Geschwindigkeitskörper zeigen gewisse Unterschiede.

Ein Teil der Effekte mag durch einzelne Sternströme bewirkt werden sowie durch eine verschieden hohe Beimischung von Schnelläufern; außerdem treten mit abnehmender Helligkeit Auswahleffekte auf zugunsten höherer Geschwindigkeit. Weiterhin sind die beobachteten Effekte zum Teil von einander abhängig: z. B. muß sich bei zunehmender Streuung die einseitige Begrenzung durch die Entweichgeschwindigkeit der Milchstraße bemerkbar machen und eine Verschiebung des Schwerpunktes entgegen der Rotationsrichtung bewirken. Nur ein Ausschnitt dieser Fragen soll hier untersucht werden.

Für unser Problem der Sternentstehung interessiert am meisten die Streuung der räumlichen Geschwindigkeit. Mit welcher kinetischen

Energie sind die Sterne bei ihrer Entstehung versehen? Ist es nötig anzunehmen, daß Sterne verschiedener Masse mit verschiedener Geschwindigkeit entstehen? Als nächstes interessiert speziell die Streuung der z-Komponente (senkrecht zur galaktischen Ebene), da sie Aufschluß gibt über die Dicke der Schicht, über die sich die betrachteten Sterne verteilen. Nur so können wir eine über z integrierte Leuchtkraftfunktion erhalten, die wir im folgenden Kapitel benötigen werden. Im Anschluß an die Schichtdicke stellen wir noch die Frage, ob auch die Sterne größerer Schichtdicke ursprünglich in der relativ dünnen Schicht des interstellaren Gases entstanden sind oder nicht.

B.I. a) Die Streuung der räumlichen Geschwindigkeiten

1. Wir beginnen mit einem mehr qualitativen Hinweis. In Abb. 15 tragen wir über dem Spektraltyp (log T linear) zunächst die Streuung σ nach PARENAGO (1951) auf:

$$\sigma = \sqrt{\sigma_1^2 + \sigma_2^2 + \sigma_3^2},$$

wobei die σ_i die drei Achsen des jeweiligen Geschwindigkeits-Ellipsoides sind. Damit ist σ die quadratisch gemittelte Raumgeschwindigkeit, bezogen auf den jeweiligen Schwerpunkt der betrachteten Gruppe. PARENAGO verwendet ein sehr umfangreiches Material von 1312 Sternen. Wegen der schwer zu überblickenden Gefahr von Auswahleffekten wollen wir für die detaillierteren Analysen dieses Kapitels die Sterne der Sonnenumgebung (innerhalb 20 pc) des Kataloges von GLIESE (1956, 1957) vorziehen. Aus Tab. 9 der Arbeit [GLIESE (1956)] entnehmen wir für 537 Sterne der Hauptreihe die dem Betrag nach gemittelten Raumgeschwindigkeiten V, wobei wir sämtliche 237 M-Sterne verschiedener Herkunft zusammengefaßt haben. Gemeinsamer Bezugspunkt aller Gruppen ist hierbei der Schwerpunkt der McCormick-Sterne.

Trotz verschieden ausgewählten Materials, verschiedener Bezugspunkte und verschiedener Arten der Mittelung zeigen die beiden Kurven σ und V qualitativ das gleiche Verhalten: anfangs eine deutliche Zunahme der Geschwindigkeit mit wachsendem Spektraltyp, von G ab jedoch etwa konstante Geschwindigkeit.

Diese Art der Abhängigkeit vom Spektraltyp (und somit von der Masse $\mathfrak{M}$) macht es nicht sehr wahrscheinlich, daß dies die Geschwindigkeiten sind, mit denen Sterne verschiedener Masse entstehen. Vielmehr deutet der Knick gerade zwischen F und G und die Konstanz oberhalb von G darauf hin, daß wir es mit einem Alterseffekt zu tun haben. Zur Veranschaulichung zeichnen wir in Abb. 15 ein durchschnittliches Alter τ von Sternen verschiedenen Typs ein. Zum Beispiel können Sterne vom Typ A5 höchstens $1.2 \cdot 10^9$ Jahre alt sein (Tab. 4), werden also im Mittel halb so alt sein, d. h. etwa $6 \cdot 10^8$ Jahre. Von etwa F6 an ist eine konstante Altersmischung zu erwarten mit rund $\tau =$ halbes Weltalter.

Die Abb. 15 möchten wir als ein starkes Argument dafür betrachten, daß die Abhängigkeit der Geschwindigkeits-Streuung als ein Alterseffekt gedeutet werden kann. Dabei bleibt zunächst noch offen, ob die Geschwindigkeiten der Sterne anfangs klein sind und sich mit zunehmendem Alter vergrößern, oder ob die Sterne in der Vergangenheit mit größeren Geschwindigkeiten entstanden sind als heute.

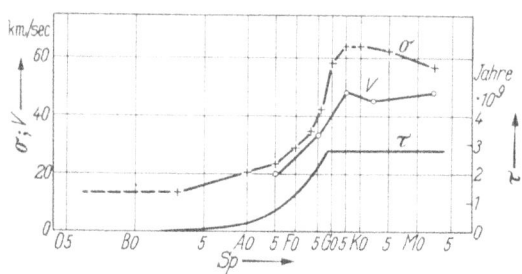

Abb. 15. Geschwindigkeit und mittleres Alter τ von Sternen in Abhängigkeit vom Spektraltyp. σ = Streuung der Raumgeschwindigkeiten nach PARENAGO. V = gemittelte Beträge der Raumgeschwindigkeit nach GLIESE

2. Für eine genauere Analyse ist in Abb. 16 das Hertzsprung-Russell-Diagramm in eine Reihe einzelner Felder unterteilt nach den folgenden Gesichtspunkten.

Feld 8 sollte nur Sterne enthalten, die garantiert sehr alt sind. Es umfaßt noch den Riesenast von M 67 (Alter $4.6 \cdot 10^9$ Jahre), ist dagegen abgesetzt gegen die Riesenäste jüngerer Sternhaufen sowie gegen die Streuung der normalen Hauptreihe. Das mittlere Alter der Sterne in Feld 8 muß somit $\geq 4.6 \cdot 10^9$ Jahre sein und soll auf $5 \cdot 10^9$ Jahre festgelegt werden. Feld 8 enthält 15 Sterne des Glieseschen Kataloges.

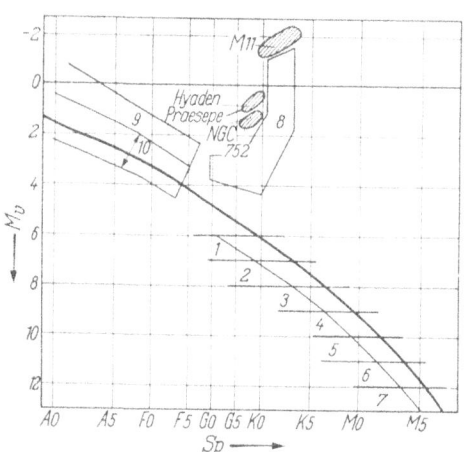

Abb. 16. H-R-Diagramm mit Feldereinteilung zur Altersbestimmung

Feld 9 enthält Sterne verschiedenen Alters, die deutlich genug von der Hauptreihe abgezweigt sind, so daß jedem einzelnen Stern nach Tab. 4 ein Alter zugeordnet werden kann. Feld 9 enthält 17 Sterne; davon haben 11 Sterne eine Leuchtkraftklasse höher als V, die restlichen 6 Sterne haben Klasse V oder keine Angabe.

Feld 10 enthält drei Sterne mit einer höheren Leuchtkraftklasse als V, weswegen wir ebenfalls ein Abgezweigtsein von der Hauptreihe annehmen und ihnen ein Alter zuordnen können. Die restlichen 26 Sterne des Feldes 10 liegen noch zu dicht an der ursprünglichen Hauptreihe. Bei ihnen läßt sich nur ein Maximalalter angeben, das sie an ihrer Stelle der Hauptreihe höchstens erreicht haben können. Da sie im Mittel etwa halb so alt sein werden, haben wir stets das halbe Maximalalter als das Alter τ eines solchen Sternes eingesetzt. Weil wir letztlich nur Mittelwerte über Gruppen betrachten, dürfte diese Näherung ausreichen.

Die 46 Sterne der Felder 9 und 10 wurden dann gemeinsam in 6 Altersgruppen eingeteilt und die Geschwindigkeiten in jeder Gruppe gemittelt. Das Ergebnis besprechen wir im Zusammenhang mit den restlichen Feldern.

3. Bevor wir die Felder 1—7 behandeln, müssen wir noch eine Entscheidung treffen über den Nullpunkt der Geschwindigkeiten. Um in jeder Altersgruppe deren eigenen Schwerpunkt zu benutzen, reicht das Material nicht aus. Wir wählen deshalb einen für alle gemeinsamen Nullpunkt, wobei wir uns vergegenwärtigen, daß eine Nullpunktsverschiebung sich nur quadratisch auf die Streuung addiert und folglich nicht viel ändert. Beträgt z. B. die Streuung einer Gruppe 30 km/sec um ihren Schwerpunkt, und benutzen wir statt dessen einen um 10 km/sec verschobenen Nullpunkt, so erhöht sich die Streuung auf $(30^2 + 10^2)^{1/2} =$ 31.6 km/sec, was im Rahmen unserer Untersuchung überhaupt nichts ausmacht.

Da es bei den jüngeren Sternen wegen ihrer geringeren Streuung der Geschwindigkeiten noch am ehesten auf Genauigkeit ankommt, wählen wir deren Schwerpunkt als allgemeinen Nullpunkt. Für 37 Sterne zwischen A0 und F2 erhalten wir als Komponenten der Sonnenbewegung die abgerundeten Werte:

Komponente	km/sec	Richtung
u	—10	weggerichtet vom gal. Zentrum
v	+10	galaktische Rotation
w	+6	galaktischer Nordpol

(103)

Dies sind etwa die gleichen Werte, wie VYSSOTSKY und JANSSEN (1951) sie vorschlagen für die Abweichung der Sonnenbewegung von der galaktischen Kreisbahn.

4. *Felder 1—7*. Eine weitere Möglichkeit, eine Gruppe von extrem alten Sternen aus dem HR-Diagramm auszusondern, besteht — wie uns scheint — in folgendem. Wenn es richtig ist, daß die schwereren Elemente in Sternen produziert werden und im Verlaufe der Sternentwicklung zwischen dem Stadium des Roten Riesen und dem des Weißen Zwerges wieder an die interstellare Materie abgegeben werden, so sollten die alleraltesten Sterne noch keine (oder zumindest sehr wenig) schwere Elemente enthalten. Nach einer Arbeit von REIZ (1954) sollten aber Sterne, deren Gehalt an schweren Elementen zu vernachlässigen ist,

links von der normalen Hauptreihe liegen, bzw. unterhalb der Hauptreihe, und zwar um 1—1$^1/_2$ Größenklassen.

In Abb. 16 haben wir die dick eingezeichnete Hauptreihe um eine Größenklasse parallel nach unten verschoben. Unterhalb dieser Grenzlinie liegen die Felder 1—7, in denen bei Gültigkeit unserer Voraussetzungen nur extrem alte Sterne liegen sollten. Um Alterseffekte durch Sternentwicklung auszuschließen, haben wir erst bei $M = 6$ begonnen. Unterhalb von $M = 13$ sind nur noch wenige Sterne im Katalog enthalten.

Es zeigt sich, daß die Sterne in den Feldern 1—7 im Mittel weit höhere Geschwindigkeiten besitzen als die Sterne der Hauptreihe, in Übereinstimmung mit Abb. 15, wo auch die älteren Sterne höhere Geschwindigkeiten zeigen. Um zu prüfen, ob dieser Effekt reell ist, müssen wir ihn von störenden Effekten trennen. Den von uns gesuchten Effekt (höhere Geschwindigkeit der Sterne, die links der Hauptreihe liegen) wollen wir mit *Lageeffekt* bezeichnen. In gleicher Richtung wirkt ein Störeffekt, den wir hier mit *Helligkeitseffekt* bezeichnen wollen; schwächere Sterne sind schwerer zu entdecken und fallen meist nur dann auf, wenn sie hohe Eigenbewegungen haben. Daher sollten Sterne unterhalb der Hauptreihe im Mittel höhere Geschwindigkeiten zeigen als Hauptreihensterne gleichen Spektraltyps. In umgekehrter Richtung wirkt ein zweiter Effekt, den wir *Güteeffekt* nennen wollen: Sterne, deren Parallaxe irrtümlich zu groß bestimmt wurde, liegen erstens unterhalb der Hauptreihe und erhalten zweitens zu kleine Eigenbewegungen. Drittens zeigte sich, daß Sterne mit *Emissionsspektrum* erstens keinen Lageeffekt zeigen und zweitens ganz allgemein (auch auf der Hauptreihe) kleinere Geschwindigkeiten besitzen als sonst gleiche normale Sterne [GLIESE (1956)]. Deshalb wurden in der folgenden Untersuchung alle Sterne mit Emission weggelassen.

Zunächst der Güteeffekt. GLIESE hat die Sterne seines Kataloges in sechs Güteklassen eingeteilt; wir testen den Güteeffekt in Tab. 13. Mit V ist bezeichnet die dem Betrag nach gemittelte Raumgeschwindigkeit, und n ist die jeweilige Anzahl von Sternen. Zum Vergleich geben wir die mittlere Geschwindigkeit der Hauptreihensterne im gleichen Helligkeitsbereich mit

$$\bar{V}_{HR} = (46 \pm 2) \text{ km/sec} \tag{104}$$

an.

Wir sehen aus Tab. 13, daß der von Klasse a bis c recht große Lageeffekt (plus eventuellem Helligkeitseffekt) von Klasse d ab durch den Güteeffekt völlig kompensiert worden ist. Daher werden wir im folgenden nur die Güteklassen a, b, c verwenden, obwohl die Anzahl verwendbarer Sterne sich damit auf 16 verringert.

In Tab. 14 sind Lage- und Helligkeitseffekt getrennt. Die Spalte „links" enthält alle Sterne der Güteklassen a, b, c in den links der Hauptreihe liegenden Feldern 1—7. Die Spalte „Alle Sterne gleicher Helligkeit" umfaßt sowohl die Sterne der Hauptreihe, als auch die links von ihr liegenden. Der Helligkeitseffekt sollte sich bei *allen* Sternen zeigen durch eine Zunahme von V in Richtung abnehmender Helligkeit. Bis einschließlich Feld 4 ist der Helligkeitseffekt nicht zu bemerken; von da ab werden die Geschwindigkeiten V zwar höher, jedoch nicht sehr viel, und keinesfalls

Tabelle 13. *Abhängigkeit der mittleren Raumgeschwindigkeit V in den Feldern 1—7 von der Güte der Parallaxenbestimmung*

Güte-klasse	Wahrsch. F. der abs. Helligkeit von	bis	V km/sec	n
a	0ᵐ00	0ᵐ08	102 ± 31	8
b	0.09	0.15		
c	0.16	0.25	109 ± 31	8
d	0.26	0.35	47 ± 9	10
e	0.36	0.50	39 ± 10	14
f	über 0.50			

vergleichbar mit dem deutlichen Lageeffekt, der sich in jeder Zeile aus dem Vergleich „Alle Sterne" — „links" ergibt. Es erscheint somit statthaft, trotz des vorhandenen Helligkeitseffektes, die Felder 1—7 alle zusammenzufassen. Im Mittel über die 16 Sterne erhalten wir

$$V_{links} = (106 \pm 21) \text{ km/sec}. \tag{105}$$

Bei den 16 Sternen sind 4 als Unterzwerge bezeichnete mit enthalten, die für sich allein $V_{sd} = (84 \pm 19)$ km/sec ergeben und somit anscheinend keine Sonderstellung einnehmen.

Aus dem Vergleich zwischen (104) und (105) schließen wir auf einen deutlich vorhandenen und relativ gut gesicherten Lageeffekt: die Geschwindigkeiten der Sterne unterhalb der Hauptreihe sind um mehr als einen Faktor 2 größer als die von gleichhellen Hauptreihensternen. Ihre Lage unterhalb der Hauptreihe läßt nach REIZ (1954) auf verschwindenden Metallgehalt schließen. Falls die Metalle erst durch Sternentwicklung erzeugt (bzw. angereichert) werden, so sollten dies die ältesten Sterne überhaupt sein. Da der Riesenast bei geringem Metallgehalt steiler verläuft [HOYLE, SCHWARZSCHILD (1955)] als das Feld 8 unserer Abb. 16, so sollten die Sterne der Felder 1—7 noch älter sein als die des Feldes 8. Wir schätzen ihr Alter auf etwa $5.4 \cdot 10^9$ Jahre.

Tabelle 14. *Trennung von Lage- und Helligkeitseffekt*

Feld	links		Alle Sterne gleicher Helligkeit	
	V km/sec	n	V km/sec	n
1	214	2	42	48
2	185	1	48	50
3	67	3	43	79
4	61	2	46	65
5	282	1	61	44
6	44	3	38	20
7	85	4	57	8

5. Wir fassen in Tab. 15 alle Ergebnisse zusammen, die sich im Anschluß an die Feldeinteilung der Abb. 16 ergeben haben. Mit τ ist das mittlere Alter jeder Gruppe bezeichnet, n ist die Anzahl der Sterne. Mit u, v, w bezeichnen wir die dem Betrag nach gemittelten Komponenten der Geschwindigkeit um ihre durch Gl. (103) gegebenen Nullpunkte, und V ist die dem Betrag nach gemittelte räumliche Geschwindigkeit. Wir haben stets die gemittelten Beträge dem quadratischen Mittel vorgezogen, um den Einfluß der wenigen sehr schnellen Sterne nicht überwiegen zu lassen. Die angegebenen Fehlergrenzen sind die wahrscheinlichen Fehler.

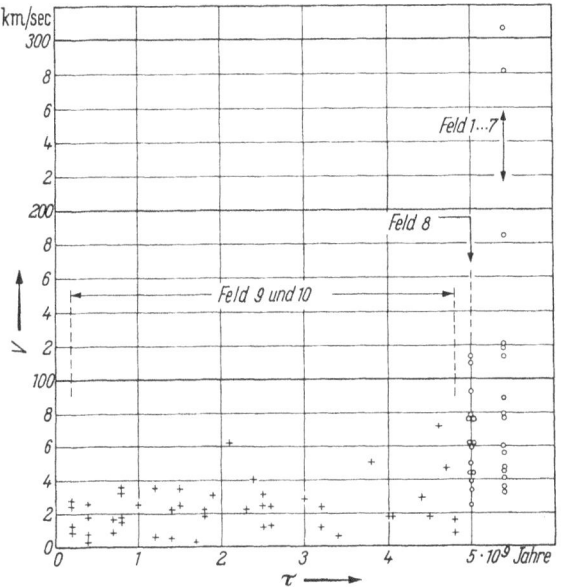

Abb. 17. Raumgeschwindigkeit und Alter von 78 einzelnen Sternen

In Abb. 17 sind alle 77 Sterne und die Sonne eingezeichnet. Mit wachsendem Alter τ nehmen die Geschwindigkeiten anfangs zu bis etwa $\tau = 2 \cdot 10^9$, bleiben fast konstant bis $\tau = 4.8 \cdot 10^9$, und erst bei den beiden ältesten Gruppen setzt ein plötzlicher, steiler Anstieg ein. Es ist noch darauf hinzuweisen, daß der Anstieg der Geschwindigkeiten nicht etwa nur durch das Hinzukommen einiger Schnelläufer bewirkt wird: lassen wir die drei extrem schnellen Sterne der ältesten Gruppe weg, so erhalten wir $V = (70 \pm 8)$ km/sec, also immer noch weit mehr als z. B. Gl. (104).

6. Als Deutung dieses Alterseffektes der Geschwindigkeiten möchten wir zwei verschiedene Ursachen vorschlagen. Erstens: ähnlich wie auch A. M. JOHNSON (1957b) nehmen wir an, daß der langsame, flache Anstieg

Die Streuung der räumlichen Geschwindigkeiten

Tabelle 15. *Alter τ und mittlere Geschwindigkeit der Sterne*

Feld	τ		u	v	w	V
	10^9 Jahre	n	km/sec			km/sec
9 und 10	0.30	8	10.1±3.2	4.8±1.3	4.1±1.0	13.3±3.2
	0.77	6	13.1±5.1	10.7±2.6	6.2±1.1	20.4±3.2
	1.5	11	12.8±3.3	8.9±2.9	7.3±2.1	20.2±3.5
	2.4	8	21.5±5.1	12.5±4.3	9.0±2.1	28.5±5.6
	3.5	7	17.8±5.5	7.3±2.1	7.7±1.9	21.9±5.5
	4.6	6	20.0±8.5	17.5±6.3	11.7±4.1	27.2±10.0
8	5.0	15	40.0±7.3	37.4±7.1	15.3±2.6	63.9±6.9
1—7	5.4	16	66±15	63±19	23±4	106±21

der Geschwindigkeit mit dem Alter durch Energieaustausch der Sterne mit größeren Stern- oder Gaswolken zustande kommt, wie SPITZER und SCHWARZSCHILD (1951) vorgeschlagen haben.

Unter Berücksichtigung der differentiellen Rotation der Milchstraße finden SPITZER und SCHWARZSCHILD (1953), daß die Pekuliargeschwindigkeiten der Sterne mit dem Alter t entsprechend der Formel

$$V(t) = V_0 \left(1 + \frac{t}{t_E}\right)^{1/3} \qquad (106)$$

anwachsen sollten. Wählen wir als ursprüngliche Geschwindigkeit $V_0 = 10$ km/sec, so erhalten wir die beste Übereinstimmung, wenn wir für die Zeitskala des Energieaustausches $t_E = 2.1 \cdot 10^8$ Jahre einsetzen. Aus den bei SPITZER und SCHWARZSCHILD angegebenen Formeln erhalten wir damit Wolkenkomplexe von rund 10^6 Sonnenmassen.

In Abb. 18 haben wir die Werte der Tab. 15 logarithmisch aufgetragen. Die durchgezeichnete Kurve entspricht der Formel (106). Wir sehen, daß sich der langsame Anstieg im Bereich $0.3 \leq \tau \leq 4.5 \cdot 10^9$ Jahre gut durch die Annahme des Energieaustausches erklären läßt.

Zweitens: Der steile Anstieg für die ganz alten

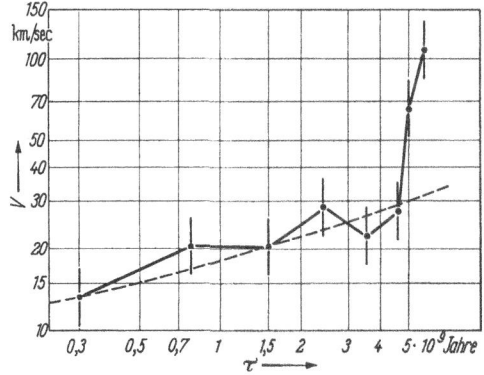

Abb. 18. Die dem Betrag nach gemittelte Raumgeschwindigkeit V von acht Altersgruppen. Die gestrichelte Linie zeigt die Geschwindigkeits-Zunahme mit wachsendem Alter durch Energieaustausch mit größeren Gaswolken nach der Theorie von SPITZER und SCHWARZSCHILD (1953). Die beiden ältesten Gruppen zeigen vermutlich den Abfall der Anfangsturbulenz der Milchstraße während der ersten 10^9 Jahre

Sterne läßt sich nicht in dieser Weise deuten, da sonst die Geschwindigkeiten innerhalb der letzten 10^9 Jahre um einen Faktor 4 zugenommen haben müßten. Wir müssen daher annehmen, daß die ganz alten Sterne bereits bei ihrer Entstehung derart hohe Geschwindigkeiten besessen haben. Da der eigentliche Prozeß der Sternentstehung damals wohl kaum sehr viel anders gewesen sein wird als heute, so müssen wir die hohen Geschwindigkeiten bereits der damaligen interstellaren Materie der Milchstraße zuschreiben. Wir kommen somit zu dem Schluß, daß die großen Geschwindigkeiten der ganz alten Sterne den Bewegungszustand im Frühstadium der Milchstraße wiedergeben, als deren ursprüngliche Turbulenz noch nicht ganz abgeklungen war. (Kugelhaufen und Halo-Sterne wären dann kurz zuvor, bei noch höherer Anfangsturbulenz, entstanden.)

Zur Kontrolle vergegenwärtigen wir uns, daß die Turbulenz einer sich selbst überlassenen Gasmasse mit der Zeitskala L/W abklingt, wobei L der Durchmesser der größten Turbulenzelemente ist und W deren Geschwindigkeit. Im Rahmen unserer groben Abschätzung mag der Hinweis genügen, daß L/W von der gleichen Größenordnung gewesen sein muß, wie die heutige Umlaufzeit der Milchstraße, die wir mit etwa $2 \cdot 10^8$ Jahren ansetzen können. Das heißt, daß eine ursprüngliche Turbulenz der Milchstraße in mehreren 10^8 Jahren abgeklungen sein sollte. In Übereinstimmung mit dieser Abschätzung lesen wir aus Abb. 18 für den Vorgang dieses Abklingens etwa 10^9 Jahre ab.

7. Zusammengefaßt erhalten wir als plausibelste Deutung unserer Ergebnisse das folgende Bild. Die ganz alten Sterne stammen aus einem Frühstadium der Milchstraße, ihre großen Geschwindigkeiten erhielten sie durch die damals noch hohe Anfangsturbulenz, die in einem Zeitraum von rund 10^9 Jahren abgeklungen ist. (Die heutige geringe Turbulenz des interstellaren Gases wird dagegen stationär aufrechterhalten durch die differentielle Rotation der Milchstraße sowie durch anisotropen Strahlungsdruck junger Sterne.) Zu späteren Zeiten entstehen alle Sterne mit relativ kleinen mittleren Geschwindigkeiten, etwa 10 km/sec, gewinnen jedoch im Laufe der Zeit Energie durch Austausch mit größeren Stern- oder Gaswolken.

Dabei darf angenommen werden, daß Sterne verschiedener Masse bei ihrer Entstehung alle die gleiche Geschwindigkeit erhalten. Eine Ausnahme hiervon bilden nach BLAAUW (1956b) die schnelleren O-Sterne.

B.I.b) z-Komponente und Schichtdicke

Für das folgende Kapitel über die zeitliche Rate der Sternentstehung benötigen wir die mittlere galaktische Schichtdicke von Hauptreihensternen als Funktion der absoluten Helligkeit. Zwar finden sich hierüber auch direkte Angaben in der Literatur, s. z. B. ALLEN (1955), doch scheint uns die genaueste Methode ihre Berechnung aus den z-Komponenten der Geschwindigkeit zu sein, da die Sterne der näheren Sonnenumgebung vollständiger bekannt sein dürften als die Sterne in größerem galaktischen Abstand. Für die Umrechnung braucht man zwar den Verlauf des Potentials als Funktion von z, und eigentlich sollte man auch den Potentialverlauf aus der Geschwindigkeits-Verteilung berechnen, was durchaus möglich ist. Wir wollen uns jedoch hier damit begnügen, das Potential aus der bei ALLEN (1955) angegebenen Gesamtdichte $\varrho(z)$ aller Sterne durch zweimalige Integration zu berechnen. Eine genauere Untersuchung soll in einer späteren Arbeit durchgeführt werden.

1. Wir entnehmen dem Glieseschen Katalog alle Sterne (mit bekannten Geschwindigkeiten) der Güteklassen a, b, c, d, die von der Hauptreihe nicht weiter als eine Größenklasse entfernt sind und deren absolute Helligkeiten oberhalb von $M = 12.5$ liegen. Ihre Anzahl ist 391.

Wir unterteilen in 14 Gruppen der absoluten Helligkeit; die erste Spalte von Tab. 16 gibt die mittlere Helligkeit jeder Gruppe an. Die zweite Spalte enthält die Anzahl der Sterne, in der dritten Spalte zeigen wir die dem Betrag nach gemittelte z-Komponente w der Geschwindigkeit mit ihrem wahrscheinlichen Fehler.

Bei einem reinen Alterseffekt sollten wir eine Zunahme der Geschwindigkeit erwarten bis zu $M = 3.75$, und Konstanz für $M \geq 4.25$. Tab. 16 bestätigt diese Erwartung. Da die Abweichungen von der Konstanz nur durch wenige schnelle Sterne bewirkt werden, zeigen wir in der vierten Spalte den Median jeder Gruppe, da allgemein der Median sehr viel weniger als der

Tabelle 16. *Die z-Komponente der Geschwindigkeit für 391 Sterne der Hauptreihe*
w = mittlere Beträge,
w_m = Median der Beträge

M_v	n	w km/sec	w_m km/sec
0.8	4	5.2 ± 1.9	5.0
2.2	13	5.5 ± 1.1	4.5
3.25	12	7.5 ± 2.0	4.5
3.75	21	10.7 ± 1.5	10.0
4.25	27	15.0 ± 2.5	12.0
4.75	28	15.3 ± 2.7	12.0
5.25	25	14.5 ± 3.1	12.0
6	56	14.0 ± 1.4	12.5
7	41	14.6 ± 1.9	12.0
8	36	13.8 ± 1.6	13.5
9	54	14.7 ± 1.4	13.5
10	33	22.8 ± 3.8	13.0
11	32	16.9 ± 3.2	11.0
12	9	12.5 ± 2.3	13.0

Mittelwert von der zufälligen Anwesenheit weniger „Außenseiter" abhängt. Die Konstanz ist jetzt völlig befriedigend, was wir als eine gute Bestätigung des angenommenen Alterseffektes bezeichnen möchten.

Die Verteilung der z-Komponenten streut weit weniger stark als die der beiden anderen Komponenten: nur 19 Sterne mit $w \geq 40$ km/sec sind vorhanden, nur 5 Sterne mit $w \geq 60$, und die höchste Geschwindigkeit beträgt 80 km/sec. Die Art der Verteilung soll jedoch hier nicht weiter untersucht werden.

2. Als nächstes berechnen wir die Geschwindigkeit $w(z)$, die ein Stern bei $z = 0$ haben muß, um gerade bis zur Höhe z fliegen zu können. Nennen wir P das Potential der Milchstraße, und vernachlässigen wir

alle Ableitungen außer der nach z, so erhalten wir aus

$$\frac{d^2 P(z)}{dz^2} = 4\pi G \cdot \varrho(z)$$

mit $\varrho_0 = \varrho(0) = 5.8 \cdot 10^{-24}$ g/cm^3 [SCHMIDT (1956)] die einfache Beziehung:

$$w(z) = 9.4 \left[\int_0^z dz \int_0^z dz\, \varrho(z)/\varrho_0 \right]^{1/2} \quad \begin{array}{l} z \text{ in } 100 \text{ pc} \\ w \text{ in km/sec} \end{array} \quad (107)$$

Tab. 17 zeigt einige der so berechneten Werte.

Tabelle 17. *Sterne der Geschwindigkeit w fliegen bis zur Höhe z*

z	50	75	100	150	200	250	300	400	500	600	700	pc
w	3.3	4.9	6.5	9.6	12.6	15.4	18.1	23.0	27.4	31.5	35.3	km/sec

3. Im Hinblick auf den Abschnitt über die Leuchtkraftfunktion müssen wir noch folgende Frage beantworten. Gegeben sei die Anzahl Φ von Sternen (einer bestimmten Helligkeit) pro pc^3 in Sonnenumgebung sowie die Verteilung $f(w)\,dw$ der Beträge w ihrer Geschwindigkeiten senkrecht zur galaktischen Ebene. Wie groß ist dann ihre Gesamtzahl Φ_z in einem Zylinder (von 1 pc^2) senkrecht zur Ebene?

Nennen wir $\Theta(w)$ die Flugdauer von $z = 0$ bis zum Umkehrpunkt, so ist

$$\Phi_z = \Phi \int_0^\infty w \cdot 2\,\Theta(w) \cdot f(w)\,dw. \quad (108)$$

Das rechts stehende Integral hat die Dimension einer Länge und entspricht etwa der mittleren Schichtdicke der Sterngruppe. Durch eine Reihenentwicklung, auf die wir hier verzichten wollen, ließ sich zeigen, daß wir nur Fehler von wenigen Prozent erhalten, wenn wir

$$\Phi_z = 2\,\bar{z} \cdot \Phi \quad (109)$$

setzen, mit

$\bar{z} = z(\bar{w})$ nach Tab. 17, und $\bar{w} = \int_0^\infty w f(w)\,dw$.

Tabelle 18. *Die mittlere Schichtdicke $\bar{z}$*

M_v	$\bar{w}$	$\bar{z}$
0.8	5.2 km/sec	78 pc
2.2	5.5 km/sec	83 pc
3.25	7.5 km/sec	115 pc
3.75	10.7 km/sec	168 pc
≥ 4.25	15.4 km/sec	250 pc

Aus Tab. 16 erhalten wir in dieser Weise die Umrechnungsfaktoren $\bar{z}$ der Tab. 18, die das Verhältnis der in Sonnenumgebung beobachteten Leuchtkraftfunktion zu der über z integrierten LKF angeben.

B.I.c) Die Dicke der sternerzeugenden Schicht

1. Wir stellen die Frage, ob die Dicke derjenigen Schicht, innerhalb deren die Sterne entstehen, sich zeitlich verändert oder konstant bleibt. Nach Tab. 15 haben die älteren Sterne größere z-Komponenten der Geschwindigkeit und nach Tab. 17 somit auch größere Schichtdicken als die jüngeren Sterne, was zwei Grenzfälle der Deutung zuläßt.

Fall 1. Die Dicke der sternerzeugenden Schicht nimmt zeitlich ab, und jede zu irgendeiner Zeit erzeugte Gruppe von Sternen behält ihren mittleren z-Abstand unverändert bei.

Fall 2. Alle Sterne sind zu allen Zeiten innerhalb der gleichen geringen Schichtdicke entstanden, aber ihre w-Geschwindigkeiten nehmen durch Energieaustausch mit wachsendem Alter langsam zu und somit auch ihre Schichtdicke.

Im ersten Grenzfall bleibt die räumliche Dichte einer Sterngruppe in der Umgebung von $z = 0$ zeitlich konstant, im zweiten Grenzfall nimmt sie zeitlich ab.

2. Ebenso wie früher bei den räumlichen Geschwindigkeiten, so zeigt sich auch jetzt bei den w-Geschwindigkeiten, daß Fall 2 in keiner Weise den steilen Anstieg bei den ganz alten Sternen erklären kann. Die Dicke der sternerzeugenden Schicht muß in den ersten 1 bis $2 \cdot 10^9$ Jahren von etwa $z = 400$ pc auf rund $z = 100$ pc abgeklungen sein.

Wollten wir für den weiteren Verlauf auch Fall 1 annehmen, so zeigt eine Abschätzung, daß die Dicke der sternerzeugenden Schicht einige 10^9 Jahre hindurch fast konstant geblieben wäre, in den letzten 10^9 Jahren jedoch um einen Faktor 2 abgenommen hätte, was diese Deutung recht unwahrscheinlich macht. Andererseits ergibt eine logarithmische Auftragung der w-Komponenten über dem Alter wieder ein ganz ähnliches Bild wie das der Abb. 18, was zugunsten von Fall 2 spricht, und man sollte ja ohnehin annehmen, daß ein Energieaustausch mit größeren Wolken, falls er überhaupt stattfindet, dann auch die z-Komponenten mit wachsendem Alter vergrößert.

3. Für das folgende Kapitel möchten wir die Formulierung bereitstellen: a) die Dicke der sternerzeugenden Schicht hat in den ersten 10^9 Jahren mit Sicherheit stark abgenommen. b) Die mittlere Schichtdicke der später erzeugten Sterne nimmt aller Wahrscheinlichkeit nach mit der Zeit zu, d. h. die räumliche Dichte bei $z = 0$ einer in einem bestimmten Zeitintervall entstandenen Gruppe von Sternen nimmt zeitlich ab.

B.II. Die Zeitliche Rate der Sternentstehung

B.II.a) Aus der Leuchtkraftfunktion

1. Nehmen wir an, daß die Sterne bei ihrer Entstehung nach einer „ursprünglichen" Leuchtkraftfunktion (LKF) verteilt sind, so muß die

jetzt zu beobachtende LKF von ihr verschieden sein, da die älteren und massiveren Sterne die Hauptreihe bereits wieder verlassen haben. Nach den Theorien der Sternentwicklung ist die eine LKF aus der anderen zu berechnen, sobald man Annahmen macht über die zeitliche Rate $R(t)$ der Sternentstehung. Diese Zusammenhänge wurden von SALPETER (1955) untersucht und durch SANDAGE (1957) fortgesetzt. Den Anlaß dazu gab ein auffälliger Knick der beobachteten LKF in der Nähe von $M_v = 4$, s. Abb. 13. Hellere Sterne können die Hauptreihe bereits wieder verlassen haben, schwächere Sterne noch nicht.

SALPETER setzt versuchsweise $R(t) = $ const und erhält damit aus der beobachteten LKF $\Phi(M)$ eine ursprüngliche LKF $\Psi(M)$, die erstens glatt über den besagten Knick hinweggeht und zweitens der in offenen Haufen beobachteten LKF $\psi(M)$ recht ähnlich ist. SALPETER (1955), SANDAGE (1957) und V. D. BERGH (1957a) schließen hieraus, daß die Voraussetzung der zeitlich konstanten Rate der Sternentstehung eine brauchbare Näherung darstellt.

Dabei tauchte die Schwierigkeit auf, daß nach V. D. BERGH (1957c) der Vorrat der interstellaren Materie durch Sternentstehung relativ rasch aufgebraucht wird und in einigen 10^8 Jahren ganz verbraucht ist, wobei man sich eine zeitlich konstante Sternentstehung bei zeitlich abnehmendem Vorrat schwer vorstellen kann. Zur Behebung dieser Schwierigkeit schlägt V. D. BERGH einen ständigen Nachschub interstellarer Materie vor, die im Zentrum der Milchstraße durch die Entwicklung alter Sterne frei wird. Wir werden später ausführlich hierauf eingehen.

2. Gegen die bisherige Behandlung des Problems haben wir zwei Einwände zu machen. Erstens: bisher wurde von allen Bearbeitern implizit vorausgesetzt, daß sich die mittleren z-Abstände der Sterne seit ihrer Entstehung nicht verändert haben. Nach den Ergebnissen des vorigen Kapitels ist es dagegen viel wahrscheinlicher, daß die Abstände durch Energieaustausch zeitlich zunehmen. Dann dürfte nur die *über z integrierte* LKF verwendet werden.

Zweitens: die v. d. Berghsche Materiebilanz beschränkt sich auf eine relativ dünne galaktische Schicht. Es wird weder danach gefragt, wo und woraus die Sterne größeren Abstandes entstanden sind, noch wird das Gas berücksichtigt, das diese Sterne im Laufe ihrer Entwicklung wieder abgeben. Beide Einwände gelten unabhängig voneinander.

3. Wir nennen Φ_z die in Gl. (109) definierte, über z integrierte beobachtete LKF. Es ist also $\Phi_z(M)$ die Anzahl von Sternen im Intervall $M \pm 1/2$ innerhalb eines Zylinders senkrecht zur galaktischen Ebene von 1 pc² Querschnitt am Ort der Sonne. Weiterhin sei $\Psi(M)$ die auf 1 Sonnenmasse normierte ursprüngliche LKF, von der wir, wie auch die bisherigen Bearbeiter, annehmen wollen, daß sie zeitlich und räumlich konstant

sei. Mit $R(t)\,dt$ bezeichnen wir die Gesamtmasse der Sterne (in Sonnenmassen pro Jahr und pc²), die innerhalb des Zylinders im Zeitintervall $t \ldots t + dt$ entstanden sind. Dann ist, mit $T = 5.6 \cdot 10^9$ Jahre,

$$\Phi_z(M) = \Psi(M) \int_{T-\tau(M)}^{T} R(t)\,dt, \qquad (110)$$

wobei wir mit $\tau(M)$ dasjenige Alter bezeichnen, das Sterne der Helligkeit M maximal auf der Hauptreihe erreichen können.

Wir wollen nun so vorgehen, daß wir nicht wie bisher eine Annahme über $R(t)$ machen, sondern umgekehrt eine Hypothese über $\Psi(M)$ benutzen und $R(t)$ berechnen. Im Sinne einer Arbeitshypothese nehmen wir an, daß Feldsterne und Haufensterne ursprünglich nach der gleichen LKF entstehen, wir benutzen also für $\Psi(M)$ die (durch 1000 dividierten) Werte $\psi(M)$ der Tab. 7. Nennen wir

$$F(M) = \frac{\Phi_z(M)}{\psi(M)} = \frac{2\bar{z}\,\Phi(M)}{\psi(M)}, \qquad (111)$$

so ist dann

$$R(T - \tau) = \frac{dF}{d\tau}, \qquad (112)$$

da wir $F(M)$ bei bekanntem $\tau(M)$ auch als $F(\tau)$ schreiben können.

4. Wir benutzen $\Phi(M)$ nach SANDAGE (1957), $\bar{z}(M)$ nach Tab. 18, $\psi(M)$ nach Tab. 7 und $\tau(M)$ nach Tab. 4. Das Ergebnis ist dargestellt in Abb. 19. Die Rate der Sternentstehung ist zwar in den letzten $4 \cdot 10^9$ Jahren etwa konstant geblieben, war jedoch anfangs rund um einen Faktor 10 größer. Dabei sind bereits nach $1 \cdot 10^9$ Jahren die Hälfte aller Sterne entstanden.

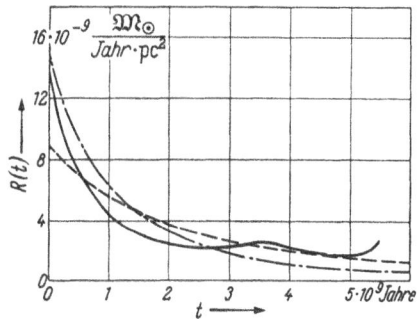

Abb. 19. Die zeitliche Rate der Sternentstehung. ——— nach (112) aus der beobachteten Leuchtkraftfunktion. - - - - - $a = 2 \cdot 10^9$ Jahre, -·-·-·- $a = 1 \cdot 10^9$ Jahre, nach (114) aus dem theoretischen Ansatz. Alle drei Kurven sind auf gleiche Fläche normiert. Die gegenwärtig vorhandene Gasmenge beträgt: 4% bei $a = 1 \cdot 10^9$ und 13% bei $a = 2 \cdot 10^9$

Die nach unserer Meinung unbedingt nötige Integration über z liefert somit eine Rate der Sternentstehung, die anfangs stark abgenommen hat und weiterhin etwa konstant geblieben ist mit im Mittel

$$R(t) = 2.1 \cdot 10^{-9} \text{ Sonnenmassen/Jahr} \cdot \text{pc}^2. \qquad (113)$$

B.II.b) Theoretischer Ansatz

1. Wir fragen nun, ob wir auch bei dieser Rate der Sternentstehung einen Nachschub an Gas annehmen müssen oder nicht. Als Gegentest

berechnen wir $R(t)$ probeweise aus derjenigen Voraussetzung, die theoretischerseits am einfachsten ist: $R(t)$ sei zu allen Zeiten der jeweils vorhandenen Gasmenge G proportional gewesen

$$R(t) = \frac{1}{a} G(t), \qquad (114)$$

wobei $G(t)$ die im Zylinder von 1 pc² Querschnitt vorhandene Gasmenge ist, die nicht von außen nachgeliefert werden soll. Die Konstante a hat die Dimension einer Zeit.

Wir betrachten eine zur Zeit $\tau = 0$ entstandene Gruppe von Sternen der LKF ψ, und nennen $A(\tau) d\tau$ denjenigen Bruchteil ihrer ursprünglichen Masse, der im Zeitintervall $\tau \ldots \tau + d\tau$ infolge Sternentwicklung wieder als Gas abgegeben wird. Zur Durchführung berechnen wir $A(\tau)$ aus Tab. 7. Mit Gl. (114) ist dann

$$\frac{dG(t)}{dt} = a \frac{dR(t)}{dt} = - R(t) + \int_0^t R(t-\tau) \cdot A(\tau) d\tau. \qquad (115)$$

Führen wir noch die Funktionen ein

$$H(t) = e^{t/a} R(t) \quad \text{und} \quad B(t) = e^{t/a} A(t),$$

so erhalten wir

$$\frac{dH(t)}{dt} = \frac{1}{a} \int_0^t H(t-\tau) \cdot B(\tau) d\tau \qquad (116)$$

als Integro-Differentialgleichung für den zeitlichen Abfall der Sternentstehung.

2. Wir haben Gl. (116) in zweiter Näherung und mit grober Schrittweite für zwei Werte von a integriert und das Ergebnis in Abb. 19 mit eingezeichnet. Wir gewinnen den Eindruck, daß sich die aus der Leuchtkraftfunktion erhaltene Rate der Sternentstehung, im Rahmen der Fehlergrenzen unserer Abschätzungen, durchaus mit der Annahme verträgt, daß die Sternentstehung stets der vorhandenen Gasmenge proportional gewesen sei. Es ist also nicht nötig, eine Nachlieferung von Gas aus dem galaktischen Kern anzunehmen.

Ist man jedoch der Meinung, daß die aus der Leuchtkraftfunktion erhaltene Konstanz von R während der letzten $3 \cdot 10^9$ Jahre in voller Schärfe reell und ernst zu nehmen sei, so müßte tatsächlich Gas nachgeliefert worden sein, z. B. vom galaktischen Kern oder Halo. Wenn wir weiterhin annehmen, dies nachgelieferte Gas stamme von der Entwicklung alter Sterne, und wenn wir diese Sterne in die Bilanz mit einbeziehen, so ergibt eine Abschätzung: in den ersten 10^9 Jahren müssen 30—40mal mehr Sterne entstanden sein, als in der folgenden Zeit. Nur dann könnte das gegenwärtig verbrauchte Gas in *voller* Höhe durch die Entwicklung der alten Sterne nachgeliefert werden.

Beide Darstellungen scheinen zur Zeit möglich; wir würden zwar die erstere vorziehen, doch läßt sich eine Entscheidung kaum treffen. Doch ist beiden Darstellungen gemeinsam, daß die Rate der Sternentstehung anfangs weit größer war als heute und während der letzten 10^9 Jahre kaum abgenommen hat.

B.II.c) Die zeitliche Entstehungsrate offener Sternhaufen

Das Ziel dieses Abschnittes ist die Berechnung der Anzahl

$$N(Sp, m_0)$$

von offenen Haufen, deren Sterne frühesten Typs innerhalb der Spektralklasse (Sp) liegen und scheinbare visuelle Helligkeiten $\leq m_0$ haben, wobei wir eine zeitliche Konstanz der Entstehung offener Haufen annehmen. Dies theoretische Ergebnis ist dann mit den Beobachtungen von TRÜMPLER (1930) zu vergleichen als Test auf die angenommene Konstanz. Im Gegensatz zur Methode SALPETERs und der des vorigen Abschnitts ist hierbei keine Annahme über eine ursprüngliche LKF nötig, dafür ist unsere Methode aber beschränkt auf die Entstehung von Sternen in offenen Haufen.

Ein weiterer Grund zu dieser Untersuchung liegt darin, daß wir die zeitliche Rate der Entstehung offener Haufen zur Beantwortung der Frage (des folgenden Abschnitts) benötigen, ob alle Sterne in offenen Haufen entstehen oder nicht.

1. Ist ein offener Haufen *vor* der Zeit τ im Abstand r (kpc) entstanden, so haben seine hellsten Sterne *jetzt* eine absolute Helligkeit $M(\tau)$, die sich aus Tab. 4 ergibt. Ist a die Absorption in mag/kpc, so erhalten wir die scheinbare Helligkeit

$$m(\tau, r) = M(\tau) + 5 \log r + 10 + a r. \tag{117}$$

Oder umgekehrt formuliert: ein Haufen, dessen hellste Sterne die scheinbare Helligkeit m haben und der τ Jahre alt ist, steht in einem Abstand $r(\tau, m)$, der zu berechnen ist, aus

$$\log r(\tau, m) + \frac{a}{5} r(\tau, m) = \frac{m - M(\tau)}{5} - 2. \tag{118}$$

Ist H die Dicke der Milchstraße für offene Haufen ($H \approx 0.1$ kpc), und sind alle $r \gg H/2$, so beträgt das zu r gehörige Volumen $\pi r^2 H$. Das heißt: die Anzahl offener Haufen des Alters τ bis zu einer Grenzhelligkeit m_0 beträgt

$$n(\tau, m) = \pi r^2(\tau, m_0) \cdot H \cdot R_H(T - \tau), \tag{119}$$

mit

$R_H(t) \, dt$ = Anzahl offener Haufen, die während der Zeit

$t \ldots t + dt$ pro kpc³ entstanden sind.

Das Alter τ ist nach Tab. 4 mit dem Spektraltyp Sp der frühesten Haufensterne verknüpft. Innerhalb einer Spektralklasse Sp gibt es also die folgende Anzahl von Haufen bis zur Grenzhelligkeit m_0:

$$N(Sp, m_0) = \int_{\tau_1}^{\tau_2} n(\tau, m_0) \, d\tau,$$

wobei τ_1 und τ_2 der Tab. 4 an den Grenzen der betrachteten Klasse Sp zu entnehmen sind. Wählen wir die Einteilung in Klassen dicht genug, so können wir annehmen, daß sich $R_H(t)$ innerhalb der Integrationsgrenzen nur wenig ändert. Wir erhalten

$$N(Sp, m_0) = \pi H R_H \{\overline{\tau(Sp)}\} \int_{\tau_1}^{\tau_2} r^2(\tau, m_0) \, d\tau. \tag{120}$$

Dabei ist $r(\tau, m_0)$ nach Gl. (118) gegeben, und aus Tab. 4 erhalten wir die Werte von $\tau_1(Sp)$, $\tau_2(Sp)$ und $M(\tau)$.

2. Zum Vergleich mit der Beobachtung machen wir nun die Annahme konstanter Entstehungsrate

$$R_H(t) = \text{const} = R_H. \tag{121}$$

Dann läßt sich Gl. (120) numerisch auswerten; das Ergebnis zeigt Tab. 19, unter drei verschiedenen Annahmen über Absorption a und Grenzhelligkeit m_0. Wir sehen, daß der Einfluß der Absorption nicht sehr entscheidend ist. Dabei haben wir so normiert, daß die Summe von B0 bis B9 mit der Beobachtung übereinstimmt.

3. Die Zahlen der letzten Spalte (Beobachtung) sind auf folgende Weise der Trümplerschen Arbeit (1930) entnommen. TRÜMPLER gibt für 100 offene Haufen Sp und m an. Bei Vollständigkeit sollte die über Sp summierte Anzahl anfangs mit $10^{0,4\,m}$ ansteigen und später etwas langsamer wegen der Absorption. Wegen der sich zeigenden Unvollständigkeit mit wachsendem m haben wir die Trümplerschen Angaben nur bis zu $m = 8.9$ benutzt. Eine grobe Abschätzung ergab, daß die benutzte Anzahl von 74 Haufen dann ebenso groß ist, als hätten wir Vollständigkeit bis etwa $m = 7$. Daher die in Tab. 19 angesetzten Werte von m_0.

Tabelle 19. *Die Typenhäufigkeit $N(Sp, m_0)$ offener Haufen nach Theorie und Beobachtung, unter Annahme konstanter Entstehungsrate*

Sp	Theorie			Beobachtung
	$m_0 = 7.0$ $a = 0.85$	$m_0 = 6.5$ $a = 0.65$	$a = 0.00$	
O 5 bis O 9	7.4	8.7	12.9	10
B 0, 1, 2	11.4	12.5	15.6	15
3, 4, 5, 6	25.7	25.6	25.1	23
7, 8, 9	16.8	15.8	13.3	16
A 0, 1, 2	15.3	13.9	11.1	9
3, 4, 5, 6	16.1	14.2	11.0	0
7, 8, 9	8.3	7.3	5.5	0
F 0, 1, 2	7.7	6.7	5.0	1
3, 4, 5, 6	6.7	5.8	4.3	0

In Tab. 20 zeigen wir die Trümplerschen Angaben, die wir nach Sp und m sortiert haben. Die letzte Zeile ist die Summe bis zu der gewählten Grenze $m \leq m_0 = 8.9$, die wir unter ,,Beobachtung" in Tab. 19 eintrugen.

Tabelle. 20 *Beobachtete Typenhäufigkeiten offener Haufen*

Sp = Spektraltyp
m = scheinb. vis. Helligkeit } der frühesten Sterne eines Haufens.

Längs der unteren eingezeichneten Grenzlinie ist die Flächendichte der hellsten Haufensterne 5 mal größer als die Flächendichte gleichheller Feldsterne, auf der oberen Linie 15 mal.

Sp m	O	B			A			F		Summe
	5 6 7 8 9	0 1 2	3 4 5 6	7 8 9	0 1 2	3 4 5 6	7 8 9	0 1 2	3 4 5 6	
2.0—2.9		1	1							2
3.0—3.9		1	1		1					3
4.0—4.9	2 2	2	1		1					8
5.0—5.9	1	111	3	1						8
6.0—6.9	21		113	12	1					12
7.0—7.9		31	2121	51	1					17
8.0—8.9	11	1 12	1131	132	41			1		24
9.0—9.9		1 1	2	11	31					10
10.0—10.9		1		114	1					8
11.0—11.9			1	121	1	1				7
12.0—12.9				1						1
Summe $m \leq 8.9$	10	15	23	16	9	0	0	1	0	74 / 100

4. Gelegentliche Diskussionseinwände legten es nahe, die Abhängigkeit der Entdeckungswahrscheinlichkeit offener Haufen von Spektraltyp und Helligkeit etwas ausführlicher zu behandeln. Der Einfluß der (mit zunehmendem Typ abnehmenden) Helligkeit der hellsten Haufensterne wurde bei Ableitung der Formel (120) bereits quantitativ berücksichtigt, da wir in Gl. (117) nur Haufen bis zu einer Grenzhelligkeit m_0 erfaßt haben.

Ein zweiter Einwand bezieht sich auf die Flächendichte der Haufensterne. Ihr Verhältnis zur Flächendichte gleich heller Feldsterne nimmt zu den späten Typen hin ab, wodurch sich die Entdeckungswahrscheinlichkeit vermindert. Dieser, aus der Verschiedenheit und der Krümmung der Leuchtkraftfunktion von Feld- und Haufensternen herrührende Effekt wurde bisher noch nicht berücksichtigt.

Zur Veranschaulichung der Verhältnisse zeigen wir in Abb. 20 die Flächendichte σ von Feld- und Haufensternen (in Sternzahl pro Quadratgrad und Größenklassenintervall). Die Daten für Feldsterne stammen von ALLEN (1955). Für Haufensterne haben wir einen „mittleren" offenen Sternhaufen vorausgesetzt mit einer Leuchtkraftfunktion nach Tab. 7, einer Masse von 1500 Sonnenmassen nach Gl. (91) oder anders gesagt mit 85 Sternen heller als $M_v = 2.5$, und mit einem Durchmesser von 3.5 pc nach BECKER (1951).

Ein Haufen, der z. B. $2.7 \cdot 10^9$ Jahre alt ist, beginnt bei F0 (Tab. 4). Liegt er innerhalb der Ebene der Milchstraße und in 100 pc Entfernung,

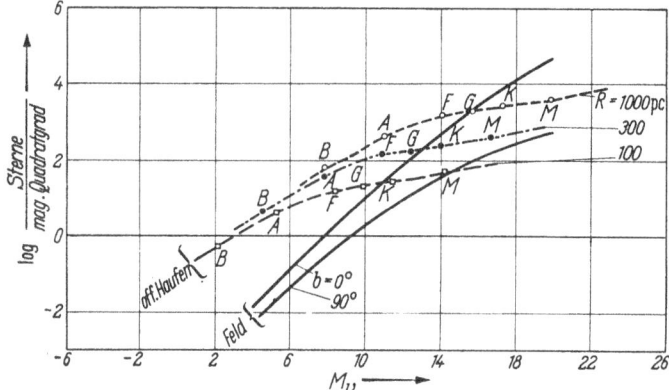

Abb. 20. Flächendichte der Feldsterne (der Helligkeit $m \pm {}^1/_2$) und Flächendichte eines offenen Haufens mittlerer Sternzahl und Größe in drei verschiedenen Entfernungen R. b = galaktische Breite

so ist die Flächendichte seiner hellsten Sterne nach Abb. 20 gerade 10mal so groß, wie die der gleichhellen Feldsterne,

$$\frac{\sigma \text{ Haufen}}{\sigma \text{ Feld}} = Q = 10,$$

er ist somit noch recht auffällig. Seine Hauptreihe dürfte sich etwa bis G7 verfolgen lassen, von da ab überwiegt die Flächendichte der Feldsterne. Liegt der gleiche Haufen jedoch in 1000 pc Entfernung, so ist das Verhältnis Q der Flächendichten nur noch 2.3 bei F0, der Haufen würde nur noch als zufällige Verdichtung des Feldes betrachtet werden. Ein jüngerer Haufen dagegen würde auch in 1000 pc Entfernung sich noch deutlich vom Feld abheben. Die Hauptreihe wäre etwa bis F9 zu verfolgen. Unter anderem geht aus der Abbildung auch hervor, daß man im Normalfall die Hauptreihe nie bis zu sehr späten Typen hin verfolgen kann. Nur in Ausnahmefällen wird dies möglich: extrem massereiche Haufen, hohe galaktische Breite (z. B. M67), Dunkelwolke als Hintergrund.

Sind m und Sp eines Haufens bekannt, so läßt sich seine Entfernung angeben und somit seine Lage in Abb. 20, wenn wir wieder die mittleren Werte von Masse und Radius voraussetzen. Auf ähnliche, aber numerische Weise haben wir für eine Anzahl von Punkten der Tab. 20 das

Verhältnis Q der Flächendichten berechnet und die Höhenlinien $Q = 5$ und $Q = 15$ interpoliert (die Absorption wurde mit $a = 1$ mag/kpc eingesetzt). Das Ergebnis ist in Tab. 20 eingezeichnet. Daß die Linie $Q = 5$ die Beobachtungen gerade umrandet, sehen wir als eine gute Bestätigung für die vorgeschlagene Behandlung der Entdeckungswahrscheinlichkeit an. Weiterhin meinen wir, daß sich innerhalb von $Q = 15$ durch eine sorgfältige Durchmusterung (z. B. des Mt. Palomar-Atlas) noch Vollständigkeit erreichen lassen sollte.

5. Betrachten wir die Lage der von uns benutzten Haufen in Tab. 20, sowie den Verlauf der beiden Q-Linien, so kommen wir zu folgendem Schluß: oberhalb der benutzten Grenzhelligkeit $m = 8.9$ ist die beobachtete Verteilung der Typenhäufigkeit durch die Variation der Entdeckungswahrscheinlichkeit noch nicht oder nur unwesentlich beeinflußt. Speziell das scharfe Abbrechen der Häufigkeiten bei A 1 kann hierdurch *nicht* erklärt werden.

Nachdem wir in dieser Weise die benutzten Beobachtungsdaten gerechtfertigt haben, kehren wir zu Tab. 19 zurück. Der Vergleich zwischen Beobachtung und Theorie zeigt volle Übereinstimmung bis etwa A 1; von A 2 ab fehlen die Beobachtungen fast ganz, während nach der Theorie bis F 6 etwa noch 30 Haufen folgen sollten.

Aus der Übereinstimmung bis A 1 schließen wir:

Die zeitliche Rate der Entstehung offener Haufen
war konstant (bis auf einen Faktor von etwa 1.5)   (122)
während der letzten $5 \cdot 10^8$ Jahre.

Drücken wir R_H in Sternhaufen pro Zeit und pro Flächeneinheit der galaktischen Ebene aus, so erhalten wir aus den in Tab. 19 nötigen Normierungsfaktoren für die verschiedenen Annahmen der Absorption:

a	R_H
0.00 mag/kpc	$1.9 \cdot 10^{-13}$ Haufen/Jahr · pc²
0.65 mag/kpc	$2.7 \cdot 10^{-13}$ Haufen/Jahr · pc²
0.85 mag/kpc	$2.9 \cdot 10^{-13}$ Haufen/Jahr · pc²

Für den folgenden Abschnitt wollen wir mit einem runden Wert von

$$R_H = 3 \cdot 10^{-13} \text{ Haufen/Jahr · pc}^2 \quad (123)$$

rechnen.

Zufolge der in Tab. 20 auffälligen Tendenz zum „Abrunden" der Spektraltypen werden auch bei A0 einige Haufen registriert sein, die eigentlich nach A 1 gehörten. Wir setzen daher das Abbrechen der Typenhäufigkeit bei A 1.5 an, wozu nach Tab. 4 ein Alter von $5 \cdot 10^8$ Jahren gehört. Hieraus ziehen wir den bereits in Gl. (102) benutzten Schluß:

Die mittlere Auflösungszeit offener Haufen beträgt $5 \cdot 10^8$ Jahre.   (124)

Die Unsicherheit dieser Angabe schätzen wir auf einen Faktor 1.5 ein.

6. Ebenso wie die Entstehungsrate der Feldsterne ist also auch die Entstehungsrate der Haufensterne während der letzten Zeit konstant gewesen. Das Ergebnis für die Haufensterne benötigt weniger Voraussetzungen als das der Feldsterne, ist dafür jedoch wegen der Auflösung der Haufen auf einen weit kürzeren Zeitraum begrenzt.

Bezüglich der in der Einleitung dieses Kapitels gestellten Frage lautet unsere Antwort zusammengefaßt: Zwar war die Sternentstehung im Frühstadium der Milchstraße weit häufiger als heute, doch sind die erst kürzlich entstandenen Sterne und Sternhaufen keineswegs nur als vereinzelte Nachzügler zu betrachten. Seit einigen 10^9 Jahren läuft der Prozeß der Sternentstehung nahezu konstant.

B.III. Sternentstehung einzeln oder in Haufen?

Für die Ansätze einer physikalischen Theorie der Sternentstehung wäre es äußerst wichtig zu wissen, ob die Sterne normalerweise einzeln entstehen oder ob etwa alle Sterne in offenen Haufen (oder Assoziationen) entstanden sind. Da wir um so eher gravitations-stabile Massen erhalten, je größer diese Massen sind, so wäre die Entstehung in Haufen am leichtesten zu verstehen.

Eine Behandlung dieser Frage liegt bereits von ROBERTS (1957) vor, der zu dem Schluß kommt, daß vermutlich die O- und frühen B-Sterne nur in Haufen und Assoziationen entstehen, daß dagegen für die Gesamtzahl der übrigen Sterne etwa ein Faktor 100 fehlt. Gegen die Durchführung von ROBERTS machen wir zwei Einwände. Erstens schätzt ROBERTS die mittlere Anzahl von Sternen pro Haufen mit 200 ein, was um rund eine Zehnerpotenz zu niedrig ist, falls unsere Annahme einer allgemeingültigen ursprünglichen LKF zutrifft; und *nur* unter dieser Annahme könnten alle Sterne in Haufen entstanden sein.

Zweitens betrachtet ROBERTS die Gesamtmasse der Milchstraße und die Gesamtdauer der Zeit, für beides reichen jedoch unsere Kenntnisse nicht aus. Wir beschränken daher unsere Untersuchung sowohl auf die über z integrierte Sonnenumgebung, als auch auf die letzten 10^9 Jahre.

B.III.a) Offene Haufen

1. Zur Beantwortung der Frage, welcher Bruchteil aller Sterne in offenen Sternhaufen entsteht, gehen wir aus von der Entstehungsrate offener Haufen, die sich in Gl. (123) zu $R_H = 3 \cdot 10^{-13}$ Haufen/Jahr · pc² ergab. Multiplizieren wir dies mit der mittleren Haufenmasse $\mathfrak{M}_H = 1500$ Sonnenmassen von Gl. (91), so erhalten wir die in offenen Haufen erzeugte Sternmasse von $4.5 \cdot 10^{-10}$ Sonnenmassen/Jahr · pc². Dies haben wir nun zu vergleichen mit der in Gl. (113) erhaltenen Rate der

Sternentstehung von $R = 2.1 \cdot 10^{-9}$ Sonnenmassen/Jahr $\cdot$ pc². Für den Bruchteil q aller Sterne, die in offenen Haufen entstehen, erhalten wir somit

$$q = \frac{\mathfrak{M}_H R_H}{R} = 0.215 \; . \tag{125}$$

Unser Ergebnis lautet demnach: rund $^1/_5$ aller Sterne sind in offenen Sternhaufen entstanden.

2. Wären *alle* Sterne in offenen Haufen entstanden, und beträgt die Auflösungszeit offener Haufen etwa $5 \cdot 10^8$ Jahre, so sollte es keine (oder nur sehr wenige) Sterne früher als etwa A0 außerhalb von offenen Haufen geben. Daß dies nicht zutrifft, ist ein weiteres Argument gegen die ausschließliche Entstehung in Haufen.

B.III.b) Assoziationen

Bezüglich der Assoziationen ist die uns interessierende Frage weit schwerer zu beantworten als bei den offenen Sternhaufen, da wir für die Massen der Assoziationen kaum Anhaltspunkte besitzen. Um zu weiteren Untersuchungen dieser Frage *anzuregen*, wollen wir trotzdem eine Abschätzung durchführen, die wir jedoch wegen ihrer Unsicherheit hier nur kurz skizzieren wollen.

1. Falls alle Sterne in Assoziationen entstehen, so dürfte man diejenigen Sterne, für die $\tau(Sp) \ll t_A$ gilt, nur innerhalb von Assoziationen antreffen. Dabei ist $\tau(Sp)$ das Alter, das ein Stern des Typs Sp höchstens erreichen kann, und t_A ist die mittlere Auflösungszeit von Assoziationen, falls letztere nicht stabil sind, worauf wir zum Schluß noch eingehen wollen.

Aus Radius und beobachteten Geschwindigkeiten [Gl. (128)] lassen sich Auflösungszeiten von knapp 10^7 Jahren abschätzen, dieses Alter können Sterne vom Typ B1 gerade noch erreichen. Merklich frühere Typen sollten somit nur in Assoziationen (bzw. offenen Haufen) vorkommen, während spätere Typen, nach Auflösung ihrer Assoziation, noch die ihrem Typ entsprechende Zeitspanne als Feldsterne weiterleben.

Dies scheint auch der Fall zu sein. Nach BLAAUW (1956a) befinden sich so gut wie alle O-Sterne (sowie auch alle Sterne der Leuchtkraft-Klassen I und II) in Assoziationen, oder können zumindest einer Assoziation als „Ausreißer" zugeordnet werden. Hieraus können wir schließen, daß jedenfalls die O-Sterne nicht einzeln entstehen.

2. Zur Beantwortung der gleichen Frage für die anderen Sterne müßten wir die LKF der Assoziationen kennen. Nachdem alle O-Sterne in Assoziationen (bzw. offenen Haufen) entstehen, so würden genau dann *alle* Sterne in Assoziationen entstehen, wenn die LKF der Assoziationen gleich der ursprünglichen LKF der Feldsterne wäre.

Aus einer Reihe von Angaben der Literatur der letzten Jahre haben wir für 7 Assoziationen eine gemeinsame mittlere LKF bestimmt, die jedoch auf den Bereich $-7 \leq M_v \leq -1$ beschränkt ist. Innerhalb der Streuung stimmt sie einigermaßen gut mit der in Tab. 7 benutzten LKF $\psi(M)$ überein. Betrachten wir wie bisher $\psi(M)$ als brauchbare Näherung für die ursprüngliche LKF und nehmen wir die Übereinstimmung ernst, so hieße das: auch die Sterne bis B5 sind alle in Assoziationen entstanden.

Unter Benutzung von Gl. (125) läßt sich also zusammenfassend nur sagen: Unter der Voraussetzung, daß alle Sterne, wo sie auch immer entstehen, nach dem gleichen Massenspektrum verteilt sind, kann man darauf schließen, daß etwa $^4/_5$ aller Sterne in Assoziationen entstehen und der Rest in offenen Sternhaufen. Im Bereich O5 bis B5 scheint sich die Voraussetzung zu bestätigen. Enthalten die Assoziationen dagegen eine geringere Anzahl der weniger massiven Sterne, so müßte für die Sterne mit $M_v \geq 0$ eine gesonderte Entstehung von Einzelsternen angenommen werden.

3. Falls die obige Voraussetzung erfüllt ist, läßt sich die Masse der Assoziationen abschätzen. Wir benutzen eine LKF, die für die helleren Sterne aus dem Mittel über 7 Assoziationen stammt, und weiterhin mit der LKF der Tab. 7 identisch ist. Nach der gleichen Methode wie bei der Massenabschätzung offener Haufen haben wir aus den beobachteten Sternzahlen die Massen von 16 Assoziationen der Liste von MORGAN, WITHFORD und CODE (1953) berechnet. Im Mittel erhalten wir 2200 Sonnenmassen/Assoziation. Einen ähnlichen Wert ergeben 5 Assoziationen, über die sich ausführlichere Angaben in der Literatur finden. Weiterhin wollen wir mit dem runden Wert rechnen:

$$\mathfrak{M}_A = 2000 \text{ Sonnenmassen/Assoziation.} \tag{126}$$

4. Als nächstes interessiert noch die zeitliche Entstehungsrate von Assoziationen. Die obige Liste enthält 14 Assoziationen, deren früheste Sterne in O5...O8 liegen, innerhalb eines mittleren Abstandes von 1.8 kpc. Nach Tab. 4 können wir ihnen ein Alter von $3 \cdot 10^5 ... 1.5 \cdot 10^6$ Jahren zuschreiben. Hieraus berechnet sich die zeitliche Rate R_A der Entstehung von Assoziationen zu $1.14 \cdot 10^{-12}$ Ass/(Jahr · pc²). Die 13 restlichen Assoziationen der Liste liegen in O9 und B0, die Altersgrenzen dieser Gruppe betragen $1.5 \cdot 10^6 ... 4.6 \cdot 10^6$ Jahre, die Entfernungen reichen bis 1.0 kpc. Für die zweite Gruppe erhalten wir aus diesen Werten: $1.34 \cdot 10^{-12}$ Ass/(Jahr · pc²). Wir haben also auch bei den Assoziationen eine konstante Entstehungsrate in beiden Altersgruppen. Im Mittel ergibt sich somit

$$R_A = 1.2 \cdot 10^{-12} \text{ Ass/(Jahr} \cdot \text{pc}^2). \tag{127}$$

3. Zum Schluß stellen wir noch die Frage, welche Massen die Assoziationen haben müßten, um gravitationsstabil zu sein. Nennen wir r den Radius einer Assoziation und v die mittlere Geschwindigkeit ihrer Sterne bezüglich ihres Schwerpunktes, so erhalten wir im Mittel über 7 Assoziationen für v und über 5 Assoziationen für r die Werte:

$$\bar{v} = 5.6 \text{ km/sec} \quad \text{und} \quad \bar{r} = 53 \text{ pc}. \tag{128}$$

Falls die Assoziationen stabil wären, erhalten wir für die Masse als Abschätzung:

$$\mathfrak{M} = \frac{r v^2}{G} = 4 \cdot 10^5 \text{ Sonnenmassen}. \tag{129}$$

Multipliziert mit der Entstehungsrate (127) der Assoziationen ergäbe dies jedoch 300mal mehr Sterne, als zur Erklärung der Feldsterne nötig ist, falls die Masse der Assoziationen im wesentlichen aus Sternen besteht oder sich zumindest irgendwann in Sterne verwandelt. Diese Möglichkeit entfällt somit. Unser Ergebnis lautet:

Um stabil zu sein, müßten die Assoziationen mindestens 300mal mehr Gas als Sterne enthalten. (130)

Wir machen darauf aufmerksam, daß diese Formulierung nicht von unserer Abschätzung der LKF oder der Masse abhängt. Sie basiert allein auf Gl. (129), (127) und (113).

Nicht auszuschließen wäre dagegen folgende Darstellung: Die Assoziationen enthalten $4 \cdot 10^5$ Sonnenmassen an Gas und sind zunächst stabile Gebilde. Nur $1/300$ der Masse verwandelt sich in Sterne, deren Strahlungsdruck dann das restliche Gas wieder zerstreut, wozu die Energie weniger O-Sterne bereits ausreicht. Anschließend laufen auch die Sterne auseinander, die Assoziation löst sich auf.

B.IV. Sterne und interstellare Materie

Nachdem wir den Ort und die zeitliche Rate der Sternentstehung behandelt haben, bleibt noch das Material, aus dem die Sterne sich bilden, zu diskutieren übrig. Wenn unsere Vorstellung, daß die Sterne aus der interstellaren Materie entstehen, richtig ist, dann müssen wir erwarten, daß wir die jungen Sterne noch mit ihrem Ausgangsmaterial vergesellschaftet vorfinden. Zur Prüfung dieser Frage behandeln wir im Abschnitt a) die räumliche Verteilung der jüngeren Sterne und des interstellaren Mediums. Das letztere wird nun nicht nur aus ursprünglichem — d. h. vor der Entstehung der ersten Sterne schon vorhanden gewesenem — Material bestehen, sondern auch einen Anteil enthalten, der bereits von sich hinreichend rasch entwickelnden Sternen zurückgeliefert wurde. Wie wir in Abschnitt A.I.a gesehen haben, müssen die massiven Sterne ja einen wesentlichen Teil ihrer Masse erst abgeben, ehe sie einen Endzustand hinreichend langer Lebensdauer erreichen können. Die sich hieraus ergebende notwendige zeitliche Rate der Massenrücklieferung sowie eine Abschätzung der Prozesse, durch die die Sterne Masse verlieren können, behandeln wir im Abschnitt b) „Massenabgabe von Sternen an das interstellare Medium". Mit dieser Massenabgabe ist zugleich die Möglichkeit verbunden, daß ein Teil der Sterne sich aus Material bildet, das bereits einmal Bestandteil eines hinreichend massiven Sternes gewesen ist. Da jeder dieser Sterne einen Teil seines Wasserstoffes in Helium umgewandelt hat, wird die chemische Zusammensetzung der zurückgelieferten Materie bezüglich des Heliumgehaltes von der ursprünglichen Ausgangsmaterie abweichen. Neben dem Heliumgehalt ist aber auch der Gehalt an Metallen, wie wir gesehen haben, für das Schicksal eines Sternes sehr entscheidend, und manche Effekte lassen sich nur mit einem grundsätzlichen Unterschied im Metallgehalt (z. B. zwischen den Sternen der Population I und denen der Population II) erklären. Die bisherige Ansicht, daß die Häufigkeit der schwereren Elemente auf Prozesse zurückzuführen sei, die nur am Anfang der Welt unter völlig anderen Bedingungen als den heute herrschenden ablaufen konnten, hat sich in jüngster Zeit dadurch etwas geändert, daß eine Reihe von Möglichkeiten entdeckt wurden, nach denen unter gewissen Bedingungen auch in den Sternen selbst schwerere Elemente als Helium aufgebaut werden können. Diese Möglichkeiten und einige sich daraus ergebende Fragen der Änderung der chemischen Zusammensetzung werden wir in Abschnitt c), „Erzeugung schwererer Elemente in Sternen", behandeln.

Nach dem bisher Gesagten wird verständlich sein, daß sich eine größere Zahl von Fragen ergeben wird, deren Beantwortung eine vollständige, über die Zeit integrierte Bilanz des gesamten Materieumsatzes der Milchstraße voraussetzt, ja eigentlich sogar die Kenntnis der Geschichte jedes einzelnen Atomes unseres galaktischen Systems erfordert. Wir müßten dazu eine detaillierte Entwicklungsgeschichte der einzelnen Teile der Milchstraße angeben und dabei notwendigerweise eine Reihe von weiteren Parametern in das Problem hineinbringen, die bei dem heutigen Stand unserer Kenntnisse mit einem immer höheren Grad an Unsicherheit behaftet wären, z. B. den Massenaustausch zwischen dem Kern der Milchstraße und ihren äußeren Teilen [c. f. van den Bergh (1957c)]. Stattdessen werden wir im letzten Abschnitt d) versuchen, den Punkt zu bezeichnen, bis zu dem sich die auftretenden Fragen mit einem Minium an Voraussetzungen beantworten lassen, ohne daß die innere Konsistenz der sich dabei ergebenden Folgerungen unter ein der Unsicherheit der gemachten Voraussetzungen entsprechendes Maß absinkt.

B.IV.a) Die räumliche Verteilung von Sternen frühen Typs und interstellarer Materie

1. Die O-Sterne

Daß die O-Sterne eng mit dem Vorkommen interstellarer Materie verknüpft sind, gilt schon seit langem als grundlegender Tatbestand. Das gleichzeitige Vorhandensein von Sternen sehr früher Typs und interstellarer Materie in Form von Gas, Staub und neutralem Wasserstoff dient geradezu als Definition eines „Spiralarmes" [c. f. HILTNER und IRIARTE (1955)]. Es hat jedoch den Anschein, als ob nicht nur die O-Sterne selbst gemäß ihrer Entwicklung einen exceptionellen Charakter besitzen, sondern auch die Art und der Ort ihrer Entstehung. BLAAUW (1956b) gibt an, daß 25% aller O- bis B0-Sterne Geschwindigkeiten > 30 km/sec besitzen und ferner, daß mit zunehmender Geschwindigkeit der Prozentsatz an Doppel- und Mehrfachsystemen unter diesen Sternen abnimmt. So sind 80% der Sterne mit Geschwindigkeiten zwischen 20—35 km/sec und alle Sterne mit > 35 km/sec Einzelsterne, während der normale Prozentsatz für Sterne vom Typ A—K 50% beträgt [ALLEN (1955)]. Nimmt man noch die schon in A.III.d und e behandelten Fakten über die Expansion von Assoziationen und ihre „Ausreißer" hinzu, so gewinnt man den Eindruck, daß in Regionen besonders hoher interstellarer Dichte bevorzugt Sternsysteme mit positiver Energie entstehen, wodurch sich die Produkte der Sternerzeugung dieser Regionen über ein größeres Volumen verbreiten. Übrigens ist die Existenz z. B. des „Ausreißerpaares" AE Aurigae und μ Columbae, wenn wir es als ein Doppelsternsystem mit positiver Energie auffassen, ein wichtiger Hinweis darauf, daß das Problem der Entstehung von Doppel- und Mehrfachsternen wohl eher eine Frage der Sternentstehung selbst als der späteren Entwicklung der Sterne ist.

2. Die Korrelation der räumlichen Verteilung junger Sterne mit der interstellaren Materie

Den besten Überblick über diese Frage gibt im Augenblick wohl eine Arbeit von VAN RHIJN (1957). Dieser hat die räumliche Verteilung von drei Sterngruppen untersucht, nämlich der Draperschen Spektralklassen A0—A5, der Klasse A0 gesondert und der Riesen der Klassen K0—K5. Die Sterne dieser drei Gruppen wurden nach ihrem Abstand $|z|$ von der galaktischen Ebene unterteilt und die ermittelte Sterndichte $\bar{\Delta}$ (in Einheiten der Dichte in der Sonnenumgebung) jeder dieser Untergruppen mit den Werten für die Dichte des neutralen Wasserstoffes verglichen, die die 21 cm-Messungen des Observatoriums Leiden ergaben. Dabei wurde zum Vergleich der maximale Wert H_m der Wasserstoffdichte auf einer zur galaktischen Ebene senkrechten Geraden durch den betrachteten Punkt zwischen der galaktischen Breite $b = \pm\ 10°$ benutzt. Für die Sterne in der Nähe der Milchstraßenebene wurde auch der Vergleich mit der Wasserstoffdichte H_2 in dem betrachteten Punkt selbst durchgeführt. Das Material wurde nach den Werten von H noch einmal in zwei Gruppen geteilt, um die Gradienten $d\varDelta/dH$ zu erhalten. Wir geben in der Abb. 21 das Resultat van Rhijns nach seinen Tab. 12 und 13. Bis $|z| = 50$ pc zeigt sich eine deutliche Korrelation der Dichte der A-Sterne und der des interstellaren Wasserstoffes, die für $|z| > 100$ pc schwächer wird und für noch größere Abstände von der galaktischen Ebene ganz verschwindet. Für die K-Riesen dagegen ergibt sich überhaupt keine Korrelation. Da der Anstieg bei den A-Sternen maximal nur knapp die doppelte Streuung beträgt, hat sich van Rhijn dadurch von der Realität des Effektes überzeugt, daß er das Material nach galaktischer Länge noch einmal in fünf Gruppen teilte, die alle das gleiche Resultat ergaben.

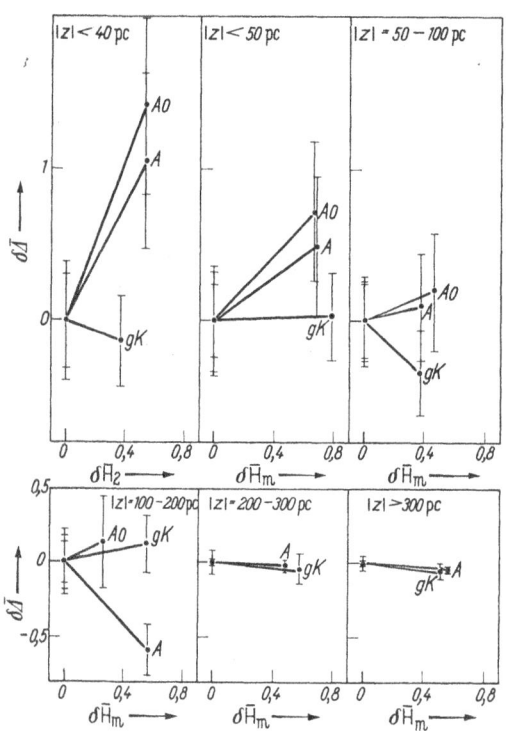

Abb. 21. Relation zwischen der räumlichen Dichte dreier Sterngruppen und der Dichte des neutralen Wasserstoffes nach van Rhijn (1957) l. c. Tab. 12 u. 13

Ein anderes sehr umfangreiches Material über Sterndichten stammt von McCuskey (1956). Van Rhijn (1957, Tabelle 14) hat auch dessen Ergebnisse in der gleichen Weise mit den Leidener Messungen verglichen. Das Ergebnis zeigt Abb. 22a und b. Zunächst fällt auf, daß die Zunahme der Sterndichte mit der interstellaren Dichte für die B8-A0-Sterne schwächer ist als für die A2-A5-Sterne. Nach dem im vorigen Abschnitt Gesagten läßt sich immerhin verstehen, daß die sehr frühen Typen zufolge ihrer höheren Raumgeschwindigkeiten eine etwas schwächere Korrelation mit ihrem Ursprungsort zeigen. Schwerer zu verstehen ist aber, daß sich für die G8-K3-Riesen nach McCuskeys Material im

Gegensatz zu VAN RHIJNs Resultat eine durchaus deutliche Korrelation vor allem mit $\bar{H}_m$ ergibt. Eine Aufklärung dieser Diskrepanz wäre für unser Problem sehr wichtig, erscheint aber zur Zeit nicht möglich. VAN RHIJNs Resultat würde der Meinung entsprechen, daß die K-Riesen Endstadien der Sternentwicklung darstellen, während MCCUSKEYs Ergebnis eher darauf hindeutet, daß sie gleichzeitig mit den frühen Typen entstehen.

Der gleichen Arbeit VAN RHIJNs entnehmen wir schließlich noch eine Korrelation zwischen der photographischen Absorption pro 200 pc in Größenklassen und der Dichte des neutralen Wasserstoffes nach seinem eigenen und MCCUSKEYs Material, die in guter Übereinstimmung zeigt, daß Staub und Gas in der Milchstraßenebene hinreichend miteinander korreliert sind (Abb. 22c).

B.IV.b) Die Massenabgabe von Sternen an das interstellare Medium

1. Für unser Problem ist neben der Rate der Sternentstehung, d. h. des Verbrauches von interstellarem Material, die Rate der Rücklieferung von Materie an das interstellare Medium durch die Sternentwicklung von zentraler Bedeutung, da sie sozusagen das „Soll" einer Gesamtbilanz darstellt. Wie wir in Abschnitt A.I.a 4 gesehen haben, muß die Masse eines Sternes kleiner sein als die Chandrasekharsche

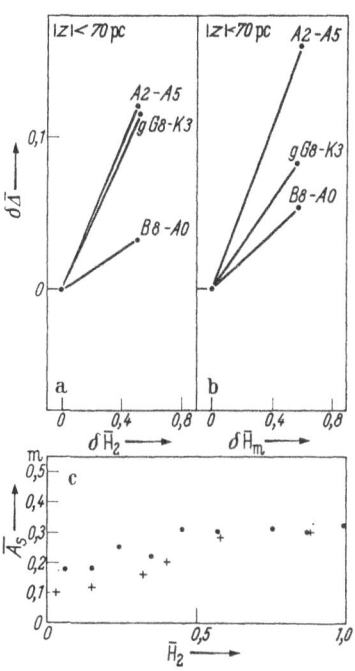

Abb. 22a—c. a) und b): Relation zwischen der räumlichen Sterndichte in Anzahl/1000 pc³ und der Dichte des neutralen Wasserstoffes für ein mit Abb. 21 vergleichbares Material von MCCUSKEY (1956). c) Relation zwischen der photographischen Absorption m/200 pc und der Dichte des neutralen Wasserstoffes nach VAN RHIJN, l. c. Tab. 14, 16, 17

Grenzmasse für relativistische Elektronenentartung, damit der Stern, nachdem er seine subatomaren Energiequellen erschöpft hat, sich zu einem überdichten Zwerg mit praktisch beliebig langer Lebensdauer kontrahieren kann. Das heißt aber, seine Endmasse darf höchstens von der Größenordnung einer Sonnenmasse sein.

Um die sich aus dieser Bedingung ergebende Rücklieferungsrate abzuschätzen, bedienen wir uns wieder der Leuchtkraftfunktion nach Tab. 7. Was wir benötigen, ist die Summe

$$\sum_{-\infty}^{3.5} \psi(M) \cdot (\mathfrak{M} - 1) = \sum_{-\infty}^{3.5} \psi \mathfrak{M} - \sum_{-\infty}^{3.5} \psi$$

über alle die Sterne, deren Masse größer als 1 $\mathfrak{M}_\odot$ und deren Lebensdauer kleiner als $5 \cdot 10^9$ Jahre ist. Dazu genügt es bis $M_v = 3$ oder 4 zu gehen; aus der Differenz der beiden letzten Spalten sehen wir, daß etwa 200 $\mathfrak{M}_\odot$ auf 1000 Sonnenmassen, d. h. 20% der Gesamtmasse aller Sterne, auf Grund unserer Bedingung wieder an das interstellare Medium zurückgegeben werden müssen. Wenn wir annehmen, daß die Gesamtmasse der Milchstraße mit $1.6 \cdot 10^{11} \mathfrak{M}_\odot$ [ALLEN (1955)] an der Sternerzeugung beteiligt ist, ergibt sich als mittlere Rate der Massenabgabe über das Weltalter in grober Näherung

$$A = \frac{0.2 \cdot 1.6 \cdot 10^{11}}{5 \cdot 10^9} \approx 6 \, \mathfrak{M}_\odot/\text{Jahr} \,. \tag{131}$$

Bei einer genaueren Rechnung müßte man natürlich vor allem berücksichtigen, daß mit abnehmender Masse und Leuchtkraft die Lebensdauer eines Sternes wächst. Für Sterne mit nicht zu großem Massenüberschuß bedeutet dies, daß sie lediglich zu Anfang ihrer Entwicklung eines einigermaßen effektiven, masseabtreibenden Prozesses bedürfen, später können sie sich in zunehmendem Maße Zeit lassen. Aus der Tab. 7 lesen wir noch ab, daß der Hauptbeitrag zu der geforderten Gesamtrate der Massenabgabe, nämlich das Maximum der Differenz ($\mathfrak{M}\,\psi - \psi$) bei $M_v = 0$ bis 1 liegt, d. h. bei den B5-A5-Sternen, deren Masse etwa zwischen 3 und 5 $\mathfrak{M}_\odot$ liegen. Für diese und die späteren Typen dürfte das oben Gesagte gelten. Für die verbleibenden etwa 0.3% aller Sterne dagegen, die — wie wieder die Differenz der beiden letzten Spalten für $M_v = -1$ zeigt — zusammen 8% der Gesamtmasse abgeben müssen, werden wir einen sehr viel effektiveren Prozeß des Massenverlustes erwarten dürfen, der in einer Katastrophe am Ende der Entwicklung oder in einer bis auf wenige Sonnenmassen hinunter wirksamen Materieabstrahlung bestehen kann. Die zugehörige Teilrate wäre

$$A = 2.6 \, \mathfrak{M}_\odot/\text{Jahr} \,. \tag{132}$$

2. Um dieses mehr theoretische Ergebnis mit der Wirklichkeit zu vergleichen, ziehen wir den entsprechenden Teil einer Massenbilanz des interstellaren Mediums von BIERMANN (1955) heran, der folgende Abschätzung gibt (die Zahlen bedeuten $\mathfrak{M}_\odot/\text{Jahr}$ und beziehen sich auf die ganze Milchstraße):

1. Rasche Massenverluste durch
   a) Supernovae 0.010
      Novae 0.001
   b) Planetarische Nebel 0.017
   c) Besondere Sterne z. B. P Cyg . . . . <0.01
                                Summe   <0.1 $\mathfrak{M}_\odot/\text{Jahr}$

2. Kontinuierliche Korpuskularemission:
   a) Wolf-Rayet-Sterne . . . . <0.05
   b) Überriesen früher als A . . . . . . ≈ 1.0
   c) Sonnenähnliche Sterne $N \cdot 10^{-12}$ . . . <0.1
   d) O- und B-Sterne u. U. . . . . $\approx (3.2)$
   e) Schnell rotierende Sterne . . . . . . ?

   Summe $1-10\ \mathfrak{M}_\odot/\mathrm{Jahr}$

Demnach spielen die eigentlichen Sternkatastrophen nur eine untergeordnete Rolle, und der Hauptbeitrag zur Materierücklieferung stammt aus der Korpuskularemission. Wenn wir deren Summe mit unseren beiden obigen Raten vergleichen, dürfen wir feststellen, daß die Übereinstimmung angesichts der beiderseitigen Unsicherheit recht befriedigend ist, bis auf den in Gl. (131) enthaltenen Beitrag der weniger massiven A—F-Sterne, der unter Umständen durch 2e gedeckt werden müßte. Dabei wäre zu bemerken, daß die auf einem zu großen Gesamtdrehimpuls beruhenden Massenverluste gerade in der ersten Entwicklungsphase von Bedeutung sein werden, die auf einer zu großen Leuchtkraft beruhende Emission dagegen während der Entwicklung noch anwachsen kann.

B.IV.c) Die Erzeugung schwererer Elemente in Sternen

Die im vorigen Abschnitt besprochene Rücklieferung von Sternmaterie an das interstellare Medium führt auf die Frage, inwieweit dadurch die chemische Zusammensetzung des Ausgangsmaterials weiterer Sternbildungen verändert wird, so daß das Schicksal eines Sternes unter Umständen davon abhängt, welcher Prozentsatz seiner Materie den Prozeß der Sternentwicklung schon durchlaufen hat und wie oft das schon geschehen ist, d. h. welcher „Generation" der Stern selbst bezüglich seines Ausgangsmaterials angehört.

1. Eine Bilanz der Erzeugung von Helium bringen wir im folgenden Abschnitt B.IV.d im Zusammenhang mit der Anzahl Weißer Zwerge.

Für das nächste, für die Sternentwicklung wichtige Element, den Kohlenstoff, ergibt sich folgendes Bild. Damit der CN-Zyklus in einem Stern anlaufen kann, muß seine Materie einen Massenanteil von etwa 0.5% Kohlenstoff als Katalysator enthalten. Das heißt aber, selbst wenn die sich rasch auflösenden Sterne nur einen Mindestheliumgehalt von 14% besitzen, würde es bereits genügen, wenn sie vor ihrer Auflösung noch etwa 10% ihres Heliums nach der von SALPETER (1952) angegebenen Reaktion (5) in Kohlenstoff umwandeln, damit in einem aus dem rückgelieferten Material bestehenden Stern der CN-Zyklus möglich ist. Dabei wird wiederum vorausgesetzt, daß alles interstellare Material schon einmal durch die Sternentwicklung gegangen ist, andernfalls müßte der Prozentsatz des umgewandelten Heliums entsprechend höher gewesen oder der nötige Kohlenstoff in der ursprünglichen Materie doch schon enthalten gewesen sein.

2. Für den Aufbau und die Entwicklung eines Sternes ist noch von besonderer Wichtigkeit sein Gehalt an leicht ionisierbaren Elementen, d. h. an Metallen der Aluminium- und Eisengruppe, und es erhebt sich die Frage, ob auch diese Elemente im Laufe der Sternentwicklung erzeugt werden können. Wenn dies der Fall wäre, würden sich die für Sterne verschiedenen Metallgehaltes aus der Beobachtung ergebenden Unterschiede [SCHWARZSCHILD (1957a) und WELLMANN (1955)] als Generationsunterschiede erklären lassen, z. B. in dem Sinne, daß die Population II die ältere und die Population I tatsächlich die jüngere ist.

Wegen der mit zunehmender Kernladungszahl anwachsenden elektrostatischen Abstoßungskräfte benötigt man zum Aufbau schwererer Elemente freie Neutronen. In den letzten Jahren sind nun eine Reihe von Prozessen gefunden worden, die bei Temperaturen um 10^8 Grad Neutronen liefern und unter Bedingungen, wie sie beim Auslaufen der Salpeterreaktion im Stern herrschen, tatsächlich eine Synthese höherer Elemente ermöglichen [CAMERON (1955); BURBIDGE, BURBIDGE und FOWLER (1955); BURBIDGE, BURBIDGE, FOWLER und HOYLE (1957)].

Die Vorstellung einer kontinuierlichen Synthese schwerer Elemente im Sterninnern stößt jedoch an zwei Stellen auf unüberwindliche Hindernisse: 1. Die beobachtete Häufigkeit der Eisen-Nickel-Gruppe kann nicht erklärt werden, und die der höheren Elemente nur, wenn das Eisen mit der beobachteten Häufigkeit zu Beginn der Synthese schon vorhanden ist. Diese Schwierigkeit sucht man nach HOYLE (1954) dadurch zu umgehen, daß man eine instantane Erzeugung des Eisens während eines Supernovaausbruches annimmt. 2. Die α-Instabilität der Elemente oberhalb des Wismut verhindert auf jeden Fall die Entstehung der natürlichen radioaktiven Elemente, z. B. des Urans, durch einen fortgesetzten stetigen Neutroneneinfang. Auch diese Schwierigkeit kann nach BURBIDGE, CHRYSTY und FOWLER (1956) dadurch behoben werden, daß während eines Supernovaausbruches für eine kurze Zeit von 10—100 sec ein praktisch unerschöpflicher Neutronenstrom im Stern vorhanden ist, in dem der Neutroneneinfang sehr viel schneller erfolgt als die β-Zerfälle. Hierdurch können (unter bestimmten Voraussetzungen bezüglich der relativen Häufigkeit des Wasserstoffes und des Eisens zu der der übrigen Elemente) in etwa 7 sec aus dem Eisen die Elemente in der Gegend des Californium 254 synthetisiert werden.

Die Anregung zu dieser Vorstellung gab die von BAADE (1945) entdeckte Tatsache, daß die zur Population II gehörenden Supernovae vom Typ I in ihren Lichtkurven übereinstimmend einen linearen Abfall zeigen, dem eine exponentielle Abnahme der Leuchtkraft mit einer Halbwertzeit von 55 ± 1 Tagen entspricht. Mit der gleichen Halbwertzeit spaltet das Cf^{254}. Ebenfalls passen würden aber auch die Halbwertzeiten, mit denen das Beryllium-Isotop Be^7 und das Strontium-Isotop

Sr^{89} zerfallen. Diese Möglichkeiten schließen die Autoren mit dem Hinweis aus, daß dann die beobachtete Häufigkeit der Zerfallsprodukte Li^7 und Y^{89} hundertmal größer sein müßte. Dagegen kann man mit BIERMANN [vgl. TEMESVÁRY (1957)] einwenden, daß das so entstandene Lithium beim nächsten Durchgang der Materie durch das Sternstadium wieder eliminiert würde. Da sowohl diese instantane Erzeugung des Eisens wie die des Urans außer von einer Reihe kernphysikalischer Bedingungen noch von den uns nur ungenügend bekannten Vorgängen beim Ausbruch einer Supernova abhängt, müssen wir darauf verzichten, die Erzeugungsrate selbst abzuschätzen.

Immerhin gibt es einen unwiderlegbaren Beweis dafür, daß in Sternen schwerere Elemente wirklich erzeugt werden können, nämlich die von MERILL (1959) in den Spektren einiger S-Sterne entdeckten Linien des instabilen Technetium-Isotops Tc^{99}, das durch Neutroneneinfang aus Molybdän entsteht und mit einer Halbwertzeit von $2.4 \cdot 10^5$ Jahren in Ruthenium zerfällt. Gerade die Kürze dieser Halbwertzeit, die zwingend beweist, daß das Isotop im Stern erzeugt werden muß, bereitet andererseits einige Schwierigkeiten, wenn man annimmt, daß die Umwandlungen in der Kernregion des Sternes erfolgen. Der Stern müßte dann durch Konvektion oder rasche Zirkulation ganz durchmischt werden, damit das Technetium noch während seiner Lebensdauer an die Oberfläche gelangen kann.

Die wahrscheinliche Lösung dieses Dilemmas liegt in einer anderen, ebenfalls von FOWLER und dem Ehepaar BURBIDGE (1955) entwickelten Vorstellung, wonach eine Synthese höherer Elemente auch möglich ist durch Kernreaktionen an der Oberfläche von Sternen unter der Einwirkung einer im Magnetfeld des Sternes beschleunigten Höhenstrahlung. Den Ausgangspunkt für diese Vorstellung bildete die Tatsache, daß in den Spektren einiger Apm-Sterne bestimmte Elemente wie Si, Mn, Cr, Eu und Pb mit einer um eine bis drei Größenordnungen größeren Häufigkeit beobachtet werden.

Man hofft nun, nicht nur die anormalen Häufigkeiten in den Apm-Sternen, sondern auch das Vorkommen des Technetiums in den S-Sternen, ja unter Umständen sogar die Erzeugung des Urans durch solche Prozesse erklären zu können. Ein kritischer Beitrag von BIERMANN (1956) zu dieser Frage zeigt, daß die Entstehung der benötigten Höhenstrahlung unter weniger exceptionellen Bedingungen möglich ist als nach der Vorstellung der erstgenannten Autoren, wenn sie nämlich in größerer Höhe, in den äußersten Schichten der Sternatmosphäre entsteht. Um die zur Erklärung der beobachteten Häufigkeiten nötigen Umwandlungen zu erhalten, müßte die Intensität des auf den Stern zurückkommenden Teiles dieser Strahlung etwa 1—10% der Wärmestrahlung entsprechen. Das bleibt immer noch eine recht harte

Anforderung. Die damit erhaltene Erzeugungsrate würde für den betrachteten Stern von etwa 4 $\mathfrak{M}_\odot$ Masse und 2 $R_\odot$ Radius $10^{9.8}$— $10^{10.8}$ angelagerte Nukleonen pro cm² und sec oder 10^{10} g erzeugte schwere Elemente pro sec betragen. Das ergibt über eine Entwicklungszeit von $10^{16.2}$ sec (vgl. Tab. 1) $10^{-7}\mathfrak{M}_\odot$ so erzeugter schwerer Elemente, während der Stern in der gleichen Zeit 3 $\mathfrak{M}_\odot$ abstrahlen muß. Danach spielen diese Kernumwandlungsprozesse an der Sternoberfläche für die chemische Zusammensetzung der von dem einzelnen Apm-Stern an das interstellare Medium zurückgelieferten Masse kaum eine Rolle, geschweige denn für die Gesamtbilanz der schweren Elemente.

B.IV.d) Bilanz der Weißen Zwerge und des Heliums

Nach den im ersten Kapitel dargestellten Theorien der Sternentwicklung wird der Energiehaushalt der Sterne im wesentlichen durch die Verbrennung von Wasserstoff zu Helium gedeckt, und die Weißen Zwerge wurden als Endprodukte der Entwicklung betrachtet. Bisher blieb jedoch offen, ob *alles* vorhandene Helium und *alle* Weißen Zwerge durch Sternentwicklung entstanden sind oder nicht. Die Milchstraße könnte bereits vor der Sternentstehung einen wesentlichen Prozentsatz Helium enthalten haben; desgleichen blieb offen, ob ein wesentlicher Teil der Weißen Zwerge bei der Sternentstehung direkt entstanden ist, was z. B. für Sterne unterhalb von 0.25 Sonnenmassen durch MASANI (1956) wahrscheinlich gemacht wurde (oberhalb von 0.25 Sonnenmassen können nach MASANI jedoch nur normale Sterne entstehen).

Einer Anregung von L. BIERMANN (private Mitteilung) folgend, wollen wir diese Fragen näher untersuchen, zumal sich hierbei ein weiteres (und von Kapitel B.II. völlig unabhängiges) Argument dafür ergibt, daß die Mehrzahl der Sterne relativ früh entstanden sein muß.

Die Durchführung unserer Abschätzung basiert auf den folgenden Voraussetzungen:

    a) Alle abgestrahlte Energie stammt aus dem Umsatz von Wasserstoff in Helium.
    b) Alles vorhandene Helium stammt aus Sternentwicklung.   (133)
    c) Jeder Stern oberhalb von 1.2 Sonnenmassen wird schließlich zum Weißen Zwerg.
    d) Alle Weißen Zwerge stammen aus Sternentwicklung.

Nach den Theorien der Sternentwicklung ist a) eine brauchbare Näherung und auch c) läßt sich begründen. Voraussetzung d) dürfte vermutlich richtig sein; dagegen ist b) zunächst eine reine Arbeitshypothese, die erst durch die Abschätzung selbst zu prüfen ist.

Unter diesen Voraussetzungen läßt sich nun eine „Dreierbilanz" aufstellen: Leuchtkraft — Heliumgehalt — Anzahl Weißer Zwerge. Würden wir z. B. den zeitlichen Verlauf $L(t)$ der gesamten Leuchtkraft der Milchstraße kennen, so ließe sich sowohl der heutige Prozentsatz Y an Helium berechnen als auch die Anzahl N der Weißen Zwerge. Da uns jedoch die Kenntnis von $L(t)$ fehlt, so beschränken wir uns auf zwei Grenzfälle:

Modell A: die zeitliche Rate der Sternentstehung war zu allen Zeiten gleich ihrem heutigen Wert [Gl. (113)]: (134)
$R(t) = \text{const.} = 2.1 \cdot 10^{-9} \mathfrak{M}_\odot / \text{Jahr} \cdot \text{pc}^2$.

Modell B: alle Sterne sind anfangs entstanden.

Bei Modell A können wir in genügender Näherung auch die Gesamtleuchtkraft als zeitlich konstant ansetzen, da die sich zeitlich vergrößernde Anzahl von Sternen geringer Helligkeit wenig zur Gesamtleuchtkraft beiträgt, während die jeweils vorhandene Anzahl heller Sterne konstant bleibt. Bei Modell B ließe sich zwar die zeitliche Abnahme von $L(t)$ berechnen, doch wäre der dazu nötige Arbeitsaufwand durch die Unsicherheit der zu verwendenden Zahlen nicht gerechtfertigt. Der einfachere Weg ist hier die Berechnung von Y und N aus der in Tab. 7 gegebenen Massenverteilung in offenen Sternhaufen, die wir, wie bisher, als ursprüngliche Verteilung bei Sternentstehung benutzen.

Die Massen der Weißen Zwerge setzen wir, unabhängig von der ursprünglichen Masse des Sternes, mit

$$\mathfrak{M}_{WZ} = 0.7 \text{ Sonnenmassen} \qquad (135)$$

an, ein Stern der ursprünglichen Masse $\mathfrak{M}$ gibt also während seiner Entwicklung die Masse $(\mathfrak{M} - 0.7)$ wieder an die interstellare Materie ab. Den Bruchteil des Heliums dieser zurückgegebenen Masse nennen wir Q, und nach den Theorien der Sternentwicklung dürften wir, vor allem im Hinblick auf die massiveren Sterne, etwa mit

$$Q \approx 0.5 \qquad (136)$$

rechnen.

1. Wir prüfen zunächst Voraussetzung b), da sich zeigt, daß sich dies unabhängig von den beiden Modellen (134) durchführen läßt.

Modell A:

Die Gesamtmasse aller jemals erzeugten Sterne beträgt

$$\mathfrak{M}_* = R \cdot T, \qquad (137)$$

und der Bruchteil an Sternen der Helligkeit $M \pm 1/2$ ist $\psi(M)$ von Tab. 7. Ist $\tau(M)$ wieder das maximale Entwicklungsalter von Sternen der ursprünglichen Helligkeit M nach Tab. 4, so sind alle die Sterne zu Weißen Zwergen geworden, die zur Zeit $t \leq T - \tau(M)$ entstanden sind. Damit ist die Anzahl heute vorhandener Weißer Zwerge

$$N = \mathfrak{M}_* \int_{-\infty}^{3.5} \psi(M) \left\{1 - \frac{\tau(M)}{T}\right\} dM \qquad (138)$$
$$= (0.088 - 0.037) \mathfrak{M}_* = 0.051 \mathfrak{M}_*, \mathfrak{M}_* \text{ in Sonnenmassen}$$

wobei wir noch offen lassen können, ob wir die Gesamtheit der Milchstraße meinen oder die Sonnenumgebung.

Die Menge des an die interstellare Materie zurückgegebenen Heliums berechnet sich in gleicher Weise zu

$$\mathfrak{M}_{He,z} = \mathfrak{M}_* Q \int_{-\infty}^{3.5} \psi(M) \cdot (\mathfrak{M} - 0.7) \cdot \left(1 - \frac{\tau}{T}\right) dM = 0.089 \mathfrak{M}_*. \qquad (139)$$

Schätzen wir mit ähnlichen Formeln noch die Menge des Heliums in noch vorhandenen Sternen ab, so erhalten wir insgesamt für die Masse des erzeugten Heliums:

$$\mathfrak{M}_{He} = 0.13 \mathfrak{M}_*. \qquad (140)$$

Wir nennen $\mathfrak{M}_T$ die insgesamt vorhandene Materie (Gas + Sterne) und $Y_T = \mathfrak{M}_{He}/\mathfrak{M}_T$, setzen Gl. (140) in Gl. (138) ein und erhalten

$$N = 0.39 \frac{\mathfrak{M}_T}{\mathfrak{M}_\odot} \cdot Y_T.$$

Zum Vergleich mit der in Sonnenumgebung beobachteten Anzahl Weißer Zwerge sollten wir berücksichtigen, daß sie sehr lichtschwach sind. Wir dürfen daher nicht mit der Gesamtzahl aller Sterne vergleichen, sondern am besten nur mit Hauptreihensternen gleicher Helligkeit. Wählen wir im Hinblick auf die Angaben von GLIESE (1956, Tab. 3) das Intervall $10 \leq M_v \leq 13$, so ergibt Tab. 7:

$$N_{HR} = \mathfrak{M}_* \int_{10}^{13} \psi(M) dM = 0.556 \mathfrak{M}_*. \qquad (141)$$

Etwa $^3/_4$ von $\mathfrak{M}_*$ ist in den weniger massiven Sternen „eingefroren", nur der Rest könnte mehrfach durch Sterne gelaufen sein. Andererseits ist der heutige Restbestand interstellaren Gases klein. Somit kann $\mathfrak{M}_*$ weder wesentlich größer, noch wesentlich kleiner sein als $\mathfrak{M}_T$; wir benutzen $\mathfrak{M}_* \approx \mathfrak{M}_T$ und erhalten

$$N = 0.70 N_{HR} \cdot Y_T. \qquad (142)$$

Nach GLIESE sind innerhalb von 5 pc Abstand von der Sonne $N = 5$ Weiße Zwerge beobachtet, sowie $N_{HR} = 18$ Hauptreihensterne im

Intervall $10 \leq M \leq 13$. Somit sollte sein:

$$Y_T = \frac{5}{0.70 \cdot 18} = 0.40. \qquad (143)$$

Modell B:

Sind alle Sterne anfangs entstanden, so ist

$$N = \mathfrak{M}_* \int_{-\infty}^{3.5} \psi(M)\, dM = 0.088\, \mathfrak{M}_*, \qquad (144)$$

und die Menge des zurückgegebenen Heliums ist

$$\mathfrak{M}_{He,z} = \mathfrak{M}_* Q \int_{-\infty}^{3.5} \psi(M) \cdot (\mathfrak{M} - 0.7)\, dM = 0.111\, \mathfrak{M}_*. \qquad (145)$$

Schätzen wir wieder das noch in Sternen befindliche Helium ab, so erhalten wir als Summe:

$$\mathfrak{M}_{He} = 0.165\, \mathfrak{M}_*, \qquad (146)$$

und statt Gl. (142) erhalten wir nun

$$N = 0.95\, N_{HR}\, Y_T. \qquad (147)$$

Mit $N = 5$ und $N_{HR} = 18$ erhalten wir schließlich für Modell B:

$$Y_T = 0.30. \qquad (148)$$

Vergleich:

In bezug auf die Bilanz „Weiße Zwerge — Helium" sehen wir, daß sich die beiden Modelle nicht wesentlich unterscheiden. Der aus der beobachteten Anzahl von Weißen Zwergen berechnete Heliumgehalt ist, verglichen mit dem beobachteten Heliumgehalt, von gleicher Größenordnung und dürfte eher zu hoch als zu niedrig sein. In Anbetracht der durch die geringe Anzahl von nur 5 Weißen Zwergen bedingten Unsicherheit formulieren wir unser Ergebnis:

> Der aus der Anzahl Weißer Zwerge berechnete Heliumgehalt läßt darauf schließen, daß zumindest ein wesentlicher Teil, vielleicht auch die (149) Gesamtmenge des heutigen Heliums durch Sternentwicklung entstanden ist.

2. Als nächstes wollen wir die Anzahl der Weißen Zwerge direkt berechnen und mit der Beobachtung vergleichen. Nach Gl. (134) gehen wir für *Modell A* von der heute beobachteten Rate R der Sternentstehung aus. Multiplizieren wir R mit dem Weltalter T und dividieren wir durch eine mittlere Schichtdicke von $2\bar{z} = 500$ pc, so erhalten wir die zu erwartende Dichte aller jemals erzeugten Sterne in Sonnenumgebung in $\mathfrak{M}_\odot/\text{pc}^3$. Wir multiplizieren mit dem innerhalb von 5 pc

befindlichen Volumen von 524 pc³ und erhalten $\mathfrak{M}_*$. Setzen wir dies in Gl. (138) ein, so ergibt sich:

$$N = 0.56 \,. \tag{150}$$

Bei *Modell B* müssen wir dagegen auf die beobachtete Anzahl von Sternen gleicher Helligkeit normieren. Aus Gl. (144) und (141) erhalten wir

$$N = 2.9 \,. \tag{151}$$

Der Vergleich mit der beobachteten Anzahl $N = 5$ von Weißen Zwergen innerhalb von 5 pc zeigt, daß wir Modell B den Vorzug geben müssen.

3. Es folgt nun die direkte Berechnung von Y zum Vergleich mit der Beobachtung. Bei *Modell A* gehen wir von der Voraussetzung aus, die Gesamtleuchtkraft sei zu allen Zeiten die gleiche wie heute gewesen. Für die Erzeugung von Helium durch den CN-Zyklus gilt dann

$$L \cdot T = K \, \mathfrak{M}_{H_e} \quad \text{mit} \quad K = 6 \cdot 10^{18} \, \text{erg/g} \,. \tag{152}$$

Wir dividieren durch die Gesamtmasse $\mathfrak{M}$ und erhalten für den heutigen totalen Heliumgehalt (in Sternen und Gas)

$$Y_T = 0.05 \, \frac{L}{\mathfrak{M}} \,, \tag{153}$$

wobei L und $\mathfrak{M}$ in Sonneneinheiten zu messen sind. In Modell A gibt uns also das Masse-Leuchtkraftverhältnis direkt den Heliumgehalt an.

Für die gesamte Milchstraße und zwei weitere Sb-Nebel erhalten wir aus den bei ALLEN (1955) angegebenen Werten für Masse und Helligkeit:

$$\begin{array}{l} \phantom{\text{Milchstraße }} \mathfrak{M}/L \\ \hline \text{Milchstraße } 10.7 \\ \text{M 31} 12.3 \\ \text{M 81} 13.9 \end{array} \Bigg\} \text{in Sonneneinheiten.} \tag{154}$$

Setzen wir $\mathfrak{M}/L = 10.7$ in Gl. (153) ein, so erhalten wir für die Milchstraße

$$Y_T = 0.005^1 \,. \tag{155}$$

Für die Sonnenumgebung gehen wir aus von der über z integrierten Leuchtkraftfunktion des Abschnittes B.II.a (plus dem dort nicht enthaltenen Anteil an Riesen und Überriesen); wir überschlagen hier die weiteren Einzelheiten und erhalten schließlich

$$\mathfrak{M}/L = 1.69 \,, \tag{156}$$

---

[1] Dieser Wert ist zu klein um einen Faktor 20 — 40. Inzwischen ist eine ähnliche Abschätzung von BURBIDGE (1958) erschienen, mit etwa gleichem Resultat. BURBIDGE schließt hieraus ebenfalls, daß die Energieerzeugung der Milchstraße anfangs weit größer gewesen sein muß als jetzt. Er diskutiert allerdings nicht die Möglichkeit, daß bereits anfangs der Hauptteil des Heliums vorhanden gewesen sein könnte. Erst durch unseren Vergleich mit den Weißen Zwergen läßt sich diese Möglichkeit ausschalten.

woraus wir
$$Y_T = 0.03 \tag{157}$$
erhalten.

Für *Modell B* könnten wir aus Gl. (146), indem wir wieder $\mathfrak{M}_* = \mathfrak{M}_T$ setzen, auf
$$Y_T = 0.165 \tag{158}$$
schließen. Um zu zeigen, an welchen Stellen die Unsicherheiten unserer Abschätzung liegen, wollen wir jedoch etwas genauer vorgehen. Die größte Unsicherheit steckt vermutlich in dem benutzten Wert von $Q = 0.5$, der den Bruchteil des Heliums in dem bei Sternentwicklung zurückgegebenen Gas angibt. Statt 0.165 schreiben wir daher ausführlicher: $(0.222\, Q + 0.054)$; der zweite Summand ist das noch in Sternen befindliche Helium. Weiterhin kann bei der Sternentstehung ein Restbestand an Gas übriggeblieben sein; wir nennen G den Bruchteil übriggeblieben Gases, bezogen auf die Gesamtmenge aller Materie. Dann ist
$$Y_T = (1 - G)\,(0.222\, Q + 0.054). \tag{159}$$
Schließlich ist zu bedenken, daß ja eigentlich nicht Y_T zu beobachten ist, sondern der Heliumgehalt an der Oberfläche von relativ jungen Sternen. (Auch die Sonne könnte in unserem Modell B durchaus *nach* dem Gros der Sterne entstanden sein.) Wir sollten somit den Heliumgehalt Y_G des interstellaren Gases berechnen:
$$Y_G = \mathfrak{M}_{He,z}/\mathfrak{M}_{Gas}.$$
Nach Gl. (145) ist
$$\mathfrak{M}_{He,z} = 0.222\, Q\, \mathfrak{M}_*$$
und weiterhin ist
$$\mathfrak{M}_{Gas} = 0.222\, \mathfrak{M}_* + G\, \mathfrak{M}_T,$$
woraus sich
$$Y_G = Q\,\frac{1-G}{1+3.5\,G} \tag{160}$$
ergibt.

Wie zu erwarten, ist Y_G sehr viel empfindlicher als Y_T in bezug auf den Restbestand G des Gases. Der heutige Anteil des Gases an der Gesamtmasse dürfte nach neueren Abschätzungen [v. D. HULST, MÜLLER und OORT (1954); v. D. HULST und RAIMOND (1956); SCHMIDT (1956)] nur etwa 12% betragen, somit sollte auch $G \ll 1$ gewesen sein. Mit $Q = 0.5$ ergibt sich für die verschiedenen Annahmen über G:

G	Y_T	Y_G
0.0	0.17	0.50
0.05	0.16	0.40
0.1	0.15	0.33
0.2	0.13	0.24

(161)

Von seiten der Beobachtung läßt sich Y nur schwer festlegen. Nach UNSÖLD (1948) ist $Y = 0.41$, während z. B. von Miss UNDERHILL (1953) etwa $Y = 0.15$ angegeben wird. Wir sollten also mit $0.20 \leq Y \leq 0.40$ rechnen, was recht gut zu unserem in Gl. (161) dargestellten Ergebnis paßt, keineswegs dagegen zu Modell A.

4. Wir fassen zusammen. Erstens: der aus der Anzahl Weißer Zwerge berechnete Heliumgehalt weist darauf hin, daß zumindest ein wesentlicher Teil des vorhandenen Heliums durch Sternentwicklung entstanden ist. Zweitens: sowohl die Anzahl Weißer Zwerge als auch der heutige Heliumgehalt lassen sich dann (und nur dann) verstehen, wenn die Mehrzahl der Sterne relativ früh entstanden ist.

In unserer Abschätzung ist nicht enthalten, daß die Materie zum Teil mehrfach durch den Prozeß der Sternentwicklung hindurchgegangen sein kann. Eine genauere Rechnung, die auch diesen Effekt berücksichtigt, ist in Vorbereitung, dürfte jedoch an unseren bisherigen Ergebnissen nichts Wesentliches ändern.

B.V. Zusammenfassung

Wir hatten uns die Aufgabe gestellt, eine Reihe von Voruntersuchungen durchzuführen, um die Unterlagen und den äußeren Rahmen einer Theorie der Sternentstehung soweit als möglich festzulegen. Wir fassen unsere Ergebnisse nochmals zusammen, weisen auf mögliche Verbesserungen hin und erwähnen einige weitere, noch ungeklärte Probleme.

B.V. a) Ergebnisse

Die Verschiedenheit der Geschwindigkeitsstreuungen von Sternen verschiedenen Spektraltyps ließ sich durch zwei Alterseffekte gut erklären: die ganz alten Sterne sind zu einer Zeit entstanden, als die Anfangsturbulenz der Milchstraße noch nicht abgeklungen war, was etwa 10^9 Jahre benötigt hat. Von da ab entstehen alle Sterne, unabhängig vom Typ, mit Geschwindigkeiten von rund 10 km/sec (vermutlich bilden sehr massive Sterne eine Ausnahme), die Geschwindigkeiten wachsen jedoch im Laufe der Zeit langsam an durch Energieaustausch mit großen Stern- oder Gaswolken.

Die Dicke der sternerzeugenden Schicht hat ebenfalls anfangs schnell abgenommen, blieb während der letzten $3 \cdot 10^9$ Jahre jedoch konstant. Durch das Anwachsen der Geschwindigkeit (z-Komponente) nimmt die räumliche Dichte von Sternen, die in einem bestimmten Zeitintervall entstanden sind, mit der Zeit ab.

Daraus folgt, daß die bisherige Ableitung der zeitlichen Rate $R(t)$ der Sternentstehung falsch ist. Es ergibt sich: $R(t)$ war anfangs rund eine Zehnerpotenz größer als heute und ist erst in den letzten $3 \cdot 10^9$

Jahren nahezu konstant geblieben. Mit diesem Ergebnis ist die Vorstellung vereinbar, daß R stets der jeweils vorhandenen Gasmenge etwa proportional gewesen ist, auch ohne Nachlieferung von Gas aus dem galaktischen Kern. Mindestens die Hälfte aller Sterne der Sonnenumgebung sind in den ersten 10^9 Jahren entstanden.

Auch die Betrachtungen des heutigen Heliumgehaltes und der Häufigkeit der Weißen Zwerge führen zu dem Schluß, daß die Mehrzahl der Sterne relativ früh entstanden sein muß.

Die räumliche Korrelation junger Sterne mit der Dichte des interstellaren Gases deutet auf die Gebiete hoher Dichte (Spiralarme) als Ort der Sternentstehung hin. Dabei entstehen rund $^1/_5$ aller Sterne in offenen Haufen; alle restlichen Sterne entstünden genau dann in Assoziationen, wenn deren Leuchtkraftfunktion gleich der ursprünglichen Leuchtkraftfunktion der Feldsterne wäre. Diese Annahme ist mit den vorliegenden Beobachtungen der helleren Sterne ($M \leq -1$) verträglich und gibt rund 2000 Sonnenmassen/Assoziation.

Die Gesamtbilanz der durch Sternentwicklung geforderten Materieabgabe stimmt mit einer Abschätzung der Beobachtungsdaten von BIERMANN überein. Eine Reihe von Möglichkeiten der Erzeugung schwererer Elemente im Sterninneren und an der Oberfläche wurden diskutiert.

B.V.b) Mögliche Verbesserungen

Einige unserer Untersuchungen lassen sich verbessern oder weiter fortführen. So würde z. B. eine genauere Diskussion des Potentialverlaufes senkrecht zur galaktischen Ebene eine verbesserte Bestimmung von $R(t)$ ermöglichen, und die Bilanz des Heliumgehaltes sollte neu aufgestellt werden, unter Berücksichtigung des Mehrfachdurchlaufes von Materie durch den Prozeß der Sternentwicklung, am besten im Zusammenhang mit dem „theoretischen Ansatz" von Abschnitt B.II.b.

Weiterreichende Beobachtungen der Leuchtkraftfunktion von Assoziationen könnten die Frage klären, ob man zur Annahme der Entstehung von einzelnen Sternen (außerhalb von Assoziationen und Haufen) gezwungen ist oder nicht.

Wenn auch die Gesamtbilanz der Materieabgabe bei Sternentwicklung in Abschnitt B.IV.b ein positives Ergebnis hatte, so ist doch noch ungeklärt, wie im einzelnen die Sterne gegebener Masse $\mathfrak{M}$ ihren Materieüberschuß $\mathfrak{M} - 1$ loswerden.

B.V.c) Nicht behandelte Probleme

Auf eine weitere Anzahl für die Sternentstehung wichtiger Probleme konnte im Rahmen dieser Arbeit nicht eingegangen werden. So sollte z. B. eine vollständige Theorie der Sternentstehung in der Lage sein, das Massenspektrum der entstehenden Sterne zu berechnen und somit eine „ursprüngliche Leuchtkraftfunktion" zu liefern.

Über die Verschiedenheit der chemischen Zusammensetzung von Sternen verschiedener Population bzw. verschiedener Geschwindigkeit (siehe Abschnitt B.I.a 4) sind von seiten der Beobachtung in den letzten Jahren zahlreiche Untersuchungen erschienen, deren Diskussion an anderer Stelle erfolgen soll.

Weiterhin ist zu klären, ob man nach den Aussagen von BLAAUW tatsächlich gezwungen ist, für O-Sterne eine größere Geschwindigkeitsstreuung bei ihrer Entstehung anzunehmen als für die weniger massiven Sterne, während wir im Gegensatz hierzu die übrigen Unterschiede der Geschwindigkeiten in Abschnitt B.I. als reine Alterseffekte deuten konnten.

Auch die Frage, warum bei den schnelleren O-Sternen ein geringerer Anteil an Doppelsternen als sonst vorhanden ist, sowie die Behandlung der Entstehung von Doppelsternen überhaupt, sollte in einer vollständigen Theorie der Sternentstehung enthalten sein, und eine sorgfältige Sichtung des Beobachtungsmaterials im Hinblick auf dieses Ziel wäre zu wünschen.

C. Physikalische Theorien zur Sternentstehung

C.I. Einleitung

In diesem Kapitel wollen wir versuchen zu zeigen, wie unter den heutigen Bedingungen, oder unter Bedingungen, die den heutigen ähnlich sind, eine Sternentstehung möglich ist. Es werden dabei die vorhandenen Ansätze zu einer physikalischen Theorie der Sternentstehung diskutiert werden. Wir beschränken uns aber auf Arbeiten, die die interstellare Materie als Quelle einer Sternentstehung oder -verjüngung annehmen, behandeln also nicht Auffassungen der Materieerzeugung, nach denen immer ein ganzer Stern auf einmal aus Nichts entstehen könnte; sie würden über den Rahmen dieser Arbeit hinausgehen.

Bei der Sternverjüngung handelt es sich um die Frage, ob die heute als jung zu bezeichnenden Sterne ursprünglich alte Sterne sind, die durch die Aufsammlung von interstellarer Materie soviel Wasserstoff gewonnen haben, daß ihre Entwicklung wieder wie die eines jungen Sterns abläuft. Der Gedanke der Aufsammlung geht auf EDDINGTON zurück. Er fand den durch diesen Effekt bedingten Massezuwachs jedoch klein verglichen mit dem Masseverlust durch die Strahlung des Sterns. In neuerer Zeit ist der Aufsammlungseffekt unter veränderten Voraussetzungen neu berechnet worden mit dem Ergebnis einer größeren Aufsammlung.

Zur Erklärung der Beobachtungen würde es genügen, wenn der Prozeß der Verjüngung nur bei einem kleinen Teil aller Sterne wirkungsvoll wäre. Die Zahl nämlich der als jung zu bezeichnenden Sterne — also hauptsächlich O- und B-Sterne — ist verglichen mit der aller Sterne gering.

Die Frage nach der Herkunft der Sterne würde durch diesen Prozeß jedoch nicht beantwortet, denn es bleibt offen, wie die sich verjüngenden Sterne, und alle anderen, entstanden sind. Es ergäbe sich aber die Möglichkeit, die Entstehung der Sterne in eine viel frühere Entwicklungsstufe unseres Milchstraßensystems zu verlegen und dadurch die Freiheit zu gewinnen, quantitativ von den heutigen vielleicht sehr verschiedene Bedingungen einführen zu können. Es ist daher wichtig zu sehen, wie wirkungsvoll dieser Prozeß sein kann. Der Abschnitt II wird eine genaue Diskussion des Verjüngungsprozesses bringen. Wir werden sehen, daß er nur unter extremen Bedingungen wirksam sein kann und nicht ausreicht, die Zahl der heute jungen Sterne zu erklären.

Dieses Ergebnis führt uns dazu, die zweite Möglichkeit, die Entstehung von Sternen durch Kondensation interstellarer Materie, ausführlich zu betrachten. JEANS war der erste, der versuchte, sie quantitativ zu fassen. In den letzten Jahren hat sie mehr und mehr mathematische Gestalt gewonnen, und obgleich bisher keine direkten Beobachtungen dafür vorliegen, so sprechen doch Beobachtungen wie die von sehr jungen Sternhaufen, bei denen ein Teil der Sterne noch im Kontraktionsstadium ist und die Hauptreihe noch nicht erreicht hat, stark dafür, daß diese Möglichkeit in der Natur verwirklicht ist.

Sehen wir von Strahlungsdruck, Drehimpuls und Magnetfeldern vorerst ab — wir werden in der genauen Diskussion auf sie eingehen —, so ist der Entstehungsprozeß im Prinzip sehr einfach. Die Eigengravitation der interstellaren Materie stellt eine Kraft dar, die versucht, Verdichtungen innerhalb der interstellaren Materie zu erzeugen. Die bei diesen Verdichtungen entstehenden Druckgradienten des Gases dagegen wirken im Mittel der Gravitation entgegen, versuchen also die entstehenden Verdichtungen wieder aufzulösen. Hat die Gravitation aber einmal die Überhand gewonnen, so kann unter ihrem Einfluß in den meisten Fällen die Verdichtung fortschreiten, weil das sich verdichtende Gas durch Ausstrahlung Energie nach außen abgibt und dadurch die Gesamtenergie des Systems weiter abnimmt. Dieser Prozeß kann fortschreiten, bis die Temperatur im Innern Werte erreicht, bei denen die Kernreaktionen einsetzen, also eine neue Energiequelle erschlossen wird. Die Entstehung des Sterns ist beendet. Auf der Hauptreihe im HR-Diagramm angekommen, folgt der zum Teil stürmischen Entstehungsperiode im allgemeinen eine damit verglichen langsame und lange Entwicklungsphase.

Einleitung

Wir sehen, die Frage, ob und unter welchen Bedingungen eine Sternentstehung möglich ist, wird weitgehend in der ersten Phase der Entstehung entschieden: es ist die Frage, in welchen Fällen die Gravitation die ihr entgegenwirkenden Kräfte überwinden kann. Die Zeitskala der gesamten Entstehung dagegen und die Frage, mit welcher Masse ein Stern schließlich die Hauptreihe erreicht, lassen sich nicht durch Betrachtung der ersten Phase allein entscheiden. Hier werden Abklingen der Turbulenz, Strahlungstransport und -druck sowie Magnetfelder und Drehimpulstransport eine entscheidende Rolle spielen. Unser einfaches Bild des Entstehungsprozesses bedarf also wesentlicher Erweiterungen durch Berücksichtigung dieser Vorgänge.

Magnetfeld und Drehimpuls der anfänglichen Gasmasse wirken im allgemeinen der Verdichtung entgegen, das Magnetfeld durch seinen Druck, der Drehimpuls infolge der durch ihn auftretenden Rotation des Gesamtsystems. Kann der größte Teil des Drehimpulses im Laufe der Verdichtung nicht abgeführt oder, wenn das gesamte System in Untersysteme zerfällt, in Bahndrehimpuls umgewandelt werden, so wird die Kontraktion und damit die Sternentstehung in diesem Stadium unterbrochen. Die Drehimpulsabführung ist daher für den weiteren Entstehungsvorgang eine der wichtigsten Fragen. Wir werden sehen, daß über die Möglichkeiten der Umwandlung des Drehimpulses in Bahndrehimpuls oder der Abführung durch turbulente Reibung hinaus gerade das Magnetfeld, das einerseits die Entstehung des Sterns erschwert, andererseits den weiteren Enstehungsvorgang erleichtern kann, indem es unter geeigneten Bedingungen die Abführung des Drehimpulses übernehmen kann.

Die Turbulenz erschwert ebenfalls durch ihren Druck die Entstehung, sie kann aber, wie schon erwähnt, zur Drehimpulsabführung beitragen. Im Unterschied zum Magnetfeld ist bei ihr der Drehimpulstransport mit einem Materietransport verbunden.

Der Strahlungstransport wird wichtig, wenn die Verdichtung soweit fortgeschritten ist, daß die Wolke optisch dicht wird. Die Größe des Transportes wird, zusammen mit der Größe der Drehimpulsabführung, die Zeitskala der Kontraktion bestimmen.

Ob der Strahlungsdruck der in dem entstehenden Stern erzeugten Strahlung in irgendeiner Entstehungsphase eine bedeutende Rolle spielt, ist bisher ungeklärt. Zu Beginn des optisch Dichtwerdens kann er den Gasdruck mitunter erheblich übertreffen. Da die Kräfte aber vom Gradienten abhängen und dieser nur durch eine bisher nicht vorliegende genauere Modellrechnung abzuschätzen ist, bleibt offen, ob dieses Überwiegen zu mechanischen Wirkungen führt. Eine mögliche mechanische Wirkung könnte im Abblasen eines Teiles der sich verdichtenden Materie bestehen. In den späteren Entstehungsphasen kann der Strahlungsdruck im Mittel nur bei sehr massereichen Sternen mit 100 Sonnenmassen oder mehr den Gasdruck überwiegen. Wenn solche Sterne die Hauptreihe erreichen, die Kernprozesse also einsetzen, führt dieses Überwiegen zu Instabilitäten in ihrem inneren Aufbau. Ob das Überwiegen des Strahlungsdruckes bei solchen Sternen schon vor ihrem Erreichen der Hauptreihe zu Instabilitäten führt, ist wiederum eine offene Frage.

Abschnitt III und IV werden die Darstellung und Diskussion der Entstehungsphasen bringen, soweit darüber Arbeiten vorhanden sind. Die überwiegende Zahl der Arbeiten gilt der ersten Entstehungsphase. In Abschnitt IV wollen wir auf die Frage der Magnetfelder und des Drehimpulses gesondert eingehen.

C.II. Sternverjüngung durch Aufsammlung

Aus Beobachtungen wissen wir, daß ein gewisser Teil der gesamten Masse unseres Milchstraßensystems in Form von interstellarer Materie vorhanden ist, sich also zwischen den Sternen befindet. Durch ihr starkes Gravitationsfeld und ihren geometrischen Querschnitt könnten die Sterne — wir sehen vorerst von anderen Kräften ab — einen Teil dieser Materie aufsammeln.

EDDINGTON (1926), der als erster diesen Prozeß betrachtete, machte die Annahme, die mittlere freie Weglänge der Teilchen der interstellaren Materie sei so groß, daß für den Aufsammlungsprozeß die Stöße der Teilchen untereinander vernachlässigt werden können. Bewegt sich der Stern durch ein solches Medium, so ist sein Massezuwachs pro Zeiteinheit nach EDDINGTON

$$\frac{dM}{dt} = \frac{2\pi G M R \varrho}{V} ; \qquad (162)$$

hierin ist G die Gravitationskonstante, M und R sind Masse und Radius des Sterns, ϱ die Dichte der interstellaren Materie, und V ist die Geschwindigkeit des Sterns relativ zur interstellaren Materie. Diesen Massegewinn vergleicht EDDINGTON mit dem Masseverlust, der mit der Energieabgabe des Sterns durch seine Leuchtkraft verbunden ist. Für eine Dichte von 1 H/cm^3 und eine Relativgeschwindigkeit von 1 km/sec stellt sich der Massegewinn mehrere Zehnerpotenzen kleiner als der Verlust heraus, ist also völlig zu vernachlässigen, selbst unter für die Aufsammlung günstigeren Bedingungen.

In neuerer Zeit haben HOYLE und LYTTLETON (1939, 1940) den Aufsammlungsprozeß neu berechnet unter der von EDDINGTON abweichenden Annahme, die Stöße der interstellaren Teilchen untereinander seien nicht zu vernachlässigen. Diese Annahme wurde nahegelegt durch die beobachtete Kopplung von jungen Sternen mit dichten interstellaren Gaswolken in den Spiralarmen. Die Gasdichte der Wolken kann mitunter erheblich oberhalb des von EDDINGTON angenommenen mittleren

Wertes liegen. Dadurch und durch ihre weitere Verdichtung, die das Gas bei seiner Annäherung an den Stern während der Aufsammlung erfährt, kann die mittlere freie Weglänge der Teilchen vergleichbar mit dem Durchmesser des Sterns werden.

Die interstellare Materie bildet dann ein Gas, das den Gleichungen der Hydrodynamik gehorcht, und in dem sich der Stern bewegt oder ruht. Dies führt, wie leicht zu sehen ist, zu erheblich größeren Aufsammlungsraten. Gasteilchen, die vorher durch ihren Drehimpuls in bezug auf das Sternzentrum verhindert wurden, die Sternoberfläche zu treffen, können nun durch Stöße Drehimpuls verlieren und dadurch zur Oberfläche gelangen.

Der für die Aufsammlung günstigste Fall ist der des in der Wolke *ruhenden* Sterns. BONDI (1952) hat eine exakte Lösung dieses Problems gegeben. Er erhält als Aufsammlungsrate

$$\frac{dM}{dt} = \frac{2\pi \beta_0 G^2 M^2 \varrho}{a^3}, \qquad (163)$$

wobei a die Schallgeschwindigkeit des Gases in großer Entfernung vom Stern ist und β_0 eine Konstante, die je nach der Zahl der Freiheitsgrade der Gasteilchen zwischen 0.5 und 2.24 (für Isothermie) liegt.

Eine exakte analytische Lösung für den Fall einer *Relativgeschwindigkeit* des Gases zum Stern ist wegen der außerordentlichen mathematischen Schwierigkeiten bisher nicht gegeben worden. HOYLE und LYTTLETON (1939, 1940) und später BONDI und HOYLE (1944) haben unter Vernachlässigung der thermischen Geschwindigkeit gegenüber der Relativgeschwindigkeit, aber mit Berücksichtigung des Drehimpulsverlustes durch Stöße der Teilchen, eine Näherungsformel abgeleitet. Darüber hinaus hat DODD (1953) für einige einfache Fälle das Problem numerisch gelöst. Das Ergebnis der Näherung, zusammen mit der numerischen Lösung und dem Grenzfall verschwindender Relativgeschwindigkeit (163) legt es nahe, als analytische Näherungslösung des allgemeinen Falles die Formel

$$\frac{dM}{dt} = \frac{2\pi \beta G^2 M^2 \varrho}{(a^2 + V^2)^{3/2}} \qquad (164)$$

zu vermuten. BONDI (1952) hat diese Formel vorgeschlagen. Sie geht für $a^2 \ll V^2$ in die Näherung von BONDI und HOYLE, und für $a^2 \gg V^2$ in die exakte Lösung [Gl. (163)] über. Dabei ist β eine Konstante der Größenordnung 1. Übereinstimmung mit den numerischen Ergebnissen für isothermes Gas erhält man, wenn $\beta = 2.24$ gewählt wird. Die Formel (164) dürfte daher für einen Wertbereich von a und V, der die im interstellaren Raum vorkommenden Werte umfaßt, eine gute Näherung sein, sofern die zu ihrer Herleitung gemachten Voraussetzungen zutreffen.

Wir werden später sehen, inwiefern diese Voraussetzungen zu eng sind und wie sie erweitert werden müssen.

Bevor wir darauf eingehen, wollen wir sehen, zu welchen Ergebnissen Gl. (164) führt. Eine direkte Anwendung von Gl. (164) über eine große Zeitspanne hinweg ließe aber die Bremsung außer acht, die der relativ zur Wolke bewegte Stern durch die Aufsammlung erfährt. DODD und MCCREA (1952) haben diesen Effekt berechnet, und MCCREA (1953) hat in einer umfangreichen Arbeit die mit Einschluß dieses Effektes sich ergebende Aufsammlung berechnet und diskutiert. Aus seiner Arbeit entnehmen wir folgendes Beispiel: Ein Stern mit 2 Sonnenmassen bewege sich mit einer Relativgeschwindigkeit von 5 km/sec in eine interstellare Gaswolke mit der Dichte $2.5 \cdot 10^{-21}$ g/cm^3 hinein; das Gas der Wolke verhalte sich bei der Aufsammlung isotherm; sein Molekulargewicht sei 1.4 und seine Temperatur 50°K. Dann braucht der Stern $2.6 \cdot 10^8$ Jahre, um in der Wolke zur Ruhe zu kommen, in dieser Zeit hat er eine Entfernung von 990 pc zurückgelegt. Um seine Masse zu vervielfachen, würde er nach dem Zur-Ruhe-Kommen $8.3 \cdot 10^6$ Jahre benötigen, so daß er zur Verjüngung insgesamt etwa $2.7 \cdot 10^8$ Jahre brauchen würde. Hätte er eine anfängliche Relativgeschwindigkeit von 2 km/sec anstelle von 5 km/sec gehabt, so würde bei sonst gleichen Bedingungen seine Verjüngung in etwa $2.8 \cdot 10^7$ Jahren geschehen können, und in diesem Fall hätte er 32 pc zurückgelegt.

Dichten von der Größe 10^{-21} g/cm^3 kommen nun aber nach einer Tabelle von BOK (1946) nur in den sogenannten "large globules" vor, die andererseits nur Durchmesser von etwa 0.5 pc haben, also viel zu klein sind, um einen Stern in ihnen zur Ruhe kommen zu lassen. Die mittleren und großen Wolken mit Dimensionen von 5 bis zu 100 pc haben andererseits nur mittlere Dichten von etwa 10^{-22} g/cm^3. Da die Verjüngungszeit umgekehrt proportional zur Dichte ist, würden sich damit für unsere Beispiele Verjüngungszeiten von $6 \cdot 10^8$ bis $6 \cdot 10^9$ Jahren ergeben. Beachten wir noch, daß so geringe Relativgeschwindigkeiten wie 2 km/sec sehr selten sein dürften (der beobachtete mittlere Wert liegt zwischen 5 und 10 km/sec), so ergibt sich, daß die Aufsammlung nur in sehr seltenen Fällen ein wirksamer Prozeß sein kann.

BIERMANN (1955) und SCHLÜTER (1955) haben darauf hingewiesen, daß die Voraussetzungen, unter denen die bisher erwähnten Aufsammlungsformeln abgeleitet wurden, zu eng sind. Die Strahlung eines Sternes erzeugt, wie wir durch Arbeiten von STRÖMGREN (1939) wissen, eine dem Spektraltyp des Sternes und der Dichte des ihn umgebenden interstellaren Gases entsprechende HII-Region, die im Mittel eine Temperatur von 10000°K hat [SPITZER (1950)]. Bei O- und B-Sternen haben die HII-Regionen Radien von 10 bis 150 pc. Durch die Aufheizung des Gases, die in vielen Fällen sogar zu einer Expansion der HII-Region führt, muß die Aufsammlung erheblich herabgesetzt werden. Es ist schwer vorzustellen, wie bei einer expandierenden HII-Region noch eine wesentliche Aufsammlung stattfinden kann. MESTEL (1954) hat daraufhin diesen Effekt berechnet. Nach ihm könnten nur Sterne mit Massen kleiner als eine obere Grenzmasse noch eine wirkungsvolle Aufsammlung beginnen; das spätere Anwachsen der Masse würde die Aufsammlung dann nicht unterbrechen, weil die Ausbildung einer vollen HII-Region durch den einlaufenden Teilchenstrom verhindert werden würde. Für eine Dichte des interstellaren Gases von 10^3 H/cm^3 mit einer Temperatur von 100°K beträgt die Grenzmasse 1.75 Sonnenmassen. Geringere Dichte ergibt eine kleinere Grenzmasse. Sterne mit größerer Masse können ihre volle HII-Region ausbilden, und die Aufsammlung ist dann vernachlässigbar klein.

Die beobachtete Korpuskularstrahlung bei einer Reihe von massereichen Sternen ist eine weitere Tatsache, die die Aufsammlung erschwert [BIERMANN (1955)]. Danach sieht es so aus, als würden diese Sterne eher ihre Masse durch Abblasen verringern, als sie durch Aufsammlung vergrößern. Zumindest aber wirkt die Korpuskularstrahlung der Aufsammlung entgegen.

Ferner ist auf stellare Magnetfelder hingewiesen worden [MENZEL (1955)], die eine Aufsammlung, soweit sie senkrecht zu den Feldlinien liefe, erschweren können.

Darüber hinaus sind in den letzten Jahren Beobachtungen gemacht worden, die zeigen, daß junge Sterne in Gruppen und Haufen auftreten, die auch dynamisch jung sein müssen, siehe Kapitel A.III. Weiterhin enthalten einige dieser Sternhaufen eine große Anzahl von Sternen, die sich noch in der Phase der Kontraktion befinden, was ebenfalls zu der Annahme sehr geringen Alters zwingt, siehe Kapitel A.II. Auf zwanglose Weise lassen sich diese Beobachtungen nur durch die Annahme einer *Neuentstehung* von Sternen aus interstellarem Gas deuten, nicht aber mit Hilfe der Annahme einer Sternverjüngung durch Aufsammlung von Gas.

Wir werden so zu dem Ergebnis geführt: Die Aufsammlung mag in einzelnen, seltenen Fällen von Bedeutung sein. Sie kann jedoch nicht als ein Mechanismus angesehen werden, durch den die Mehrzahl der jungen Sterne erklärt werden könnte.

C.III. Sternentstehung durch Kondensation

C.III.a) Temperaturen des interstellaren Gases

Ausgangspunkt fast aller theoretischen Arbeiten über Verdichtungen interstellarer Materie sind die grundlegenden Untersuchungen von

STRÖMGREN und SPITZER zur Temperatur des interstellaren Gases. Wir gehen aus diesem Grunde kurz auf sie ein.

Das im interstellaren Raum vorhandene Gas kann, wie STRÖMGREN (1939) gezeigt hat, mit guter Näherung in HI- und HII-Gebiete eingeteilt werden. In den HI-Regionen ist der Wasserstoff neutral, in den HII-Regionen dagegen vollständig ionisiert. Die Übergangsgebiete zwischen HI und HII sind klein, verglichen mit den Gebieten selbst.

Der interstellare Raum ist kein Strahlungshohlraum. Zur Temperaturberechnung müssen daher die einzelnen Energieprozesse, die die mittlere kinetische Energie der Gasteilchen bestimmen, betrachtet werden. SPITZER (1948, 1949) hat diese Untersuchungen durchgeführt. HII-Regionen finden wir in der Umgebung früher Sterne, vom Typ O bis A. Die hohen Oberflächentemperaturen und die damit verbundene starke ionisierende Strahlung ist der Grund für die HII-Gebiete. Je nach Spektraltyp und Dichte des den Stern umgebenden Gases haben die HII-Regionen nach STRÖMGREN (1939, 1948) Radien von 0.1 bis 150 pc. Dem Energiegewinn durch die Sternstrahlung steht als hauptsächlicher Energieverlust der sich durch Stoßanregung von OII- und OIII-Linien durch freie Elektronen ergebende gegenüber. Die Emission der verbotenen Linien kann beobachtet werden [ALLER (1956b)]. Aus der mittleren kinetischen Energie der Teilchen bei Gleichheit von Energiegewinn und -verlust ergibt sich eine Temperatur von 10000° K.

In den HI-Regionen liefert die dort vorhandene interstellare Strahlung und die Höhenstrahlung einen Energiegewinn. Als Energieverlust konkurrieren eine Reihe von Prozessen: unelastische Stöße der Elektronen an Ionen der schweren Elemente, unelastische Stöße der H-Atome an Staubteilchen, Anregung von Rotationsbanden der H_2-Moleküle durch H-Atome. Es ergibt sich für die kinetische Temperatur eines HI-Gases ein Wert von 20—100° K, je nach Staubgehalt und chemischer Zusammensetzung [SPITZER (1949, 1950); EBERT (1955a)]. Aus Beobachtungen an der 21 cm-Linie folgt der im Mittel etwas größere Wert von 100° K [v. DE HULST, MULLER, OORT (1954)].

Aus dem Temperaturunterschied zwischen HI- und HII-Region ergeben sich interessante dynamische Effekte. In Abschnitt III.e wird darauf eingegangen werden.

C.III.b) Jeanssches Kriterium und ähnliche Instabilitätsbedingungen

JEANS (1926) hat als erster quantitativ die Frage untersucht, welche Dichtestörungen in einer Gasmasse instabil sind. Er nimmt als ungestörten Zustand eine durch den Raum homogen verteilte Gasmasse der Dichte ϱ_0 an. Sie soll sich bis ins Unendliche erstrecken und keine Geschwindigkeiten haben. Dann ist kein Ortspunkt in dieser Verteilung ausgezeichnet, und JEANS schließt, daß dann auch das Gravitationspotential eine nicht vom Ort abhängige Konstante Φ_0 sein muß (dazu siehe Bemerkung S. 299).

Unter diesen Annahmen gilt, wenn es sich um eine infinitesimale Störung handelt und die Abweichungen vom ungestörten Zustand durch ein δ bezeichnet werden,

$$\frac{\partial}{\partial t}(\delta v) + \frac{1}{\varrho_0}\operatorname{grad}(\delta p) + \operatorname{grad}(\delta \Phi) = 0$$
$$\frac{\partial}{\partial t}(\delta \varrho) + \varrho_0 \operatorname{div}(\delta v) = 0 \qquad (165)$$
$$\Delta\,\delta\Phi - 4\pi G\,\delta\varrho = 0\,.$$

v bezeichnet die Geschwindigkeit, p und ϱ Gasdruck und -dichte, Φ und G Gravitationspotential und -konstante. Δ ist der Laplace-Operator. Die erste der drei Gleichungen geht aus der Bewegungsgleichung hervor, die zweite aus der Kontinuitäts- und die dritte aus der Poisson-Gleichung. Zwischen p und ϱ gelte die Beziehung

$$\frac{\delta p}{p_0} = \gamma\,\frac{\delta \varrho}{\varrho_0}\,. \qquad (166)$$

Für Isothermie ist $\gamma = 1$, für Adiabasie gleich dem Verhältnis der spezifischen Wärmen. Aus Gl. (165, 166) läßt sich die Störungsdifferentialgleichung

$$\frac{\partial^2}{\partial t^2}(\delta \varrho) - \gamma\,\frac{p_0}{\varrho_0}\Delta\,\delta\varrho - 4\pi G\,\varrho_0\,\delta\varrho = 0 \qquad (167)$$

ableiten. Als Lösung findet man für kugelsymmetrische Störungen

$$\delta\varrho = \frac{A}{r}e^{\alpha t + i\lambda r}, \qquad (168)$$

wenn t die Zeit, r der Abstand und A, α und λ Konstanten sind. Der kleinsten Wellenlänge für Instabilität entspricht das größte λ bei reellem α.

Aus Gl. (167, 168) ergibt sich

$$\lambda_{max}^2 = \frac{4\pi G\,\varrho_0^2}{\gamma\,p_0}, \qquad (169)$$

und damit als kleinste instabile Masse, wenn wir Isothermie annehmen,

$$M = 9.8\,\varrho_0^{-1/2}\left(\frac{\Re\,T_0}{\mu G}\right)^{3/2}, \qquad (170)$$

wo T_0 die Temperatur, μ das Molekulargewicht und $\Re$ die allgemeine Gaskonstante bedeuten. Je größer die Dichte und je kleiner die Temperatur, desto kleiner die instabile Masse.

Setzt man die uns heute bekannten Werte für ϱ_0 und T_0 ein, so ergeben sich Massen von der Größe $10^3\,M_\odot$. Sind Sterne nach diesen Bedingungen entstanden, so müssen sie stets als *Sternhaufen* entstanden sein, denn einzelne Sterne mit so großer Masse sind weder beobachtet worden, noch wären sie stabil, wie sich aus dem inneren Aufbau sehr massereicher Sterne ergibt [SCHWARZSCHILD und HÄRM (1958)]. Die gesamte instabile Masse muß also während der Verdichtung in Untersysteme zerfallen. Dieser Prozeß wird in Abschnitt III.d beschrieben werden.

Durch die v. Weizsäckerschen Arbeiten (1948, 1951) über die Turbulenz des interstellaren Gases und durch viele Beobachtungen [BOK (1954)], aus denen eindeutig eine Wolkenstruktur des interstellaren Gases mit großen Dichteunterschieden hervorgeht [SCHLÜTER, SCHMIDT, STUMPFF (1953)], ist deutlich geworden, daß die von JEANS angenommene unendliche, homogene Dichteverteilung eine zu starke Vereinfachung der wirklich vorliegenden Verhältnisse darstellt. Von BONNOR (1957) ist außerdem darauf hingewiesen worden, daß die Annahme einer unendlichen Ausdehnung der konstanten Dichteverteilung nicht zu konstantem Gravitationspotential führt, sondern dieses unbestimmt läßt. Oder mit anderen Worten: die Verteilung mit konstanter Dichte und konstantem Gravitationspotential ist nur für $\varrho_0 = 0$ ein eindeutig definierter Gleichgewichtszustand.

Aus diesen Gründen ist etwa gleichzeitig und unabhängig voneinander von BONNOR (1957) und EBERT (1957) die Frage untersucht worden, wie sich das Jeanssche Kriterium abändert, wenn als Ausgangszustand keine homogene Dichteverteilung, sondern die einer isothermen sich im Gleichgewicht befindenden Gaskugel genommen wird. Hat sich nämlich ein Gebiet des interstellaren Gases durch Turbulenz oder durch die in III.a erwähnten Temperaturunterschiede oder auf andere Weise aus seiner Umgebung herausgesondert, so wird bei nicht zu großem Gesamtdrehimpuls dieses Gebiet sich der Gestalt einer kugelsymmetrischen, isothermen Gaswolke annähern. Beide Autoren kommen zum gleichen Ergebnis. Die in diesem Fall kleinste instabile Masse bei kugelsymmetrischer Dichtestörung um das Zentrum beträgt

$$M = 3.7\, \varrho_c^{-1/2} \left(\frac{\Re T_0}{\mu G} \right)^{3/2}, \tag{171}$$

wenn ϱ_c die Mittelpunktsdichte der isothermen Gaskugel mit der Temperatur T_0 ist. Diese instabile Masse unterscheidet sich von der Jeansschen um einen Faktor 2.6 und durch die Wurzel aus dem Verhältnis der Dichte ϱ_0 zur Mittelpunktsdichte ϱ_c. Tab. 21 [EBERT (1957)] zeigt einige sich aus Gl. (171) ergebende Massen.

Nach BOK (1946) haben mittelgroße interstellare Wolken eine mittlere Dichte von etwa 10^2 H/cm³. Unter sehr günstigen Bedingungen, in denen die Mittelpunktsdichte solcher Wolken vielleicht eine Zehnerpotenz oder mehr größer als ihre mittlere Dichte und die Temperatur durch Anwesenheit von Staub sehr klein wäre, könnten O-Stern-Massen instabil werden. Wir sehen aber, daß nur in seltenen Fällen so günstige Verhältnisse vorliegen können.

Die "large globules" haben zwar Dichten von 10^3 bis 10^4 H/cm^3, dafür aber nur Massen von 5 $M_\odot$. Nur wenn ihre Temperatur $< 10°$K ist, können sie instabil sein.

Der Beginn einer Entstehung von Einzelsternen scheint nach dem obigen Kriterium in sehr seltenen Fällen möglich zu sein. Wahrscheinlicher ist aber, daß die Sterne in Gruppen und Haufen entstehen, weil der Entstehungsprozeß dann mit größeren Massen beginnen kann.

Tabelle 21. *Instabile Massen*

T_0 in °K	Mittelpunktsdichte in H/cm^3		
	1	10^2	10^4
10	$8.8 \cdot 10^2$	88	8.8
100	$2.8 \cdot 10^4$	$2.8 \cdot 10^3$	$2.8 \cdot 10^2$

$M_\odot$

Wir weisen noch auf Instabilitätskriterien hin, die aus dem Jeansschen durch Hinzunahme von Turbulenz [CHANDRASEKHAR (1954)] oder von Rotationsbewegungen [FRICKE (1954)] entstehen. Sowohl die Turbulenz durch ihren Druck, wie die Rotation durch ihre Zentrifugalkraft erschweren das Instabilwerden. Sie führen daher zu größeren als den Jeansschen Massen.

C.III.c) Konzentration von Staub durch Strahlungsdruck

Der im interstellaren Gas vorhandene Staub kann durch die Wirkung des Strahlungsdruckes eine Erhöhung seiner Konzentration erfahren. Dieser Prozeß ist von SPITZER (1941) berechnet worden. Das Absorptionsvermögen der Staubteilchen ist größer als das der Gasatome. Dadurch kann in Gebieten, in denen viel Staub ist, die mittlere Energiedichte der für den Strahlungsdruck wichtigen Strahlung geschwächt werden, verglichen mit den Gebieten, in denen kein oder nur wenig Staub ist. Der Strahlungsdruck auf die Staubteilchen wird dadurch zu einer Kraft, die versucht, die Staubkonzentration weiter zu erhöhen. Die Geschwindigkeit des Prozesses hängt von der Reibungskraft zwischen Staub und Gas, ferner von der schon vorhandenen Staubkonzentration und der Strahlungsdichte ab.

WHIPPLE (1946) hat versucht, diesen Prozeß zur Erklärung der Entstehung von O-Sternen heranzuziehen. Bei ständig wachsender Staubkonzentration innerhalb einer interstellaren Wolke muß die Materiedichte in dem Gebiet, in dem sich der Staub befindet, ebenfalls wachsen. Wie aus Gl. (170) oder (171) hervorgeht, werden dadurch immer kleinere Massen gravitationsinstabil, und bei genügend großer Staubkonzentration könnten O-Stern-Massen instabil werden. Nach WHIPPLE sind ungefähr 10^9 Jahre nötig, um die erforderliche Staubkonzentration zur Bildung eines O-Sterns zu erhalten.

Durch Untersuchungen von SAVEDOFF (1955), die an die von SPITZER und WHIPPLE anschließen, hat sich aber ergeben, daß die Reibung zwischen Gas und Staub größer ist, als früher angenommen wurde, und daß ferner die Wirkung des Strahlungsdruckes überschätzt wurde, weil die hohe Albedo der Staubteilchen, bedingt durch ihr großes Streuvermögen, nicht genügend berücksichtigt worden war.

Nach SAVEDOFF würden beim Kohlensack-Nebel unter sehr günstigen Annahmen die Staubteilchen gegenüber dem Gas eine Relativgeschwindigkeit von $6 \cdot 10^3$ cm/sec gewinnen können, wenn eine Gasdichte von 10 H/cm³ angenommen wird. Bei größerer Dichte sind die Relativgeschwindigkeiten noch kleiner. Unter Annahme sphärischer Symmetrie des Gebietes würde damit die Staubkonzentration in $2 \cdot 10^7$ Jahren um den Faktor 2.3 vergrößert werden. Selbst wenn vor Beginn des Prozesses das Massenverhältnis von Staub zu Gas schon den Wert $^1/_{10}$ hatte, ist ein Verdichtungsfaktor von 10^3 bis 10^4 nötig, um O-Stern-Massen instabil werden zu lassen. Die Zeit aber, die vergeht, bis eine interstellare Wolke mit einer anderen zusammenstößt, beträgt im Mittel einige 10^7 Jahre. Aus diesem Grund können derart große Staubverdichtungen kaum zustande kommen.

In einer anderen Hinsicht mag die Staubkonzentration für die Sternentstehung von Bedeutung sein. In III.a sahen wir, daß die Staubkonzentration die Temperatur einer H I-Region beeinflußt. Ist viel Staub vorhanden, so kann die Temperatur bis auf die ,,innere Temperatur'' der Staubteilchen, die nach V. D. HULST (1949) zwischen 20° und 50°K liegt, herabgedrückt werden.

Darüber hinaus kann vielleicht, wenn das Gebiet hoher Staubkonzentration groß genug ist, der Staub diejenige Strahlung schwächen, die für die Energiegewinnprozesse des Gases wichtig ist. Die Folge wäre eine weitere Temperaturabnahme des betreffenden H I-Gebietes. Diese Möglichkeit ist von SPITZER (1957) erwähnt worden. Geringere Temperatur bedeutet kleinere instabile Masse. In Gebieten hoher Staubkonzentration herrschen daher für die Sternentstehung günstige Bedingungen. Es wäre zu untersuchen, ob die räumliche Verteilung junger Sterne eine, über die Korrelation mit der Gasdichte hinausgehende, *direkte* Korrelation mit der Staubdichte aufweist.

C.III.d) Zerfall in Untersysteme

Die in Abschnitt III.b beschriebenen Instabilitätskriterien ergeben, wie wir sahen, instabile Massen von ungefähr 10 bis $10^4\,M_\odot$, je nach Temperatur und Dichte des Gases. HOYLE (1953) hat nun einen Prozeß vorgeschlagen, nach dem aus einer großen Masse durch aufeinanderfolgenden Zerfall in kleinere Massen ein Sternhaufen entstehen könnte. Durch diesen Prozeß würde es möglich werden, die Entstehung von

Sternen zu verstehen, die kleinere Masse haben, als die durch die erwähnten Kriterien gegebenen. Da die Mehrzahl aller Sterne Massen unterhalb 10 $M_\odot$ haben, kommt diesem Prozeß eine große Bedeutung zu. Darüber hinaus eröffnet er eine Möglichkeit, die Entstehung der beobachteten Sternhaufen zu verstehen.

Ausgangspunkt des Prozesses ist eine Masse, die größer oder gleich einer nach Gl. (170) oder (171) instabilen Masse ist. Diese kann sich durch Kontraktion unter dem Einfluß der Eigengravitation aus dem Feld des übrigen Gases herauslösen. Ihre Dichte wächst bei der Kontraktion, ihre Temperatur kann, je nach der Zusammensetzung des Gases, konstant bleiben oder ebenfalls wachsen. Solange die Wolke optisch dünn ist, wird in Gegenwart von schweren Elementen, Staub oder H_2-Molekülen das Gas sich angenähert isotherm verhalten, weil die Kompressionsenergie leicht durch diese Teilchen abgestrahlt werden kann.

Besteht das Gas aus reinem Wasserstoff ohne Wasserstoffmoleküle, und ist seine Temperatur kleiner als 10000°K, so muß zwar zuerst ein Temperaturanstieg eintreten, bei 10000°K jedoch setzt die Ionisation schnell ein, und die dadurch entstehende Energieausstrahlung durch frei-gebunden und frei-frei Übergänge verhindert ein weiteres Wachsen der Temperatur. Der weitere Verdichtungsprozeß verläuft dann ebenfalls angenähert isotherm, bis schließlich andere Verhältnisse eintreten, wenn die Wolke optisch dicht wird.

Während der Verdichtung kann ein Zerfall der Wolke in Untersysteme eintreten, weil bei konstanter Temperatur und wachsender Dichte kleinere Gebilde instabil werden, wie aus Gl. (170) zu ersehen ist.

Der Zerfall soll in n Schritten erfolgen.

1. Schritt: Zerfall der Gesamtmasse M_0 in k Massen mit der jeweiligen Masse $M_1 = M_0/k$. Dieser Zerfall ist nach Gl. (170) möglich, wenn die Anfangsdichte ϱ_0 gewachsen ist auf $\varrho_1 = k^2 \varrho_0$.

2. Schritt: Die Dichte in den Untersystemen M_1 ist weiter auf $\varrho_2 = k^2 \varrho_1 = k^4 \varrho_0$ gewachsen. Jede Masse zerfällt wieder in k Massen mit der jeweiligen Masse M_1/k.

Nach n Schritten haben wir

$$\varrho_n = k^{2n} \varrho_0, \quad M_n = k^{-n} M_0, \quad R_n = k^{-n} R_0, \qquad (172)$$

wenn R_n der mittlere Radius der M_n-ten Masse ist.

Der Zerfall in kleinere Massen muß aufhören, wenn die Temperatur mit $\varrho^{1/3}$ oder stärker anwächst (im adiabatischen Fall gilt: $T \sim \varrho^{2/3}$), wie aus Gl. (170) zu ersehen ist. Wann dies eintritt, hängt davon ab, wie die entstehende thermische Energie fortgeschafft werden kann. Wir sahen, solange die Gaswolke optisch dünn ist, bleibt ihre Temperatur angenähert konstant. Wird sie jedoch optisch dicht, so entsteht im

Inneren ein Temperaturgradient, durch den Strahlung nach außen gebracht wird.

Die Größe dieses Gradienten ist durch die bei der Kontraktion freiwerdende Energie und durch den Absorptionskoeffizienten bestimmt. Solange der durch diesen Temperaturgradienten bedingte Druckgradient der Gravitation nicht das Gleichgewicht halten kann, muß die Verdichtung fast ungehindert fortschreiten, sofern sie nicht durch Auftreten von Rotation infolge nichtabtransportierten Drehimpulses unterbrochen wird (dazu siehe Abschnitt IV). Im Falle eines Temperaturgradienten sind die Formeln (170), (171) nicht mehr gültig, weil dort Isothermie vorausgesetzt war.

Das Zerfallen in Untersysteme muß aber spätestens aufhören, wenn der Temperaturgradient einen ungefähren hydrostatischen Gleichgewichtszustand herbeiführen kann. MESTEL und SPITZER (1956), die den Zerfall für ein Gas mit hohem Staubgehalt berechnet haben, machen diese Annahme. Aus dem Virialsatz folgt nämlich, daß bei Verdichtung eines sich im Gleichgewicht befindenden Systems die mittlere Temperatur wie $\varrho^{1/3}$ ansteigt (ϱ = mittlere Dichte) und dadurch nach Gl. (170) ein weiteres Kleinwerden der instabilen Masse verhindert wird.

HOYLE (1953), der den Prozeß bei reinem Wasserstoff untersucht, wählt eine ähnliche Annahme. Der Zerfall soll aufhören, wenn der durch die Mittelpunktstemperatur von 10000°K und die Oberflächentemperatur von 4500°K (Begründung siehe seine Arbeit) bestimmte Temperaturgradient ebensoviel Energie aus der Wolke durch Strahlungstransport herausschaffen kann, wie bei der Kontraktion (mit der Zeitskala des freien Hineinfallens) frei wird.

Die Zeitskala des freien Falls ist nach dem n-ten Zerfall

$$t_n = (G \varrho_n)^{-1/2}, \tag{173}$$

und die bei der Kontraktion freiwerdende Energie ist angenähert

$$E_n = \frac{G M_n^2}{R_n}. \tag{174}$$

Ist L_n die Leuchtkraft der Gasmasse M_n, so soll der Zerfall aufhören, wenn

$$L_n t_n = E_n \tag{175}$$

ist.

Für die Leuchtkraft gilt bei Strahlungstransport die Beziehung

$$L = \frac{16 \pi\, ac\, r^2 T^3}{3 \varkappa \varrho} \frac{dT}{dr}, \tag{176}$$

wenn r die Entfernung vom Mittelpunkt, c die Lichtgeschwindigkeit, a die Strahlungskonstante und $\varkappa$ der Absorptionskoeffizient ist. Unter den vereinfachenden Annahmen

$$T_{Mittelp.} = 10\,000°\,\text{K}, \qquad T_{Oberfl.} = 4500°\,\text{K}$$
$$\frac{dT}{dr} = \frac{T_{Mittelp.} - T_{Oberfl.}}{R}, \qquad \varkappa = 10^{-1}\,P_e$$

wo P_e der Elektronendruck ist, ergibt sich aus Gl. (175), (176) zusammen mit (172) die Grenzmasse, bei der der Zerfall aufhören muß, zu 1.5 $M_\odot$. Dieser Wert ist unabhängig von der Anfangsdichte ϱ_0, aber abhängig von der zu 10 000° K angenommenen Ausgangstemperatur T_0 in Gl. (170). Die einzelnen Annahmen, die HOYLE macht, sind alle nicht zwingend. Vielleicht stellen sie im Mittel aber eine gute Abschätzung dar.

MESTEL und SPITZER erhalten für ihren Fall eine Grenzmasse von $^1/_2\,M_\odot$. Sie nehmen eine Temperatur von $T_0 = 100°\,\text{K}$ an. Beide Abschätzungen zeigen, daß der Zerfall in Untersysteme zu Sonnenmassen führen kann. Der Prozeß enthält außerdem die Möglichkeit einer Umwandlung des Drehimpulses der anfänglichen Wolke in Bahndrehimpuls der entstehenden Untersysteme. Das schwierige Problem der Drehimpulsabführung wäre hier also zum Teil gelöst. Die entstehenden Sternhaufen müßten dann als Ganzes eine Rotation zeigen.

Es bleibt aber zu beachten, daß die Umwandlung in Bahndrehimpuls erst nach Einsetzen des ersten Zerfalls der Wolke beginnen kann. Dazu muß die Wolke eine Verdichtung erfahren haben. Ist aber der Drehimpuls der anfänglichen Wolke groß, so kann er schon diese erste Verdichtung verhindern. Auch weist das Vorhandensein der Kugelsternhaufen darauf hin, daß es wahrscheinlich noch andere Möglichkeiten der Drehimpulsabführung geben muß. Wir werden in Abschnitt IV darauf eingehen.

C.III.e) Entstehung von Sternen in der Umgebung von O-Sternen

Die Umgebung von O-Sternen scheint eine für die Neuentstehung von Sternen günstige Beschaffenheit zu haben. Beobachtungen über die Kopplung von Dunkelwolken mit O-Sternen weisen darauf hin [MINKOWSKI (1949)]. Aber auch theoretische Überlegungen im Anschluß an die Beobachtungen legen den obigen Schluß nahe.

Der im Abschnitt III.a erwähnte große Temperaturunterschied zwischen HI- und HII-Region hat starke dynamische Wirkungen zur Folge. Die Tendenz zum Druckausgleich ist zugleich eine Ursache der Wolkenbildung [OORT (1954); OORT und SPITZER (1955)]. Bei völligem Druckausgleich würde das Dichteverhältnis der zwei Regionen etwa 100 oder mehr betragen. Bis zur Herstellung dieses Gleichgewichts müssen die HII-Regionen sich in starker Expansion befinden.

Dies wurde gleichzeitig von OORT (1953) sowie von BIERMANN und SCHLÜTER (1953) vorausgesagt und ist jetzt durch Beobachtungen mit

Hilfe der 21 cm-Linie durch WADE (1957) bestätigt worden (gemessen wird die Expansionsgeschwindigkeit des an die H II-Region angrenzenden H I-Gebietes). WADE beobachtete eine Geschwindigkeit von 8 km/sec. Eine so starke Expansion einer H II-Region muß zu Verdichtungsstößen innerhalb der angrenzenden H I-Gebiete führen, wodurch in diesen Gebieten die Dichte erhöht wird [KAHN (1954); SAVEDOFF und GREENE (1955)]. Darüber hinaus ist, wie SPITZER (1954) und FRIEMAN (1954) gezeigt haben, eine ebene oder kugel-symmetrische Gestalt der Grenzfläche zwischen H I- und H II-Region unter dem Einfluß einer Expansion nicht stabil. Es müssen sich Strukturen ausbilden, die die Gestalt von „Elefanten-Rüsseln" haben, d. h. lange dichte H I-Rüssel, die in das H II-Gebiet hineinragen. Der Emissionsnebel M 16 z. B. zeigt solche Strukturen.

Im weiteren Verlauf der Expansion können sich die Rüssel abschnüren, sie bilden dann H I-Gebiete, die ganz von einer H II-Region umschlossen sind, und die durch den Druck dieser Region zusammengehalten oder noch weiter verdichtet werden. Aus ihnen könnten — unter Bedingungen, die wir noch untersuchen werden — neue O-Sterne entstehen. Ein O-Stern würde dann in seiner Umgebung mehrere oder sogar viele neue O-Sterne erzeugen, die im Mittel, wegen der Expansion der H II-Region ebenfalls eine Expansionserscheinung zeigen müßten. OORT (1954) hat diesen Prozeß als Entstehungsursache von O-Assoziationen vorgeschlagen und untersucht. Expandierende O-Assoziationen sind in der Tat beobachtet worden. BLAAUW (1956) hat Expansionsgeschwindigkeiten von 15 km/sec gemessen, was in guter Übereinstimmung mit der beobachteten Expansionsgeschwindigkeit bei λ Orionis ist.

Die Weiterverdichtung eines H I-Gebietes, das von einem H II-Gebiet umgeben ist, haben EBERT (1955b) und unabhängig davon BONNOR (1956) untersucht. Betrachtet man ein ebenes oder zylinderförmiges H I-Gebiet, das unter dem Druck der H II-Region komprimiert wird, und bleibt während der Kompression die Masse pro Flächenbzw. Längeneinheit konstant und ändert sich die Temperatur des H I-Gebietes und die Gestalt der Grenzfläche nicht, so kann bei noch so großem Druck der H II-Region niemals eine Gravitationsinstabilität innerhalb des H I-Gebietes auftreten. Nur wenn während der Kompression die Temperatur des H I-Gebietes sinkt, könnte, wie MCCREA (1957) gezeigt hat, eine Instabilität im Fall des Zylinders auftreten.

Ganz anders dagegen ist das Verhalten eines kugelsymmetrischen H I-Gebietes. Setzt man, wie in den zwei vorherigen Fällen, Isothermie innerhalb des H I-Gebietes voraus, und bleibt die Temperatur und Masse während der Verdichtung konstant, so gibt es einen eindeutig festgelegten Maximaldruck, den die Gaskugel dem Druck der H II-Region entgegensetzen kann. Besitzt die HII-Region einen größeren Druck, so bringt sie das HI-Gebiet zur Gravitationsinstabilität. Innerhalb des Instabilitätsbereiches überwiegt die Eigengravitation den Druckgradienten, ganz gleich welche Dichteverteilung innerhalb der Gaskugel sich herstellen mag. Unter dem Einfluß der Eigengravitation und des Außendruckes muß eine Kontraktion der Gaskugel einsetzen.

Die Bedingung, daß der Druck der HII-Region größer als der maximal von der Gaskugel an ihrem Rand erzeugte ist, läßt sich auch als eine Bedingung für die Masse des HI-Gebietes schreiben. Instabilität muß einsetzen, wenn [EBERT (1955b)]

$$M_{HI}^2 \geq 1.4 \frac{(\Re T_{HI})^4}{\mu^4 G^3 P_{HII}} \tag{177}$$

ist. P_{HII} bezeichnet den Gasdruck der HII-Region, T_{HI} die Temperatur des HI-Gebietes und μ das Molekulargewicht, $\Re$ ist die allgemeine Gaskonstante und G die Gravitationskonstante.

Tabelle 22. *Instabile Massen*

T_{HI} in °K	Dichte der HII-Region in H/cm³			
	10^{-2}	1	10^2	
100	$9 \cdot 10^3$	$9 \cdot 10^2$	90	$M_\odot$
50	$2.3 \cdot 10^3$	230	23	
10	90	9	0.9	

Je größer der Druck der HII-Region, desto kleinere Massen können instabil werden. Nehmen wir die Temperatur der HII-Region zu 10000°K an, so ergeben sich die in Tab. 22 aufgeführten instabilen Massen.

Wir vergleichen diese instabilen Massen mit einer von BOK (1946) stammenden Tabelle über interstellare Wolken. Sie ist hier als Tab. 23 aufgeführt.

Tabelle 23. *Charakteristische Werte interstellarer Wolken*

	Small Globule	Large Globule	Intermediate Cloud	Large Cloud
$M/M_\odot$	>0.2	5	$1.3 \cdot 10^3$	$3 \cdot 10^4$
D (pc)	0.06	0.5	8	40
n (H/cm³)	>$4 \cdot 10^4$	$1.6 \cdot 10^3$	100	20

Darin bezeichnet D den Durchmesser des Objektes und n seine mittlere Dichte. Es wurde angenommen, daß das Massenverhältnis von Staub zu Gas 1 : 100 beträgt.

Für die Entstehung neuer O-Sterne kämen die "large globules" in Frage. Sie könnten instabil sein, wenn durch hohen Staubgehalt ihre Temperatur kleiner als 50°K wäre und die HII-Region eine Dichte größer als 1 H/cm³ besäße. Die "small globules" dagegen sind wegen ihrer geringeren Masse wahrscheinlich stabil, es sei denn, daß ihr Staub-Gasverhältnis sehr viel geringer ist als oben angenommen; dann wäre ihre Gesamtmasse allerdings erheblich größer als angegeben, weil nur eine untere Grenze für die Staubmasse direkt beobachtet ist.

Aus den mittleren und großen Wolken könnten O-Sterne entstehen; es wäre hierzu ein Zerfall in Untersysteme nach dem in Abschnitt III.e beschriebenen Prozeß nötig.

Abgesehen von der Frage des Drehimpulses, die später in Abschnitt IV behandelt wird, muß, damit der oben geschilderte Verdichtungsvorgang bis zum Einsetzen der Gravitationsinstabilität stattfinden kann, die HI-Wolke die Kompressionsenergie ohne wesentliche Temperaturerhöhung abstrahlen können. Eine einfache Abschätzung zeigt, daß eine solche Abstrahlung in 10^6 bis 10^7 Jahren möglich ist [EBERT (1955b)].

MCCREA (1957) hat noch gezeigt, daß die Formel (177), die mit Hilfe der Theorie der isothermen Gaskugel abgeleitet worden ist, sich aus einer Verallgemeinerung des Virialsatzes gewinnen läßt. Durch diese Ableitung erhält man eine bessere Einsicht in die Physik des Instabilitätsprozesses.

Wie verläuft nun die Weiterverdichtung im instabilen Bereich? Unter dem überwiegenden Einfluß der Eigengravitation beginnt das Gas nach innen zu laufen. Die freiwerdende Gravitationsenergie wird in kinetische und thermische umgewandelt, die thermische wiederum wird durch Strahlung abgegeben. Wegen der Empfindlichkeit der Abstrahlungsprozesse vom Ionisationsgrad kann die Temperatur, solange die Wolke noch optisch dünn ist, 10000° K kaum überschreiten (bei dieser Temperatur ist der Ionisationsgrad des Wasserstoffes etwa 0.5). Erst wenn die Verdichtung soweit fortgeschritten ist, daß die Strahlung infolge der zunehmenden optischen Dichte nicht mehr schnell genug entweichen kann, muß es zu stärkeren Temperaturerhöhungen kommen. Der Verdichtungsvorgang verlangsamt sich durch die entstehenden Temperatur- und Druckgradienten. Es ist nicht ausgeschlossen, daß in dieser Phase ein Abblasen von Materie auftreten kann, weil der Strahlungsdruck größer als der Gasdruck zu werden vermag. Bei weiterer Verdichtung sinkt der Strahlungsdruck wieder unter den Gasdruck, um wahrscheinlich erst in den letzten Phasen der Entstehung bei sehr massereichen Sternen wieder von Bedeutung zu werden.

Die Endphase der Sternentstehung haben HENYEY, LELEVIER und LEVÉE (1955) untersucht. Sie haben berechnet, auf welchem Weg sich die Sterne im HR-Diagramm bis zur Hauptreihe bewegen, wobei die Kernprozesse langsam einsetzen. Ihre Arbeit ist in Abschnitt A.II diskutiert worden. Rechnungen über den quantitativen Verlauf der Verdichtung vom Einsetzen der Instabilität bis zur Henyeyschen Phase bestehen noch nicht, befinden sich jedoch in Vorbereitung.

C.IV. Magnetfelder und Drehimpuls

C.IV.a) Einfluß von Magnetfeldern auf die Kontraktion

In den letzten Jahren hat sich die Ansicht gefestigt, daß es neben den stellaren auch interstellare Magnetfelder gibt. Beobachtungen wie die von polarisiertem Sternlicht, zusammen mit theoretischen Überlegungen, sind ein Grund dafür. Die Polarisation läßt sich verstehen durch Annahme eines die interstellaren Staubteilchen ausrichtenden Magnetfeldes [SPITZER und TUKEY (1951); DAVIS und GREENSTEIN (1951)].

Von BIERMANN und SCHLÜTER (1950) ist ferner gezeigt worden, wie durch turbulente Bewegung des interstellaren Gases die Feldstärke eines anfänglich geringen magnetischen Feldes wachsen muß — wahrscheinlich bis magnetische und turbulente Energiedichte von gleicher Größenordnung sind. FERMI (1949) hat einen Mechanismus vorgeschlagen, der die Höhenstrahlteilchen mit Hilfe interstellarer Magnetfelder auf die beobachteten Energien beschleunigt. CHANDRASEKHAR und FERMI (1953a) haben versucht, die Spiralarme durch Magnetfelder zu erklären.

Unabhängig von der Frage, wie groß die mittlere Feldstärke eines interstellaren Magnetfeldes sein müßte, um die erwähnten Beobachtungen zu erklären (meist wird eine Feldstärke von 10^{-6} bis 10^{-5} Gauß angenommen), ist von grundsätzlicher Bedeutung die Frage, wie Magnetfelder die Sternentstehung beeinflussen können. Dazu haben MESTEL und SPITZER (1956) eine Untersuchung angestellt, auf die hier eingegangen werden soll.

Zuerst wird die Bedingung der Gravitationsinstabilität geprüft. Für ein Gas mit Magnetfeld hat der Virialsatz, wenn das System im mechanischen Gleichgewicht ist, die Form [CHANDRASEKHAR und FERMI (1953b)]

$$2 E_T + E_H + E_G = 0 . \tag{178}$$

E_T bezeichnet die thermische Energie, E_H die magnetische und E_G die Gravitationsenergie des Systems. Turbulenz und Rotation sind vernachlässigt, das Verhältnis der spezifischen Wärmen ist 5/3 gesetzt (einatomiges Gas). Wird die linke Seite von Gl. (178) positiv, so expandiert das System, wird sie negativ, so kontrahiert es.

Das Magnetfeld durchsetze als "large-scale field" die Wolke und sei in die Materie „eingefroren", d. h. der magnetische Fluß bleibe bei der Kontraktion konstant:

$$H R^2 = H_0 R_0^2 . \tag{179}$$

H ist die Feldstärke und R der Radius der kugelförmig gedachten Wolke. Für die magnetische Energie folgt

$$E_H = \frac{H_0^2 R_0^4}{6 R} , \tag{180}$$

d. h. sie zeigt die gleiche R-Abhängigkeit wie die Gravitationsenergie

$$E_G = -\frac{\alpha_0 G M^2}{R}. \tag{181}$$

α_0 ist eine Konstante von der Größenordnung 1. M ist die Masse der Wolke, G die Gravitationskonstante. Überwiegt die Gravitationsenergie die magnetische Energie, so ändert sich dies während der Kontraktion nicht. Ebenso läßt sich ein Überwiegen der magnetischen Energie durch eine erzwungene Kontraktion nicht ändern.

Aus Gl. (178) ergibt sich als kleinste Masse, die dem magnetischen Druck das Gleichgewicht halten kann, wenn die thermische Energie vernachlässigt wird und ϱ die mittlere Gasdichte der Wolke ist,

$$M = 4 \cdot 10^{-3} \frac{H^3}{(\alpha_0 G)^{3/2} \varrho^2}, \tag{182}$$

unabhängig davon, wieweit die Kontraktion fortgeschritten sein mag, weil das Verhältnis H^3/ϱ^2 konstant bleibt. Dies ist ein wesentlicher Unterschied zum Jeansschen Kriterium, wo mit wachsendem ϱ (fortschreitende Kontraktion) immer kleinere Massen instabil werden.

Mit $H = 10^{-6}$ Gauß und $\varrho = 1.7 \cdot 10^{-23}$ g/cm³ folgt aus Gl. (182) als kleinste instabile Masse ungefähr 500 Sonnenmassen. Bei Berücksichtigung der thermischen und der turbulenten Energie würde sich ein noch größerer Wert ergeben.

McCrea (1957) hat zur Anwendung von Gl. (178) bemerkt, daß nur unter Vernachlässigung des von außen auf die Wolke ausgeübten Druckes diese Gleichung richtig ist. Bei Berücksichtigung des Außendruckes können wahrscheinlich auch kleinere Massen als Gl. (182) instabil werden. Eine quantitative Untersuchung darüber besteht noch nicht.

Die weiteren von Mestel und Spitzer gegebenen Überlegungen werden aber kaum wesentlich dadurch geändert werden. Wir folgen ihnen. Es ist die Frage, ob im Laufe der Kontraktion das Magnetfeld gegenüber der sich verdichtenden Materie gleiten kann, so daß die Bedingung (179) nicht mehr erfüllt ist. Bei völligem Gleiten würde die Feldstärke konstant bleiben. Die magnetische Energie der Wolke nimmt dann infolge des kleiner werdenden Radius ab, während die negative Gravitationsenergie wie früher zunimmt. Dadurch könnten im Laufe der Kontraktion kleinere Massen instabil werden, und der von Hoyle (1953) beschriebene Zerfall in Untersysteme könnte eintreten.

Am Magnetfeld fest hängt direkt nur die ionisierte Materie, nicht das neutrale Gas. Die Frage, ob das Magnetfeld aus einer H I-Wolke herausgleiten kann, ist somit die Frage, ob die Reibungskraft zwischen neutralem und ionisiertem Gas ein solches Gleiten zuläßt. Die genaue Untersuchung für einen einfachen Fall ergibt, daß sich die magnetische

Kraft, die sich durch die Deformation des Feldes bei der Kontraktion einstellt, und die Reibungskraft das Gleichgewicht halten, wenn

$$|\mathfrak{v}_H - \mathfrak{v}_i| \cong \frac{1.5 \cdot 10^3}{n_i} \text{ cm/sec} \tag{183}$$

ist, wo $\mathfrak{v}_H$ die Geschwindigkeit des neutralen, $\mathfrak{v}_i$ die des ionisierten Gases und n_i die Teilchendichte der Ionen bezeichnet. Es wurde angenommen, die Wolke habe eine Dichte von $n_H = 2 \cdot 10^4$ H/cm³ und eine Feldstärke von $H = 1.5 \cdot 10^{-4}$ Gauß (entspricht einer Feldstärke von 10^{-6} Gauß bei $n_H = 10$). Die Ionen in einer HI-Region werden von den schweren Elementen geliefert. Auf 10^4 H-Atome kommt etwa 1 Ion, wenn die Energiedichte des interstellaren Strahlungsfeldes in der HI-Wolke nicht durch Abschirmung unter den normalen Wert herabgesetzt ist.

Aus Gl. (183) sehen wir, daß sich für unser Beispiel eine Gleitgeschwindigkeit von 10^3 cm/sec ergibt, und — mit dem Radius von 1 pc — das Feld etwa 10^8 Jahre braucht, um aus der sich kontrahierenden Wolke herauszugleiten.

Der Gleitprozeß wäre unter bisher angenommenen Bedingungen also sehr langsam. MESTEL und SPITZER weisen aber darauf hin, daß bei Zunahme der Dichte der Wolke und bei Gegenwart von Staub die Energiedichte der interstellaren Strahlung innerhalb der Wolke unter den normalen Wert fallen kann. Die Ionendichte nimmt dann ab, und die Gleitgeschwindigkeit wächst. Es ist möglich, daß sich Gleitgeschwindigkeiten von 10^5 cm/sec ergeben. In diesem Fall ist der beschriebene Prozeß sehr wirksam.

Ist durch Gleiten der Betrag der Energiedichte der Gravitation groß gegen die Energiedichte des Magnetfeldes geworden, so kann der in III.d beschriebene Zerfall in Untersysteme eintreten, ganz gleich ob das Magnetfeld später wieder durch steigende Temperatur im Innern an der Materie haftet oder nicht.

C.IV.b) Drehimpulsabführung durch Turbulenz und bei rotierender Zentralmasse

Das Problem der Drehimpulsabführung ist, wie wir schon gesehen haben, eine für die Sternentstehung entscheidende Frage. Zu ihrer Lösung sind bisher drei verschiedene Mechanismen vorgeschlagen worden. Der eine wurde bereits im Abschnitt III erwähnt. Beim Zerfall einer Wolke in Untersysteme kann der Drehimpuls in Bahndrehimpuls der Untersysteme umgewandelt werden.

Eine zweite Möglichkeit wurde durch v. WEIZSÄCKER (1944), (1948) angegeben. Führt die Nichtabführung des Drehimpulses bei der Kontraktion zu einem stark abgeplatteten, rotierenden und Turbulenz enthaltenden System, so kann die turbulente Reibung einen Drehimpulstransport nach außen bewirken. Ein Teil der Masse fließt dabei mit nach außen ab. LÜST (1952) hat genaue Lösungen gegeben. Sie beziehen sich auf die Entstehung des Planetensystems. Erweiterungen auf den Fall der Sternentstehung sind bisher nicht gegeben worden.

Ein weiterer Mechanismus ist von LÜST und SCHLÜTER (1955) vorgeschlagen worden. Er geht zurück auf den Gedanken von ALFVÉN (1942), (1951), daß Magnetfelder einen Drehimpulstransport ermöglichen könnten. Es wird untersucht, wie einmal entstandene und rotierende Sterne mit Hilfe ihres Magnetfeldes ihren Drehimpuls an das interstellare Gas zurückgeben können. Es wird angenommen, der Stern habe ein Dipolfeld, dessen Achse parallel zur Rotationsachse ist. Durch die Rotation des Sterns und durch die an den weiter außen verlaufenden Feldlinien angreifende turbulente Reibung des Gases erhält das Feld eine Torsion. Diese Torsion ermöglicht einen Drehimpulstransport senkrecht zur Dipolachse. Die dadurch bedingte Abbremsung kann unter günstigen Bedingungen sehr wirkungsvoll sein. Nimmt die Gasdichte nach außen wie r^{-2} ab mit dem Wert 10^{-8} g/cm^3 an der Oberfläche des Sterns und beträgt das Magnetfeld 10^2 Gauß an der Sternoberfläche, so kann ein an der Grenze der Stabilität rotierender Stern in 10^6 Jahren den größten Teil seines Drehimpulses abgeben. Beträgt dagegen die Gasdichte 10^{-9} g/cm^3 oder weniger, so ist, bei gleichem Magnetfeld wie vorher, die Abbremsung nur noch gering.

Bei der Entstehung von Planetensystemen und in späten Phasen der Sternentstehung wird die durch diesen Prozeß bedingte Drehimpulsabführung wahrscheinlich eine wichtige Rolle spielen. Für die frühen Phasen der Sternentstehung aber können wir kaum ein Dipolfeld der sich verdichtenden Wolke annehmen, und das Modell kann auf diese Phasen daher in dieser Form nicht angewendet werden. Es bedarf dazu einer wesentlichen Modifikation.

C.IV.c) Drehimpulsabführung durch Magnetfelder bei der Sternentstehung

In einer hier folgenden eigenen Überlegung soll der Gedanke der Drehimpulsabführung durch Magnetfelder auf den Prozeß der Sternentstehung übertragen werden. Es ist für die weiteren Ausführungen nicht wesentlich, ob aus der anfänglichen Gasmasse schließlich ein Einzelstern oder ein Sternhaufen entsteht. Der Unterschied liegt dann hauptsächlich in der Größe des abzuführenden Drehimpulses, nicht aber im Mechanismus selbst.

Der Mechanismus ist gedacht als eine weitere neben den schon erwähnten Möglichkeiten einer Drehimpulsabführung. Ob und wieweit er verwirklicht ist, hängt weitgehend von der Frage ab, wie die Gestalt und die Größe des anfänglichen, die instabil werdende Gaswolke durchsetzenden interstellaren Magnetfeldes ist. Diese Frage entscheiden wir hier nicht. Es werden aber die Bedingungen angegeben, unter denen der Mechanismus wirksam ist.

Wir gehen von folgender Annahme aus: Die Richtung der magnetischen Feldlinien der anfänglichen Wolke sei im Mittel parallel zur Richtung des resultierenden Drehimpulsvektors. Ferner seien die Feldlinien in die Materie „eingefroren". In Abb. 23 ist der angenommene Feldlinienverlauf skizziert. Die einsetzende Rotation der Wolke muß nun, da die Feldlinien eingefroren sind, zu einer Torsion des außerhalb der Wolke befindlichen Magnetfeldes führen. Die Verdrillung der Feldlinien pflanzt sich in Richtung der Feldlinien nach außen fort. Sie geschieht außerdem auf Kosten des Drehimpulses der Wolke, weil zu ihrer Erzeugung ein Drehmoment nötig ist. Mit der nach außen laufenden Torsion ist somit ein Drehimpulsstrom verbunden, durch den der Drehimpuls der kontrahierenden Wolke an das interstellare Gas zurück gegeben wird.

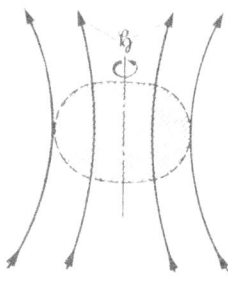

Abb. 23. Angenommener Feldlinienverlauf zu Beginn der Kontraktion und Rotation der Wolke

Die zu lösenden Gleichungen lauten, wenn die Bedingung unendlicher Leitfähigkeit schon mitverwendet wird:

$$\frac{\partial \mathfrak{H}}{\partial t} - \operatorname{rot} [\mathfrak{v} \, \mathfrak{H}] = 0 \tag{184}$$

$$\varrho \left\{ \frac{\partial \mathfrak{v}}{\partial t} + (\mathfrak{v} \operatorname{grad}) \mathfrak{v} \right\} + \operatorname{grad} p + \frac{1}{4\pi} [\mathfrak{H} \operatorname{rot} \mathfrak{H}] = 0 \tag{185}$$

$$\frac{\partial \varrho}{\partial t} + \operatorname{div} (\varrho \mathfrak{v}) = 0 \tag{186}$$

$$\operatorname{div} \mathfrak{H} = 0 \tag{187}$$

Hierin bezeichnet $\mathfrak{H}$ die magnetische Feldstärke, $\mathfrak{v}$ die Geschwindigkeit des Gases sowie ϱ und p seine Dichte und seinen Druck. Die erste Gleichung geht aus dem Induktionsgesetz hervor (durch Berücksichtigung der unendlichen Leitfähigkeit), die zweite ist die Bewegungsgleichung mit Einschluß der durch das Magnetfeld erzeugten Kraft und die dritte ist die Kontinuitätsgleichung. Das Magnetfeld muß außerdem divergenzfrei sein, was durch Gl. (187) ausgedrückt wird. Zur Ableitung der Plasma-Gleichungen siehe z. B. SCHLÜTER (1950) und (1951), ALFVÉN (1950), SPITZER (1956), COWLING (1957).

Zur mathematischen Behandlung dieser Gleichungen idealisieren wir. Das Feld soll zylindersymmetrisch sein und keine Komponente radial zur Achse haben, also nur die Komponenten H_z und H_φ in Zylinderkoordinaten, wenn z die Richtung der Achse, s den Abstand davon und φ die Richtung senkrecht zu z und s angibt. Ferner wird die Annahme gemacht, daß die bei der Verdrillung des Feldes auftretenden Geschwindigkeitskomponenten v_z und v_s (für v_φ gilt keinerlei Einschränkung) gegenüber den anderen Effekten zu vernachlässigen sind. Je langsamer sich die Rotationsgeschwindigkeit ändert, je besser trifft diese Annahme zu, wie wir später sehen werden. Es soll also gelten

$$v_s = v_z = 0, \quad \frac{\partial}{\partial \varphi} = 0 \quad \text{für alle } t$$
$$H_s = 0, \quad H_z \neq 0 \quad \text{für } t = 0. \tag{188}$$

Aus Gl. (184) zusammen mit Gl. (187) folgt dann

$$H_s = 0, \quad H_z = H(s) \quad \text{für alle } t, \tag{189}$$

d. h. es tritt im Laufe der Zeit keine s-Komponente der Feldstärke auf, und H_z ist weder von t noch von z abhängig; letzteres ist durch die Divergenzfreiheit bedingt. Aus Gl. (186) ergibt sich für die Gasdichte Zeitunabhängigkeit. Da H_z nicht von z abhängt, machen wir die Annahme, daß auch ϱ zu Beginn nicht von z abhing; dann gilt

$$\varrho = \varrho(s) \quad \text{für alle } t.$$

Weiter folgt aus Gl. (184)

$$\dot{H}_\varphi - H(s) v'_\varphi = 0, \tag{190}$$

wo der Strich die Differentiation nach z, und der Punkt die partielle Differentiation nach t bedeutet. Und aus Gl. (185) folgt

$$\dot{v}_\varphi - \frac{H(s)}{4\pi \varrho(s)} H'_\varphi = 0. \tag{191}$$

Diese zwei Gleichungen bestimmen die zeitabhängige Torsion des Feldes. Gl. (185) enthält noch zwei weitere Gleichungen, aus denen sich schließen läßt, wie weit die Annahme $v_z = v_s = 0$ gerechtfertigt werden kann. Wir werden später auf sie eingehen.

Aus Gl. (190) und (191) folgt, indem wir jeweils eine der zwei unbekannten Funktionen eliminieren,

$$\ddot{v}_\varphi - a^2 v''_\varphi = 0$$
$$\ddot{H}_\varphi - a^2 H''_\varphi = 0 \quad \text{mit } a^2 = \frac{H^2(s)}{4\pi \varrho(s)}, \tag{192}$$

d. h. die φ-Komponenten des Magnetfeldes und der Geschwindigkeit gehorchen der homogenen Wellengleichung; a ist die Alfvén-Geschwindigkeit.

Die allgemeine Lösung von Gl. (192) gewinnen wir leicht durch Einführung der Charakteristiken $x = a t - z$, $y = a t + z$. Berücksichtigen wir dann die durch Gl. (190), (191) gegebene Kopplung von v_φ mit H_φ, so erhalten wir als allgemeine Lösung

$$v_\varphi = F(x, s) + G(y, s)$$
$$H_\varphi = -(4\pi \varrho)^{1/2} \{F(x, s) - G(y, s)\}, \tag{193}$$

wo F und G zwei beliebige Funktionen zweier unabhängiger Variablen sind.

Der nächste Schritt ist die Lösung des Anfangs- und Randwertproblems. Es wird die Annahme gemacht, zur Zeit $t < 0$ sei keine Torsion des Feldes und keine Bewegung vorhanden. Zur Zeit $t = 0$ soll bei $z = 0$ eine vorgegebene zeitabhängige Rotation $V(t, s)$ einsetzen. Der magnetische Zylinder sei in Richtung der positiven z-Achse unbegrenzt, so daß die in diese Richtung laufenden Veränderungen nicht reflektiert werden. Da entlang der negativen z-Achse alles spiegelsymmetrisch zur positiven ablaufen soll, betrachten wir nur die Vorgänge auf der positiven. Die zu erfüllenden Bedingungen sind somit

$$v_\varphi = H_\varphi = 0 \text{ für } t < 0, \quad z \geqq 0, \quad s \geqq 0$$
$$v_\varphi = V(t, s) \text{ für } t \geqq 0, \quad z = 0, \quad s \geqq 0 . \tag{194}$$

Mit Hilfe der Charakteristiken, die in unserem Fall Geraden sind, erhält man aus Gl. (193), (194) die Lösung

$$\left. \begin{array}{l} v_\varphi = V\left(t - \dfrac{z}{a}, s\right) \\ H_\varphi = -(4\pi \varrho)^{1/2} V\left(t - \dfrac{z}{a}, s\right) \end{array} \right\} \text{ für } z \leqq a t ; \tag{195}$$

für $z > a t$ sind beide Komponenten Null. Das heißt, die in $z = 0$ vorhandene Rotation läuft mit der Geschwindigkeit a in Richtung der z-Achse, ohne dabei ihre Gestalt zu ändern. Mit ihr verbunden ist eine bestimmte φ-Komponente des Magnetfeldes. Aus Gl. (195) ersehen wir noch die einfache Beziehung

$$H_\varphi = -(4\pi \varrho)^{1/2} v_\varphi , \tag{196}$$

die später verwendet werden wird.

Wir sind jetzt in der Lage, den Drehimpulsstrom zu berechnen. Beachten wir, daß v_φ sich mit a in Richtung der z-Achse fortpflanzt, so folgt für den Drehimpulsstrom, der in Richtung der z-Achse durch eine senkrecht zu ihr liegenden Fläche mit dem Radius R_0 fließt,

$$D(z, t) = 2\pi \int_0^{R_0} V\left(t - \dfrac{z}{a}, s\right) a \varrho s^2 d s . \tag{197}$$

Unter der vereinfachenden Annahme, bei $z = 0$ finde eine starre Rotation statt, also

$$V(t, s) = \omega(t) \cdot s$$

und bei nicht von s abhängigem a und ϱ (durch Index 0 gekennzeichnet), folgt für den Drehimpulsstrom bei $z = 0$ die Formel

$$D(t) = \frac{\sqrt{\pi}}{4} \varrho_0^{1/2} H_0 R_0^4 \omega(t) . \tag{198}$$

Auf Grund dieser Formel berechnen wir für drei einfache Fälle die Zeit der Abbremsung einer anfänglich rotierenden Wolke. Ist J der Drehimpuls der Wolke, so muß gelten

$$\frac{dJ}{dt} + 2D(t) = 0 \,. \tag{199}$$

Der Faktor 2 entsteht durch Berücksichtigung des Abtransportes von Drehimpuls in positiver *und* negativer z-Richtung.

1. Fall: Die Wolke habe die Gestalt einer zylindrischen Scheibe mit konstanter Dichte; ihre Masse sei M, der Radius R_0 und die anfängliche Rotationsgeschwindigkeit in der Entfernung R_0 von der Achse sei V_0. Während der Drehimpulsabführung bleibe der Radius und die Masse der Wolke konstant. Dann ist $J = 2/5 \cdot M R_0^2 \omega$, und aus Gl. (198), (199) folgt durch Integration für die Zeit, nach der die Rotationsgeschwindigkeit von V_0 auf V_1 gefallen ist,

$$\tau_1 = \frac{4M}{5\sqrt{\pi}\, \varrho_0^{1/2} H_0 R_0^2} \ln \frac{V_0}{V_1} \,. \tag{200}$$

Zur leichteren physikalischen Interpretation dieses Ergebnisses schreiben wir Gl. (200) mit Hilfe der Alfvén-Geschwindigkeit [Gl. (192)] in der Form

$$\tau_1 = \frac{2M}{5\pi \varrho_0 R_0^2 a_0} \ln \frac{V_0}{V_1} \,. \tag{201}$$

Auf der rechten Seite steht, abgesehen von dem dimensionslosen Faktor $2/5 \cdot \ln(V_0/V_1)$, die Zeit, die eine Alfvénsche Welle braucht, um in dem Zylinder mit dem Radius R_0 und der Dichte ϱ_0 die Masse M zu durchlaufen. Da der Drehimpulstransport nach außen, verbunden mit einer Verdrillung der Feldlinien, gerade mit Alfvén-Geschwindigkeit erfolgt, ist einzusehen, daß die obige Zeit die Zeitskala für die Drehimpulsabführung darstellen muß.

Für die in Tab. 23 aufgeführten Objekte sind in Tab. 24 Abbremszeiten angegeben. Zu ihrer Berechnung wurden in allen drei Fällen für den magnetischen Zylinder die Werte $\varrho_0 = 10^{-23}$ g/cm³, $H_0 = 10^{-5}$ Gauß angenommen, d. h. wir machten die Annahme, daß in der unmittelbaren Umgebung interstellarer Wolken die Gasdichte größer ist als die mittlere Gasdichte in der Milchstraße, und daß mit der größeren Gasdichte auch ein etwas stärkeres Magnetfeld verbunden ist. Beide Annahmen liegen nahe, es bleibt aber die Frage offen, wieweit Beobachtungen dafür oder dagegen sprechen.

Werden diese Annahmen gemacht, so ergeben sich jedenfalls Abbremszeiten, die zum Teil von der Größenordnung der vermuteten Zeitskala der Entstehung von Sternhaufen und Einzelsternen sind.

In der Tabelle sind die Zeiten in Jahren angegeben; die letzte Spalte gibt Geschwindigkeits- und Radienverhältnisse.

Tabelle 24. *Zeiten der Drehimpulsabführung*

	Small Globule	Large Globule	Intermediate Cloud	Large Cloud	
$M/M_\odot$	>0.2	5	$1.3 \cdot 10^3$	$3 \cdot 10^4$	
τ_1	$>9.2 \cdot 10^7$	$3.3 \cdot 10^7$	$3.4 \cdot 10^7$	$3.1 \cdot 10^7$	$V_0 = 100\, V_1$
τ_2	$> 2 \cdot 10^7$	$7.2 \cdot 10^6$	$7.3 \cdot 10^6$	$6.8 \cdot 10^6$	$R_0 \gg R$
τ_3	$> 5 \cdot 10^8$	$1.8 \cdot 10^8$	$1.8 \cdot 10^8$	$1.7 \cdot 10^8$	$R_0 = 10\, R_1$

2. Fall: Bei sonst gleichen Annahmen wie im Fall 1 soll hier während der Drehimpulsabführung eine Kontraktion stattfinden. Der jeweilige Radius sei dadurch bestimmt, daß die Rotationsgeschwindigkeit $V = R\,\omega(t)$ während der Kontraktion konstant bleibt (bei Gleichheit von Zentrifugalkraft und Gravitation wächst die Rotationsgeschwindigkeit mit abnehmendem Radius; in unserem Beispiel überwiegt also die Gravitation mehr und mehr). Die Kontraktion verursacht ein Anwachsen der magnetischen Feldstärke $H \sim R^{-2}$, weil das Feld eingefroren ist. Wir nehmen an, daß mit dem Magnetfeld auch die Gasdichte im magnetischen Zylinder anwächst $\varrho \sim H$, weil die Materie sich nur in Richtung des Feldes bewegen könnte und diese Geschwindigkeit klein sein soll. Der Anfangsradius der Wolke sei R_0, ihr Endradius R_1, Anfangsfeldstärke und -dichte im Zylinder seien H_0 und ϱ_0. Die Integration ergibt als Zeit jetzt

$$\tau_2 = \frac{4M}{5\sqrt{\pi}\,\varrho_0^{1/2} H_0 R_0^2}\left(1 - \frac{R_1}{R_0}\right). \tag{202}$$

In Tab. 24 sind numerische Beispiele gegeben. Formel (202) ist nur dann eine gute Näherung, wenn die Kontraktionsgeschwindigkeit, verglichen mit der Rotationsgeschwindigkeit, klein ist, weil andernfalls die zur Herleitung gemachte Voraussetzung $v_s = 0$ nicht mehr gerechtfertigt ist. Dies gilt ebenso für den nächsten Fall.

3. Fall: Hier soll angenommen werden, daß die im 2. Fall beschriebene Kontraktion stattfindet, daß aber das Magnetfeld dabei nicht mitkontrahiert, sondern durch den im ersten Abschnitt beschriebenen Prozeß des Gleitens [MESTEL und SPITZER (1956)] seinen Anfangswert behält. Die Rotationsgeschwindigkeit sei soviel größer als die Kontraktionsgeschwindigkeit, daß in Richtung der Rotation kein Gleiten des Magnetfeldes auftritt. Die Abbremsung ist hier geringer als in den zwei früheren Fällen. Als Zeit ergibt sich

$$\tau_3 = \frac{M}{5\sqrt{\pi}\,\varrho_0^{1/2} H_0 R_0^2}\left(\frac{R_0^2}{R_1^2} - 1\right), \tag{203}$$

wenn die gleichen Bezeichnungen wie in 2 gewählt werden. Das Radienverhältnis geht empfindlich ein. Die letzte Zeile in Tab. 24 gibt wieder numerische Werte; ein Radienverhältnis von 10 : 1 wurde gewählt.

Bevor wir auf die numerischen Werte eingehen, sollen noch kurz die bisher nicht betrachteten aus Gl. (185) sich ergebenen zwei Gleichungen behandelt werden. Sie lauten

$$\varrho \, \dot{v}_z + \frac{\partial}{\partial z} \left(p + \frac{H_\varphi^2}{8\pi} \right) = 0 \tag{204}$$

$$\varrho \, \dot{v}_s + \frac{\partial}{\partial s} \left(p + \frac{H_z^2 + H_\varphi^2}{8\pi} \right) + \frac{H_\varphi^2}{4\pi s} - \frac{\varrho \, v_\varphi^2}{s} = 0 \,. \tag{205}$$

Aus Gl. (204) gewinnen wir eine Abschätzung der nach Einsetzen der Rotation auftretenden Geschwindigkeitskomponente v_z. Dazu vernachlässigen wir den Gasdruck und nehmen zeitlich konstante Dichte an. Mit Hilfe von Gl. (191) folgt dann

$$\dot{v}_z - \frac{(4\pi \varrho)^{1/2}}{H} v_\varphi \dot{v}_\varphi = 0 \tag{206}$$

und nach Integration

$$v_z = \frac{v_\varphi^2}{2a} \quad \text{mit} \quad a = \frac{H}{(4\pi \varrho)^{1/2}}, \tag{207}$$

wenn vor Einsetzen der Rotation $v_z = v_\varphi = 0$ war. Die in z-Richtung auftretende Geschwindigkeit ist also von der Größenordnung des Produktes aus Rotationsgeschwindigkeit und Verhältnis von Rotations- zu Alfvén-Geschwindigkeit. Der Ansatz $v_z = 0$ kann also gerechtfertigt werden, solange $|v_\varphi| \ll a$ ist. In unseren Beispielen ist bei Kontraktionsbeginn $a = 10$ km/sec. Sind die Anfangsrotationsgeschwindigkeiten größer als dieser Wert, so treten wahrscheinlich Verdichtungsstöße in z-Richtung, verbunden mit erheblichem Materiestrom, auf. Für die in Tab. 24 gegebenen Objekte liegen die Anfangsrotationsgeschwindigkeiten, bei denen Zentrifugalkraft und Gravitation sich das Gleichgewicht halten, bei etwa 1 km/sec, sind also geringer als die Alfvén-Geschwindigkeit.

Zur Behandlung von Gl. (205) schlagen wir einen anderen Weg ein. Wir fragen, welche Radiusänderung der Zylinder nach Einsetzen der Rotation erfahren muß, um wieder Gleichgewicht in s-Richtung herzustellen. Das Glied mit v_s kann dann gestrichen werden, und unter Beachtung von Gl. (196) geht Gl. (205) über in

$$\frac{\partial}{\partial s} \left(p + \frac{H_z^2 + H_\varphi^2}{8\pi} \right) = 0 \,. \tag{208}$$

Vernachlässigen wir noch den Gasdruck und nehmen starre Rotation und, wegen des Haftens der Materie am Feld, $\varrho \sim H$ an, so folgt für den Betrag H der Feldstärke

$$H^2(s) + \frac{4\pi \varrho_0 \omega^2 s^2}{H_0} H(s) = H_0^2, \tag{209}$$

mit $H_0 = H(0)$ und $\varrho_0 = \varrho(0)$. In Abb. 24 ist der schematische Verlauf von H wiedergegeben. Der neue Radius R_ω des Zylinders ergibt sich aus

$$2\pi \int_0^{R_\omega} H\, s\, ds = H_0 R_0^2,$$

wenn R_0 der Radius vor Einsetzen der Rotation ist.

Die Reihenentwicklung ergibt

$$\frac{R_\omega}{R_0} = 1 + \frac{\pi \varrho_0 \omega^2 R_0^2}{2 H_0^2} + \ldots, \tag{210}$$

sofern das zweite Glied der Reihe klein gegen das erste, also gegen 1 ist. Für unsere numerischen Beispiele ergibt sich $R_\omega/R_0 \approx 1.01$, wenn $\omega R_0 = 3 \cdot 10^5$ cm/sec gesetzt wird. Die Annahme $v_s = 0$ kann somit für unsere Fälle gerechtfertigt werden.

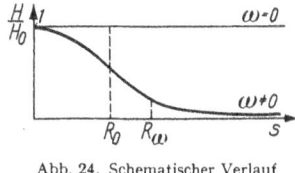

Abb. 24. Schematischer Verlauf der Feldstärke

Das durch Gl. (209) und (196) beschriebene Magnetfeld ist ein kraftfreies, rotierendes und Drehimpuls transportierendes Feld. Es hat große Ähnlichkeit mit den von SCHLÜTER (1957) entdeckten kraftfreien, Drehimpuls transportierenden Feldern.

Wir fragen zum Schluß: kann mit Hilfe des beschriebenen Mechanismus soviel Drehimpuls abgeführt werden, daß eine einmal begonnene Kontraktion einer interstellaren Wolke bis zur Bildung von Sternen oder eines Sternhaufens fortgehen kann?

Zuerst werde die Entstehung eines Haufens betrachtet. Der Beginn sei eine Wolke mit 10^3 Sonnenmassen und einem Radius von 1 pc, was einer mittleren Dichte von 10^4 H/cm^3 entspricht. Die Temperatur des Gases sei 100° K, die Feldstärke 10^{-4} Gauß (der hohen Gasdichte entsprechend). Die Dichte außerhalb der Wolke sei $\varrho_0 = 10^{-23}$ g/cm^3. Nach dem Kriterium von SPITZER und MESTEL (1956) ist diese Wolke gerade gravitationsinstabil. Das Magnetfeld sei zuerst eingefroren, bis der Radius 1/10 seines Anfangswertes erreicht. Dann gleite das Magnetfeld, bis der Radius sich erneut auf 1/10 seines Wertes verkleinert hat. Die Gasdichte der Wolke ist jetzt so groß, daß nach dem Jeanschen Kriterium Sonnenmassen instabil werden. Wir nehmen an, die Wolke zerfalle jetzt in viele Untersysteme und der noch verbliebene Drehimpuls werde dabei weitgehend in Bahndrehimpuls umgewandelt. Die Zeit für die Drehimpulsabführung bis zum Zerfall der Wolke beträgt nach unserer Formel etwa $3 \cdot 10^7$ Jahre.

Das aus Beobachtungen erschlossene Alter eines Sternhaufens läßt sich mit dieser Zeit nicht unmittelbar vergleichen. Der Nullpunkt für das Alter eines Haufens ist durch das Einsetzen der Kernprozesse oder, bei expandierenden Haufen, der Expansionsbewegung gegeben. Der Nullpunkt für die Drehimpulsabführung aber liegt beim Einsetzen der Gravitationsinstabilität oder noch davor.

Für die Bildung eines einzelnen O-Sterns aus einer anfänglichen Wolke ergeben sich bei Berücksichtigung eines Gleitens des Magnetfeldes zu große Zeiten für die Drehimpulsabführung, es sei denn, daß vor Einsetzen des Gleitens die Wolke unerwartet viel Drehimpuls abtransportiert hat [SPITZER (1958)].

C.V. Zusammenfassung

Wir wollten untersuchen, ob wir in der Lage sind, die Entstehung von Sternen unter heutigen kosmischen Bedingungen physikalisch zu verstehen. Zunächst ergab sich, daß die Aufsammlung von Materie nur unter Annahme sehr extremer Bedingungen zu einer wesentlichen Vergrößerung der Sternmasse führen kann. Für die jungen Sterne müssen wir also eine direkte Entstehung durch Kondensation annehmen.

Nach einer Modifikation des Jeansschen Kriteriums können bei normalen Werten von Temperatur und Dichte nur Massen oberhalb von 10 bis 1000 Sonnenmassen kondensieren; durch die Anwesenheit interstellarer Magnetfelder wird diese Grenze heraufgesetzt. Innerhalb einer interstellaren Wolke können somit in diesem ersten Kondensationsschritt nur Sternhaufen, höchstens noch O-Sterne, nicht dagegen einzelne Sterne normaler Masse entstehen. Auch die Berücksichtigung der Wirkung des Strahlungsdruckes auf Staubteilchen kann nichts Wesentliches hieran ändern.

Im Falle der primären Entstehung eines Sternhaufens wurde der schrittweise Zerfall in weitere Untersysteme untersucht. Dieser Zerfall ist dann nicht möglich, wenn Magnetfeld und Materie fest aneinanderhängen. Abschätzungen ergaben jedoch, daß bei Annahme günstiger Bedingungen die Materie schnell genug durch das Feld gleiten kann; der Zerfall wird schließlich dadurch gestoppt, daß in Systemen einer kleinsten Masse der Temperaturgradient ein hydrostatisches Gleichgewicht herzustellen vermag. Verschiedene Arten der Abschätzung ergaben 1.5 bzw. 0.5 Sonnenmassen. Dieser Zerfall in Untersysteme scheint also in der Lage zu sein, die Entstehung von Sternen normaler Masse in Sternhaufen zu erklären.

Die zweite Möglichkeit ist die primäre Entstehung eines O-Sterns, in dessen Umgebung sich eine schnell expandierende HII-Region ausbildet. Die umgebende HI-Region wird stark komprimiert und beschleunigt, wobei sich aus der instabilen Grenzfläche sogenannte Elefantenrüssel herauslösen, die sich zu einzelnen, dichten Wolken abschnüren und weiter kontrahieren. Die Entstehung expandierender Assoziationen kann in dieser Weise qualitativ verstanden werden.

Der Druck umgebender HII-Regionen auf eine einzelne, abgeschnürte HI-Wolke kann zu einer fortlaufenden Kontraktion bis zum Einsetzen der Gravitations-Instabilität führen, falls die Wolkenmasse größer ist als eine von Dichte und Druck abhängige Grenzmasse. Es ergibt sich auch hier, daß Sterne von Sonnenmasse sich primär nicht bilden können. Aus "large globules" könnten O-Sterne entstehen, aus normalen Wolken jedoch nur Sternhaufen. Eine Zeitabschätzung für die Abstrahlung der Energie während der Kontraktion ergab 10^6 bis 10^7 Jahre.

Wir gewinnen somit den Eindruck, daß sich nur die Bildung von O-Sternen direkt verstehen läßt, die Bildung von Sternen durchschnittlicher Masse dagegen nur in größeren Gruppen oder Haufen von rund 1000 Sonnenmassen, ganz gleich, ob sich zunächst einzelne O-Sterne bildeten oder nicht. Dies steht im Einklang mit dem Ergebnis von Abschnitt B.III., wonach $1/5$ aller Sterne in offenen Haufen entstehen und der Rest vermutlich in Assoziationen, als auch mit unseren früheren Massenabschätzungen von 1500 Sonnenmassen/offener Haufen und 2000 Sonnenmassen/Assoziation.

Eines der Hauptprobleme jeder Kontraktion (sowohl Sternhaufen als auch Sterne betreffend), die Abführung des Drehimpulses, scheint sich mit Hilfe der interstellaren Magnetfelder lösen zu lassen. In den ersten Phasen der Kontraktion kann der hier vorgeschlagene Mechanismus den Drehimpuls längs der Rotationsachse fortführen, in der letzten Phase hat sich das Magnetfeld des Sterns bereits abgeschnürt und kann zusammen mit der turbulenten Reibung den Drehimpuls in der Äquatorebene abführen. Beide Mechanismen setzen zwar keine extremen, aber doch günstige Verhältnisse voraus.

Ganz offen mußten wir dagegen die Frage lassen, welches Massenspektrum der entstehenden Sterne sich aus theoretischen Überlegungen ableiten ließe. Ihre Beantwortung ist gegenwärtig eines der brennendsten Probleme der Theorie der Sternentstehung.

Schlußbemerkung

Zur Zeit der Planung der Arbeit befanden sich die drei Verfasser gemeinsam am Max-Planck-Institut für Physik, Göttingen. Während der Fertigstellung ging S. v. HOERNER nach Heidelberg und R. EBERT auf Grund eines Stipendiums nach Princeton, USA.

Im einzelnen stammen der Teil A.I. von ST. TEMESVÁRY, jetzt Max-Planck-Institut für Physik und Astrophysik, München, die Teile A.II., A.III. und B von S. v. HOERNER, jetzt Astronomisches Rechen-Institut, Heidelberg, und der Teil C von R. EBERT, jetzt Universität Hamburg.

Die Autoren sind den folgenden Herren für zahlreiche Diskussionen und wertvolle Hinweise zu großem Dank verpflichtet: L. BIERMANN, L. DAVIS, W. FRICKE, W. GLIESE, R. KIPPENHAHN, W. LOHMANN, R. LÜST, G. MÜNCH, A. SCHLÜTER, M. SCHWARZSCHILD, L. SPITZER, B. STRÖMGREN, M. WRUBEL.

Der Fulbright-Kommission dankt R. EBERT für die Gewährung eines Stipendiums für einen Forschungsaufenthalt an der Universität Princeton, USA; während dieses Aufenthaltes wurde sein Beitrag zu der vorliegenden Arbeit geschrieben. Den Herren L. SPITZER und M. SCHWARZSCHILD möchte er seinen Dank aussprechen für viele fördernde und anregende Diskussionen und für die Ermöglichung des Aufenthaltes an dem dortigen Institut.

Literaturverzeichnis

1. ALFVÉN, H.: Ark. mat. astr. fys. Ser. A **28**, No. 6 (1942). — 2. ALFVÉN, H.: Cosmical Electrodynamics. Oxford 1950. — 3. ALFVÉN, H.: On the Origin of the Solar System. Oxford: Clarendon Press 1951. — 4. ALLEN, C. W.: Astrophysical Quantities. London 1955. — 5. ALLER, L. H.: Nuclear Transformations, Stellar Interiors and Nebulae. New York 1956a. — 6. ALLER, L. H.: Gaseous Nebulae, Intern. Astroph. Ser. Vol. 3, 1956b.

7. BAADE, W.: Astrophysic. J. **102**, 309 (1945). — 8. BECKER, W.: Sterne und Sternsysteme. 1951. — 9. BECKER, W., u. J. STOCK: Z. Astrophysik **31**, 316 (1953). 10. BERGH, S. v. D.: Astrophysic. J. **125**, 445 (1957a). — 11. BERGH, S. v. D.: Astronomical J. **62**, 100 (1957b). — 12. BERGH, S. v. D.: Z. Astrophysik **43**, 236 (1957c). — 13. BIERMANN, L.: Z. Astrophysik **18**, 344 (1939). — 14. BIERMANN, L.: Ergebn. exakt. Naturwiss. **21**, 1 (1949). — 15. BIERMANN, L.: Gas Dynamics of Cosmic Clouds, IAU Symp. 2 (1955). — 16. BIERMANN, L.: Z. Astrophysik **41**, 46 (1956). — 17. BIERMANN, L., u. A. SCHLÜTER: Z. Naturforsch. **5 A**, 237 (1950). — 18. BIERMANN, L., and A. SCHLÜTER: Cambridge Symp. on Gas Dynamics of Interstellar Clouds 1953. — 19. BLAAUW, A.: Bull. Astr. Netherl. **11**, 418 (1952). — 20. BLAAUW, A.: Astrophysic. J. **123**, 408 (1956a). — 21. BLAAUW, A.: Publ. Astr. Soc. Pacific **68**, 495 (1956b). — 22. BLAAUW, A., and W. W. MORGAN: Astrophysic. J. **117**, 256 (1953a). — 23. BLAAUW, A., and W. W. MORGAN: Bull. Astr. Netherl. **12**, 76 (1953b). — 24. BLAAUW, A., and W. W. MORGAN: Astrophysic. J. **119**, 625 (1954). — 25. BÖHM-VITENSE, E.: Z. Astrophysik **46**, 108 (1958). — 26. BOK, B. J.: Harv. Circ. **1934**, 384. — 27. BOK, B. J.: Centennial Symposium, Harvard Obs. Monograph No. 7 (1946). — 28. BOK, B. J.: Sky and Tel. **10**, 213 (1951). — 29. BOK, B. J.: Bericht d. Lüttich-Tagung 1954. (Les particules solides dans les astres.) S. 480. — 30. BONDI, H.: Monthly Notices **112**, 195 (1952). — 31. BONDI, H., and F. HOYLE: Monthly Notices **104**, 273 (1944). — 32. BONNOR, W. B.: Monthly Notices **116**, 351 (1956). — 33. BONNOR, W. B.: Monthly Notices **117**, 104 (1957). — 34. BURBIDGE, G. R.: Publ. Astr. Soc. Pacific **70**, 83 (1958). — 35. BURBIDGE, G., E. M. BURBIDGE, R. F. CHRYSTY and W. A. FOWLER: Physic. Rev. **103**, 1145 (1956). — 36. BURBIDGE, G. R., E. M. BURBIDGE and W. A. FOWLER: Astrophysic. J. **122**, 271 (1955). — 37. BURBIDGE, G. R., E. M. BURBIDGE, W. A. FOWLER and F. HOYLE: Rev. mod. Physics **29**, 457 (1957).

38. CAMERON, A. G.: Astrophysic. J. **121**, 144 (1955). — 39. CHANDRASEKHAR, S.: Stellar Structure. Chicago 1938. — 40. CHANDRASEKHAR, S.: Principles of Stellar Dynamics. Chicago 1947. — 41. CHANDRASEKHAR, S.: Astrophysic. J. **119**,

7 (1954). — 42. CHANDRASEKHAR, S., and E. FERMI: Astrophysic. J. **118**, 113 (1953a). — 43. CHANDRASEKHAR, S., and E. FERMI: Astrophysic. J. **118**, 116 (1953b). — 44. COWLING, T. G.: Magnetohydrodynamics. New York (1957). — 45. COX, A. N.: Astrophysic. J. **119**, 188 (1954). — 46. COX, A. N.: Astrophysic. J. **121**, 628 (1955). — 47. McCREA, W. H.: Monthly Notices **113**, 162 (1953). — 48. McCREA, W. H.: Monthly Notices **117**, 562 (1957). — 49. McCUSKEY, S. W.: Astrophysic. J. **123**, 458 (1956).

50. DAVIS, L., and J. L. GREENSTEIN: Astrophysic. J. **114**, 206 (1951). — 51. DELHAYE, J., u. A. BLAAUW: Bull. Astr. Netherl. **12**, 72 (1953). — 52. DODD, K. N.: Proc. Cambridge Phil. Soc. **49** (1953). — 53. DODD, K. N., and W. H. McCREA: Monthly Notices **112**, 205 (1952). — 54. McDOUGALL, J., and E. C. STONER: Philosophic. Trans. A **237**, 67 (1938).

55. EBERT, R.: Z. Astrophysik **36**, 222 (1955a). — 56. EBERT, R.: Z. Astrophysik **37**, 217 (1955b). — 57. EBERT, R.: Z. Astrophysik **42**, 263 (1957). — 58. EDDINGTON, A. S.: The Internal Constitution of the Stars, p. 391, 1926. — 59. EDDINGTON, A. S.: Der innere Aufbau der Sterne. Berlin 1927. — 60. EPSTEIN, J.: Astrophysic. J. **112**, 207 (1950).

61. FERMI, E.: Physic. Rev. **75**, 1169 (1949). — 62. FOWLER, W. A.: Mémoire de la Société Royale des Sciences de Liège, Tome XIV, 88 (1954). — 63. FOWLER, W. A., G. R. BURBIDGE and E. M. BURBIDGE: Astrophysic. J., Suppl. **2**, 167 (1955). — 64. FRICKE, W.: Astrophysic. J. **120**, 356 (1954). — 65. FRIEMAN, E. A.: Astrophysic. J. **120**, 18 (1954).

66. GAMOW, G.: Physic. Rev. **67**, 120 (1945). — 67. GLIESE, W.: Astr. Nachr. **272**, 97 (1941). — 68. GLIESE, W.: Z. Astrophysik **39**, 1 (1956). — 69. GLIESE, W.: Mitt. Astr. Rech. Inst. Heidelberg A, **8** (1957).

70. HARRISON, M. H.: Astrophysic. J. **100**, 343 (1944). — 71. HARRISON, M. H.: Astrophysic. J. **105**, 322 (1947). — 72. HASSELGROVE, C. B., and F. HOYLE: Monthly Notices **16**, 515, 527 (1956). — 73. HENYEY, L. G., R. LELEVIER and R. D. LEVEE: Publ. Astr. Soc. Pacific **67**, 154 (1955). — 74. HENYEY, L. G., R. LELEVIER and R. D. LEVEE: Evolution of main sequence stars (Vorabdruck) (1958). — 75. HILTNER, W. A., and B. IRIARTE: Astrophysic. J. **122**, 185 (1955). — 76. HOERNER, S. v.: Astrophysic. J. **125**, 451 (1957a). — 77. HOERNER, S. v.: Z. Astrophysik **42**, 273 (1957b). — 78. HOERNER, S. v.: Z. Astrophysik **44**, 221 (1958). — 79. HOROSHEWA, O. W.: Astr. J. UdSSR **33**, 880 (1956). — 80. HOYLE, F.: Astrophysic. J. **118**, 513 (1953). — 81. HOYLE, F.: Astrophysic. J., Suppl. **1**, 121 (1954). — 82. HOYLE, F., and R. A. LYTTLETON: Proc. Cambridge Phil. Soc. **35**, 405 (1939). — 83. HOYLE, F., and R. A. LYTTLETON: Proc. Cambridge Phil. Soc. **36**, 325 (1940). — 84. HOYLE, F., and M. SCHWARZSCHILD: Astrophysic. J., Suppl. **2**, 1 (1955). — 85. HULST, H. C. v. D.: Rech. Astr. de l'Obs. d'Utrecht, **12**, Part 2, Section 4 (1949). — 86. HULST, H. C. v. D., C. A. MULLER u. J. H. OORT: Bull. Astr. Netherl. **12**, 117 (1954). — 87. HULST, H. C. v. D., u. E. RAIMOND: Kontr. Ned. Ak. Nat. **65**, 157 (1956). — 88. HUMASON, M. L., N. U. MAYALL and A. R. SANDAGE: Astronomical J. **61**, 97 (1956).

89. JEANS, J. H.: Astronomy and Cosmogony, S. 345. Cambridge 1926. — 90. JOHNSON, H. L.: Astrophysic. J. **112**, 240 (1950). — 91. JOHNSON, H. L.: Astrophysic. J. **119**, 185 (1954a). — 92. JOHNSON, H. L.: Astrophysic. J. **120**, 325 (1954b). — 93. JOHNSON, H. L.: Astrophysic. J. **126**, 121, 134 (1957a). — 94. JOHNSON, H. L.: Publ. Astr. Soc. Pacific **69**, 54 (1957b). — 95. JOHNSON, H. L., and W. A. HILTNER: Astrophysic. J. **123**, 267 (1956). — 96. JOHNSON, H. L., A. R. SANDAGE and H. D. WAHLQUIST: Astrophysic. J. **124**, 81 (1956).

97. KAHN, F. D.: Bull. Astr. Netherl. **12**, 456 (1954). — 98. KING, I.: Astronomical J. **62**, 144 (1957). — 99. KIPPENHAHN, R.: Z. Astrophysik **46**, 26 (1958). —

100. KIPPENHAHN, R., ST. TEMESVÁRY u. L. BIERMANN: Z. Astrophysik **46**, 257 (1958). — 101. KOPPE, H.: Ann. Physik, 6. Folge **2**, 103 (1948). — 102. KUSHWAHA, R. S.: Astrophysic. J. **125**, 242 (1957).
103. LEDOUX, P.: Astrophysic. J. **105**, 305 (1947). — 104. LEE, T. D.: Astrophysic. J. **111**, 625 (1950). — 105. LOHMANN, W.: Z. Astrophysik **41**, 202 (1957a). — 106. LOHMANN, W.: Z. Astrophysik **42**, 114 (1957b). — 107. LÜST, R.: Z. Naturforsch. **7a**, 87 (1952). — 108. LÜST, R., u. A. SCHLÜTER: Z. Astrophysik **38**, 190 (1955).
109. MARKARJAN, B. E.: Obs. Bjurakan **11**, 1 (1953). — 110. MASANI, A.: Oss. Astr. Milano, Contr. 99 (1956). — 111. MAWRIDIS, L.: Z. Astrophysik **41**, 35 (1956). — 112. MENZEL, D. H.: Gas Dynamics of Cosmic Clouds, IAU Symp. 2, 1955. — 113. MERRILL, P.: Astrophysic. J. **116**, 21 (1952). — 114. MESTEL, L.: Monthly Notices **113**, 716 (1953). — 115. MESTEL, L.: Monthly Notices **114**, 437 (1954). — 116. MESTEL, L.: Astrophysic. J. **126**, 550 (1957). — 117. MESTEL, L., and L. SPITZER: Monthly Notices **116**, 503 (1956). — 118. MINKOWSKI, R.: Publ. Astr. Soc. Pacific **61**, 151 (1949). — 119. MORGAN, W. W., A. E. WHITFORD and A. D. CODE: Astrophysic. J. **118**, 318 (1953). — 120. MORSE, P. M.: Astrophysic. J. **92**, 27 (1940). — 121. MÜNCH, G.: Publ. Astr. Soc. Pacific **68**, 351 (1956). — 122. MÜNCH, G., and E. FLATHER: Publ. Astr. Soc. Pacific **69**, 142 (1957).
123. OORT, J. H.: Cambridge Symp. on Gas Dynamics of Interstellar Clouds. 1953. — 124. OORT, J. H.: Bull. Astr. Netherl. **12**, 455, 177 (1954). — 125. OORT, J. H., and L. SPITZER: Astrophysic. J. **121**, 6 (1955). — 126. OSTERBROOK, E. D.: Astrophysic. J. **118**, 529 (1953).
127. PARENAGO, P. P.: Astr. J. UdSSR **27**, 150 (1951). — 128. PETRIE, R. M.: IRAS Canada **51**, 177 (1957).
129. REIZ, A.: Astrophysic. J. **120**, 342 (1954). — 130. RHIJN, P. J. v.: Publ. Groningen No. 57, 59 (1957). — 131. ROBERTS, M. S.: Publ. Astr. Soc. Pacific **69**, 59 (1957). — 132. ROMAN, N. G.: Astrophysic. J. **121**, 454 (1955).
133. SALPETER, E. E.: Astrophysic. J. **115**, 326 (1952). — 134. SALPETER, E. E.: Astrophysic. J. **121**, 161 (1955). — 135. SANDAGE, A. R.: Astronomical. J. **59**, 162 (1954). — 136. SANDAGE, A. R.: Astrophysic. J. **122**, 263 (1955). — 137. SANDAGE, A. R.: Astrophysic. J. **123**, 278 (1956). — 138. SANDAGE, A. R.: Astronomical J. **125**, 422 (1957). — 139. SAVEDOFF, M. P.: Gas Dynamics of Cosmic Clouds, S. 218 IAU Symp. No. 2, 1955. — 140. SAVEDOFF, M. P., and J. GREENE: Astrophysic. J. **122**, 477 (1955). — 141. SCHLÜTER, A.: Z. Naturforsch. **5a**, 72 (1950). — 142. SCHLÜTER, A.: Z. Naturforschung **6a**, 73 (1951). — 143. SCHLÜTER, A.: Gas Dynamics of Cosmic Clouds, IAU Symp. No. 2, 1955. — 144. SCHLÜTER, A.: Z. Naturforsch. (1958). — 145. SCHLÜTER, A., H. SCHMIDT u. P. STUMPFF: Z. Astrophysik **33**, 194 (1953). — 146. SCHMIDT, M.: Bull. Astr. Netherl. **13**, 15 (1956). — 147. SCHÖNBERG, M., and S. CHANDRASEKHAR: Astrophysic. J. **96**, 161 (1942). — 148. SCHWARZSCHILD, M.: Astrophysic. J. **104**, 203 (1946). — 149. SCHWARZSCHILD, M.: Astrophysic. J. **125**, 123 (1957a). — 150. SCHWARZSCHILD, M.: Astrophysic. J. **125**, 233 (1957b). — 151. SCHWARZSCHILD, M., and S. BERNSTEIN: Astrophysic. J. **122**, 200 (1955). — 152. SCHWARZSCHILD, M., and R. HÄRM: Astrophysic. J. **128**, 348 (1958). — 153. SPITZER, L.: Monthly Notices **100**, 396 (1940). — 154. SPITZER, L.: Astrophysic. J. **94**, 232 (1941). — 155. SPITZER, L.: Astrophysic. J. **107**, 6 (1948). — 156. SPITZER, L.: Astrophysic. J. **109**, 337 (1949). — 157. SPITZER, L.: Astrophysic. J. **111**, 593 (1950). — 158. SPITZER, L.: Astrophysic. J. **120**, 1 (1954). — 159. SPITZER, L.: Physics of Fully Ionized Gases, New York 1956. — 160. SPITZER, L.: Rom Symposium 1957. — 161. SPITZER, L.: Mündliche Mitteilung, 1958. — 162. SPITZER, L., and M. SCHWARZSCHILD: Astrophysic. J. **114**, 385 (1951). — 163. SPITZER, L., and M. SCHWARZSCHILD: Astrophysic. J. **118**, 106 (1953). —

164. Spitzer, L., and J. W. Tukey: Astrophysic. J. **114**, 187 (1951). — 165. Strömgren, B.: Astrophysic. J. **89**, 529 (1939). — 166. Strömgren, B.: Astrophysic. J. **108**, 242 (1948). — 167. Strömgren, B.: Astronomical J. **57**, 65 (1952). — 168. Sweet, P. A.: Monthly Notices **110**, 548 (1950).
169. Taylor, R. J.: Astronomical J. **59**, 413 (1954a). — 170. Taylor, R. J.: Astrophysic. J. **120**, 332 (1954b). — 171. Taylor, R. J.: Monthly Notices **116**, 25 (1956). — 172. Temesváry, St.: Z. Naturforsch. **7a**, 103 (1952). — 173. Temesváry, St.: Mémoire de la Société Royale des Sciences de Liège, Tome XIV, 122 (1954). — 174. Temesváry, St.: Naturwissenschaften **44**, 321 (1957). — 175. Temesváry, St.: Nichtadiabatische Konvektionszonen (im Erscheinen). — 176. Trümpler, R. J.: Lick Obs. Bull. **14**, 154 (1930).
177. Underhill, A. B.: Liège Symposium p. 374 (1953). — 178. Unsöld, A.: Z. Astrophysik **25**, 11 (1948). — 179. Unsöld, A.: Sternatmosphären. Berlin-Göttingen-Heidelberg 1955.
180. Vitense, E.: Z. Astrophysik **32**, 135 (1953). — 181. Vogt, H.: Aufbau und Entwicklung der Sterne. Leipzig 1957. — 182. Vyssotsky, A. N., and E. M. Jenssen: Astronomical J. **56**, 58 (1951).
183. Wade, C. M.: Preprint 1957. — 184. Waren, G. W.: Astrophysic. J. **100**, 158 (1944). — 185. Weizsäcker, C. F. v.: Z. Astrophysik **22**, 319 (1944). — 186. Weizsäcker, C. F. v.: Z. Naturf. **3a**, 524 (1948). — 187. Weizsäcker, C. F. v.: Astrophysic. J. **114**, 165 (1951). — 188. Whipple, F. L.: Astrophysic. J. **104**, 1 (1946). — 189. Wrubel, M.: Handbuch der Physik LI, 1. Berlin-Göttingen-Heidelberg 1958.

Subject index

The following subject index refers — in english translation — also to the contribution written in german

Das folgende Sachverzeichnis enthält — in englischer Übersetzung — auch die Hinweise auf den deutschsprachigen Beitrag

Abundances of elements 131 (Tab.), see chemical composition
Accretion 8, 22 ff., 136 f., 161, 293 ff.
Age of associations and clusters 60, 68, 73, 126, 225 f. (Tab.), 228 f., 237, 246, 249
— according to expansion 53
— of stars 126, 136, 225 ff.
— of the world 229
Albedo of interstellar grain 5, 8, 133, 301
Alfvén velocity 313, 315
Andromeda Nebula, see M 31
Angular momentum, removal 16 f., 151, 304, 310 f.
— —, — by magnetic field 17, 311 f., 319
Associations 51 ff., 117 ff., 135 f.
—, age 60, 237, 246, 249
—, expansion 53 ff., 154 ff., 161, 245, 246, 305
—, formation 10, 95, 136 ff., 152, 161, 319
—, rate of formation 158, 271 ff.
—, luminosity function 271, 272, 289
—, main sequence 235
—, mass 272 f., 289
— with positive energy 152
—, stability 273
—, test for reality 174 f.
Associations:
 Cas-Tau 54
 I Cep 247, 249
 II Cep 54, 60, 246
 γ Cyg 60
 II (P) Cyg 60
 I Gem 60
 I Lac 53, 60, 118, 246, 247, 249
 Orion 60, 236, 237, 247, 249
 I (h) Per 60, 246
 II (ζ) Per 53, 60, 246, 249
 Sco-Cen 60
„Ausreißer", see runaway stars

B-stars in high galactic latitude 247 f.
Barnards ring 158
Binary systems 17 ff., 156, 227, 275, 290
Braking, magnetic and non-magnetic 17, 233, 315 ff.
Break in main sequence (concerning velocity dispersion) 81

Californium 280
Carbon, abundance 279
Carbon-cycle 105, 189 ff., 279
Carbon-Nitrogen abundance 189
Chemical composition of clusters 58, 74
— —, discontinuity in stellar interior 40 ff., 209, 225
— — of distant galaxies 91 ff.
— — of interstellar matter and protostars 8, 130 f.
— — of Magellanic clouds 87
— — of old stars 253
— — of population types 82, 128 f., 173 f., 255, 280
— — of stellar material 163
Circulation, see rotational mixing
Clusters see galactic and globular clusters
—, formation 301 ff., 318
— of galaxies 30
—, intergalactic 90
Collapse of H I cloud, see gravitational contraction
Color-magnitude diagram 35, 42, 44, 58, 124 ff., 135, 170
— — of individual clusters, see under NGC, M etc.
Coma cluster 58, 60, 229
Compression, adiabatic and isothermal 27
— under external pressure 95, 140 f., 144 ff., 156 f., 300 f.
Condensation of galaxies 28
— of stars and associations, see formation of stars and associations, and gravitational contraction

Subject index

Conductivity of interstellar gas 312
Contraction, see gravitational contraction
Convection in stellar interior 36, 164, 167f., 201, 213, 220
— — —, transfer of energy 201, 205
— — —, velocity 201, 206
Convective core 208ff., see convection in stellar interior
Corpuscular radiation 170, 279
Coupling between interstellar clouds and O-stars 304
— — magnetic field and matter 21
— — plasma and neutral gas 148f.
Cowling modell 192, 213, 228
Crab nebula, see M 1
Critical length for contraction of a cylinder 20
— mass for contraction and condensation 3, 7, 10f., 20, 25, 27, 96, 144, 298ff., 306, 309
— radius for contraction 11, 144, 157
Cross section for collision of atoms 149
— — for star collision 108, 155, 171

Degenerate core (electron degeneracy) 41, 47, 165, 213ff.
Diffusion velocity 5
Dipol field 311
Dissolution of clusters 70ff., 241ff., 269
Distribution of stars 171, 268
Dust, see interstellar dust and grain, interstellar matter

Eddington model 192, 213
Eddington-Sweet's circulation 218
Electron degeneracy, see degenerate core
— scattering 165f.
Elements, abundance 131 (Tab.), see chemical composition
—, synthesis (formation) 175f., 179ff., 289
Elephant's trunk 121ff., 151, 158, 305
"Eleven hour cluster" 90
Elliptical galaxies 88f.
Energy of compression 307
—, gain and loss (gas particles) 132, 297
— of gravitational contraction 188
—, kinetic, of gas atoms 150
—, magnetic 308, 309
— production in stellar interior 105, 164, 189, 279ff.

Energy of radiation 146
Equipartition of kinetic energies 81
Euphorion 193
Evolution of stars 32—50, 96, 167ff., 187—230
— — away from the main sequence 169f., 220f.
— — down the main sequence 37ff.
— — in external galaxies 84ff.
— —, final evolution phase 224ff.
— —, massive stars 49, 222f.
— — onto the main sequence, see gravitational contraction
— —, tracks of evolution 44ff., 221
Evolutionary sequences of galaxies 92
Exchange of energy 155, 242, 257, 262, 288
Expansion and rotation of stars 229
— — of associations, see associations

Filaments 15, 153
Formation of associations, clusters and interstellar clouds, see association etc.
— of binary systems 17, 156, 227, 290
— of stars (condensation) 3—31, 95, 136—161, 296—320
— —, individual or in clusters? 270f., 319
— —, rate of formation 55, 261ff., 288
Fragmentation 8, 16, 26, 28, 34, 65, 95, 151ff., 161, 301ff.
Friction of turbulence 310, 311

Galactic clusters 57—72, 270f.
(individual clusters see under NGC, M etc.)
— —, age 60, 68, 73, 237, 243, 249
— —, dissolution 70ff., 241ff., 269
— —, frequency of different types 245, 266ff.
— —, hydrogen content 58
— —, luminosity function 238
— —, mass 238ff.
— —, probability of detection 267, 268
— —, rate of formation 265ff.
— —, star formation in clusters and associations 270ff., 318, 319f.
Galactic rotation 109
— system 109ff.
Galaxies 84ff., 108ff., 135 (individual galaxies, see under NGC, M etc.)

Galaxies, clusters of galaxies 30
Gas filaments 15, 153
Gas sphere 299
Giant branch 48 (see Color-magnitude diagram)
Globular clusters 26, 48, 72 ff., 111, 244, 249
— —, formation 301 ff., 318
— —, rate of formation 265
Globules 14, 123, 151, 300, 306, 320
Grain, see interstellar dust and grain, interstellar matter
Gravitational contraction 3, 27, 32 ff., 64, 138 ff., 146 f., 156 ff., 161, 187 f., 232 ff., 297 ff., 305 f.
— —, time 32 ff., 64, 188, 201, 233
— — with interstellar magnetic field 147 f., 153 f., 308 ff.

H I-, H II-regions 120 ff., 130 ff., 136, 149, 156 ff., 295, 297, 304, 305
— — in Andromeda nebula 114
h and χ Per 58, 60, 66 f., 126, 236, 241, 244
Halo 117, 258
Heat conduction 168
— — for degenerate core 214
Helium burning process 46, 189 ff., 221
Helium in white dwarfs 282 ff., 289
Herbig-Haro objects 15, 119
Hertzsprung gap 68, 69, 97
Hertzsprung-Russell diagram, see color-magnitude diagram
— — of individual clusters, see under NGC, M etc.
High velocity stars 111 f., 126, 135, 229, 249
Homology transformations 34, 193 ff.
Horizontal branch 70, 75 ff., 90
Hyades 57, 58, 60, 68, 71, 126, 236, 241
Hydrogen content of galactic clusters 58
— convection zone, (see convection) 202 ff., 220

IC 4665: 60
IC 5146: 236
Induction law 312
Inhomogenities in chemical composition 40 ff., 167
Initial main sequence 228
Instability, see convection in stellar interior and gravitational contraction

Intergalactic clusters 90
Interstellar clouds (see H I-, H II-regions, see interstellar matter)
— —, collision with stars 171 f., 257
— —, collisions 132 f., 142 f.
— —, condensation, see gravitational contraction, formation of stars and associations
— —, expansion 305
— —, formation 5, 124
— —, radius 123 f., 295
— —, rotation 315
— dust and grain (see interstellar matter)
— —, acceleration 5
— —, chemical composition 133
— —, collision 150
— —, compression (concentration) 95, 140 f., 144 ff., 156, 300 f.
— —, formation 4
— —, grain size 113 f.
— —, growth of grain 134, 137
— —, temperature 132
— gas (see interstellar matter)
— —, kinetic energy 150
— —, large scale motion 112 f.
— —, mass 83 f.
— —, temperature 123, 131, 296 f.
— —, turbulence 95, 174, 299, 300, 308, 310
— matter (material, medium) 130 ff.
— —, bilance 262, 264, 277 ff., 289
— —, chemical composition 8, 130
— —, correlation between gas and dust 115, 123
— —, density 172
— —, distribution 112, 120 ff., 275, 276
— —, Helium content 274, 287
— —, mass 83 ff., 114 f., 136
— —, motion 120 ff., 173
— — and stars 273 ff., 289
— polarisation 134
Irregularities in gravitational field 171 f.
Isochrones 228, 229

Jeans criterion 3, 9, 32, 140, 297, 298
— —, with magnetic field 147

Kelvin-Helmholtz-contraction 32 (see gravitational contraction)
Kinetic energy, see energy

"Knee" (turn-off point) 44, 127, 170, 226, 234, 237, 248
Krüger 60 A: 219

Limiting mass of degenerate core 215
— — for fragmentation 304
Luminosity function of associations 271, 272, 289
— — of elliptical galaxies 89
— — of field stars and clusters 13, 42, 77, 238 ff.
— —, initial and observed 261 ff., 290
— —, integral over z 260, 262 f.

M 1 (Crab nebula): 130, 173
M 3: 48, 73 ff., 78, 126, 130, 220, 230
M 8: 14
M 10: 75 f.
M 11: 58, 60, 68, 241
M 13: 75
M 29: 241
M 31 (Andromeda nebula): 84 ff., 114 ff., 117, 286
M 32: 88 f.
M 33: 86
M 34: 60, 241
M 39: 60, 229, 241
M 41: 58, 60, 68, 241
M 67: 58, 60, 68 ff., 74, 76 f., 78, 83, 126, 170, 229, 241, 249
M 81: 91, 286
M 92: 73 ff., 130, 220, 230, 240
Mach number 148
Magellanic clouds 86 f.
Magnetic braking 17, 233, 315 ff.
— cylinder 314, 315
— field, interstellar 116 f., 132 f., 312
— — and contraction 19 ff., 147, 153 f., 161, 308 ff.
— —, slip of 309, 310, 316
— —, torsion 311, 312 ff.
Main sequence, see color magnitude diagram, gravitational contraction
— —, break 170, 248
— —, — of lower m.s. 234 ff.
— —, length 235
Mass function 13 f., 77
— of Galaxy 111, 114
— of galactic clusters 238 ff.
—, instable, see critical mass, limiting mass

Mass of interstellar matter 83 ff., 136
—, loss and gain 39, 42, 96, 170, 226, 264, 274, 277 ff., 293, 307
Mass-to-light-ratio in external galaxies 84, 88
Mass-luminosity relation 38, 129, 170, 195 f., 230, 286
Mass-radius relation 129, 196
Massive stars, evolution 49, 222 f.
McCormick stars 251
Mean free path for interstellar gas 293, 294
— — — for an ion 150
— — — for a photon 150
— — — for star collision 155
Mixing (in main sequence stars) 36 ff., 169, 216 ff.
Motion of stars 117 ff., see velocity of stars
Multiple stars 18, 54, 156, 275, see binary systems

„Nachzügler" 237
Neutron, capture of 280
NGC 147: 88
NGC 185: 88
NGC 205: 88 f.
NGC 457: 60, 241
NGC 663: 60, 241
NGC 752: 58, 60, 68 f., 228, 229
NGC 1664: 68
NGC 2264: 60, 62 ff., 126, 235 ff., 239, 241, 249
NGC 2362: 58, 126, 229, 236, 241
NGC 2516: 60, 241
NGC 3034: 91
NGC 3077: 91
NGC 3115: 88 f.
NGC 4174: 130
NGC 6357: 122
NGC 6530: 14, 62 ff., 236, 237
NGC 6611: 236
NGC 6940: 83
NGC 6960: 15
NGC 6992: 15
NGC 6995: 15
NGC 7209: 241
NGC 7243: 60, 241
Non adiabatic convection zone 206 f. (see convection)
Novae 225, 227

O-Associations, see Associations
O- (and B-) stars 3, 9 f., 83, 114, 192, 258, 275, 304 ff., 319 (see Association, Protostar formation etc.)
Oort's constants 119 f.
Origon, see formation
Orion nebula 118, 121, 158 f.
− − cluster 61 f., 63 ff.

Period-luminosity relation 107, 174
Perseus moving cluster 60
Perturbation method for studies of stability 298
Photoelectric absorption in stellar interior 165
Pinch effect 153
Plasma (see interstellar gas). Coupling with neutral gas 148 f.
−, density 21
−, equations 312
−, motion 21
Plejades 58, 60, 67 f., 71, 126, 229, 236, 239, 240, 241, 244
Poisson equation 154
Population types 78 ff., 135
− −, chemical composition 82, 128, 173 f., 255, 280
− − in other galaxies 82
− −, LC diagrams 125
− − and velocity dispersion 82
Potential of Galaxy 259, 260, 289
Praesepe 58, 60, 68, 229, 236, 240, 241
Proton-proton reaction 105, 189 ff.
Protostar, central density 31
−, central temperature 31
−, chemical composition 8
−, formation 3−31 (see formation of stars)
−, radius 31

Quasi equilibrium state 41

Radiation, energy 146
−, pressure 145 f., 292, 300, 301, 307
Red dwarfs 219
Refractive index 165
Rejuvenation 23, 24, 227, 290, 291, 293 ff.
Relaxation time 155 f., 242, 243
Rocket effect 157
Rotation of Andromeda nebula 114 f.
− and contraction 233, 236

Rotation and expansion 229
− of interstellar clouds 315
− and mass loss 279
− of T Tauri stars 53
Rotational mixing (meridional circulation) 37, 167 f., 216 ff.
− stability 152
RR Lyrae stars 47, 75, 76, 114, 230
Runaway stars („Ausreißer") 237, 247, 275
Russel-Vogt-theorem 188

Salpeter cyclus see helium burning process
Shockfront, shockwave 156 f., 305
Shrinkage, see gravitational contraction
Space density of stars, see distribution
− velocity, see velocity
Spectral classification, three dimensional 127
Spiral galaxies 108 ff., 135
− structure 83, 116
Stability conditions, see Jean's criterion, gravitational contraction
Star chains 15, 119, 153 f., 161
− formation, see formation
Stellar atmosphere 127
Strömgren sphere 9, 295 (see H II-region)
Structure of stars (see evolution) 218 ff.
− −, Central temperature 192
− −, non homogeneous model 43, 166
− −, shell source model 212, 215, 220 f.
− −, solar model (Schwarzschild) 219
− −, stationary models 161 ff.
− −, uniqueness of solution 188
Subdwarfs 224
Sun, model 219
Supernovae 10, 158 f., 161, 173 f., 175, 227, 280
Surface features of stars 124 ff.

T-Associations, see Associations
T-Tauri stars 14 f., 52, 119, 235, 236
Technetium 281
Temperature Dichotomy 27
Thomson scattering 163, 165
Tidal force 241, 243
Time of concentration and contraction 5, 32 ff., 64, 188, 201, 233
− of free fall (in contraction) 149

Time for large mass-increment 24
— of magnetic braking 315 f.
— of main sequence evolution 59
— for reducing a star to rest 24, 295
— for rejuvenation 295
— scale, astronomical 106 ff.
— — for hydrogen burning 190 ff., 201
Torsion of magnetic field 311, 312 ff.
Transfer of energy in stellar interior 163 f., 195, 201, 205
— of radiation in stellar interior 195
Turbulence in collapsing clouds 151 f.
— in H II-region 121 (see gravitational contraction)
—, initial and space velocity 257 f., 288
— of interstellar medium 95, 174, 299, 300, 308, 310
— and protostar formation 13
Turbulent friction 310, 311
Turn-off point (Knee) 44, 127, 170, 226, 234, 237, 248

Uniqueness of solution of stellar structure 188
Ursa maior stream 60, 241, 242, 244, 249

Velocity of expansion of H II-region 305
— of rotation of interstellar cloud 315 ff.
— of stars (dispersion and distribution) 79 ff., 171 f., 251 f., 288
— —, effect of age 252 ff., 259, 288
— —, — of apparent magnitude 254, 255
— —, — of population 82
— —, — of position in HR diagram 80 f., 120, 252 ff.
— —, ,,Güte-Effekt" 254, 255
— —, zero point 253
— —, z-komponente 80, 258 ff., 288
Virial theorem 11, 139, 240, 307
— —, with magnetic field 19, 308
Vogt-Russell theorem 188

White dwarfs 26, 216, 282, 289